Adaptability

*The Significance of Variability
from Molecule to Ecosystem*

Adaptability

The Significance of Variability from Molecule to Ecosystem

Michael Conrad

Wayne State University
Detroit, Michigan

PLENUM PRESS • NEW YORK AND LONDON

Library of Congress Cataloging in Publication Data

Conrad, Michael.
    Adaptability: the significance of variability from molecule to ecosystem.

    Includes bibliographies and index.
    1. Adaptation (Biology). I. Title. [DNLM: 1. Adaptation, Physiological. 2. Ecology.
QT 140 C754a]
QH546.C65 1983   574.5   82-24558
ISBN 0-306-41223-3

© 1983 Plenum Press, New York
A Division of Plenum Publishing Corporation
233 Spring Street, New York, N.Y. 10013

Printed in the United States of America

To Earl and Alyse, my parents,
whose books and manuscripts
shared space with my tabletop ecosystems

Preface and Acknowledgments

In order to survive and reproduce biological systems must be adapted to the specific features of their environment. They must also be adaptable, or capable of functioning in an uncertain environment. The adaptability of biological matter is one of its most striking properties.

This adaptability may manifest itself at many different levels of organization, ranging from the molecular and cellular levels to the levels of the population and the community. One type of population, for example a microbial population, may rely on culturability and control of gene expression to cope with the uncertainty of the environment, while another, of metazoan plants, may rely on genetic and developmental plasticity, or it may restrict itself to an environment which is not so uncertain. Still another population, say of metazoan animals, may rely on social organization or on behavioral plasticity mediated by its neuromuscular system. Indeed, over the broad spectrum of biological nature, one can find the most diverse mechanisms of adaptability and also the most diverse strategies for using these mechanisms.

This great diversity may invite a certain amount of pessimism as to the possibility of understanding, or even describing, the patterns of adaptability which actually exist in nature. Fortunately, however, the problem is simpler than it appears at first. This is because all the different mechanisms and modes of adaptability have one thing in common: they are all adaptations to the uncertainty of the environment. This means that we can expect all forms of adaptability, regardless of their diversity, to have some common denominator.

In this book I want to describe this common denominator and then show how it can be used to analyze patterns of adaptability in nature. In particular I want to answer four questions:

1. What is adaptability?
2. What are the major mechanisms of adaptability?
3. What are the major strategies with which biological systems use these mechanisms?

4. How do these strategies interlink in the development and evolution of the ecosystem as a whole?

The approach taken to these questions is formally quite straightforward. The first step is to describe the ecosystem as a biotic community and physical environment, each with a set of states and some probabilistic (and generally unknown) law governing the state-to-state transitions. The second step is to characterize the statistical properties of the biota in terms of suitable uncertainty (or entropy) measures and relate these to the uncertainty of the environment. This makes it possible to define adaptability and also to connect it to its mechanistic basis. Next I develop a convenient way of describing the complex organization of the biota in terms of its hierarchical (or, more precisely, compartmental) structure. This makes it possible to redescribe the statistical model of the ecosystem in hierarchical terms and therefore to consider the major factors which determine the allocation of various statistical properties to the different levels—in short, the factors which determine patterns of adaptability in nature. These are of crucial importance for the organization of the organism and for succession and evolution, both from the standpoint of the individual population and from the standpoint of the ecosystem as a whole.

The problem of adaptability also has a crucial conceptual connection to the problem of stability. The ability to cope with an uncertain environment is clearly a necessary condition for the maintenance of a relatively permanent form of organization, hence for stability. This is the reason for the connection between adaptability and ecological succession or evolution. Essentially, only those ecosystems with suitable adaptability properties have the "right to persist."

A number of deep and rather subtle issues arise. Adaptability involves the use of information about the environment. So information processing and reliability of information processing must be considered. Another fundamental connection is between adaptability and the structural and functional transformability of biological systems. Transformability turns out to be a generalization of reliability. A self-contained treatment requires a close analysis of fundamental biological concepts, such as information, complexity, efficiency, and fitness. Some questions of special importance concern the legitimate ways of using information measures in biology, the connection between energy and adaptability, and the relation between adaptability and various dynamical notions of stability, such as orbital stability and structural stability. To understand the relation between stability and complexity correctly it is necessary to understand the relation between stability and adaptability. There is also an important link between the adaptability of a system and the extent to which its dynamics is predictable.

My major objective, however, has been to use the theory developed to make testable statements about observable biological phenomena and to compare the concrete phenomena to the claims made about the adaptability structure of biological systems at different levels of organization. At the very lowest levels and at the very highest, decisions about which systems to focus on are easy to make. On the lower side I focus on genetic organizations, including the adaptability structure of individual genes and proteins. On the top side I focus on patterns of adaptability in populations, on the adaptability structure and successional development of communities, and on the long-term process of evolution. There are numerous specific physiological systems that could be considered in the zone between gene and population. Here I have attempted to state the general principles and to choose some examples which I believe are particularly illustrative. Examples include cyclic nucleotide and hormonal systems, ATP control systems, features of the immune system and the central nervous system, and basic morphological features of plants and animals insofar as they relate to the structure of adaptability. Features of some of the smallest objects in biology—such as genes—are deeply connected to features of large objects, such as communities. All biological objects are tied together by their contribution to the structure of adaptability, so it should not be surprising that a coherent account of adaptability would reveal new and interesting connections between superficially unrelated phenomena.

Needless to say, any formalism which is capable of coping with the full complexity of adaptability processes in nature must itself be complex. The formalism described in this book shares this feature. It would be impossible to arrive at correct conclusions without using a formal instrument. It is necessary to show the results in order to be in a position to say what they are. But I have in each case stated the results informally and have illustrated them with as many biological examples as practical. The informal statements are not as precise as the formal statements, but they are correct. As the philosopher Wittgenstein pointed out (in *Philosophical Investigations*) the concept of absolute precision cannot be useful. The suitable degree of precision must be chosen relative to the purposes at hand. The observation is remarkably apt for biological analysis. In the formal development I have chosen a degree of precision which I believe has been the most suitable for reaching useful conclusions.

But the theory is by no means all, or even primarily, mathematical formalism. Every biological theory is obliged to make a three-point landing, not only on the ground of mathematical self-consistency, but also on the ground of consistency with physical law, and most of all on the ground of consistency with and incorporation of fundamental biological principles and concepts. Thus the book requires some background of physical and thermodynamic ideas and of course a background of basic biology. I have tried to present this wide-ranging background in a way which indicates the breadth of connections, which is technical only on points relevant to adaptability, and which always gives a nontechnical description of these technicalities. As in the presentation of the formal structures of the theory itself, the nontechnical descriptions should make the book accessible to the reader who wants to familiarize himself with the main principles of adaptability in nature, but who would rather omit some of the details.

The possibility of elucidating these principles, of explaining as well as describing patterns of adaptability, is appealing from the naturalistic point of view. It is also appealing from the theoretical point of view because the generality of the problem is such that it is amenable to mathematical analysis. There is also a practical aspect. Many of the problems which arise in genetic engineering, medicine, agronomy, and ecological management are essentially ones of adaptability theory. The problem in these vital areas finally reduces to the problem of adequate adaptations to the uncertainty of the environment, of using these various adaptations to combat internal and external perturbations, and of interweaving them into stable forms of organization. I believe such practical applications would be best developed in the context of concrete situations. Of necessity this is a matter for the future. But in the concluding chapter I have used the principles to formulate a set of guidelines which should be applicable to a wide variety of practical situations. Living systems are evidently much more adaptable than present-day technical systems and have much greater potential for evolution and novel adaptation. Adaptability theory points up the features which underlie this. The design guidelines which it implies are simply guidelines for maintaining these features along with criteria for assessing whether they are in fact being maintained. The theory naturally extends from preeconomic ecosystems to ecosystems with a monetary economy. Here design is especially important and I therefore conclude the book with an application of the analysis to the adaptability structure of economic ecosystems.

Since adaptability involves both the functional organization of biological systems and their physiochemical constitution, it is inevitable that any adequate theory will include in its ancestry a number of lines of thought. It is necessary to abstract from a reality which dies when any of these lines becomes irrelevant. One important lineage is the theory of evolution. This is a line of thought which has its origins in biology itself and which is the source of basic notions such as adaptation, adaptability, and fitness. A second important lineage is physiology. This is the source of ideas about homeostasis and of analogies between physiological processes and technological control systems which have been at the same time fruitful and misleading. Ideas about feedback control and models of information processing which have their roots in automaton theory can, with some liberty, be placed here. A fourth important line of thought comes from physics and from irreversible thermodynamics. This provides the link between adaptability and energy–entropy processes and a conceptual underpinning which is necessary for the proper interpretation of the formalism. Two important sources of the underlying biology involve phenomena at the extreme scales of size. The large-scale source comes from studies of global ecosystems, particularly phenomena such as cycles and succession. The small-scale source is molecular biophysics and molecular genetics. Processes such as protein folding are fundamental for evolutionary adaptability. They are connected on the one hand with the virtually neutral sequence variability exhibited by some genetic structures and on the other to the topological transformability of biological structures which is the *sine qua non* for evolution by variation and natural selection.

The formalism itself has its origin in information theory and discrete systems theory. It captures enough of the reality to serve as a general and reliable instrument of deduction. Dynamical formalisms, another important lineage, are capable of mapping more detail. But I shall argue that for the questions addressed by adaptability theory they abstract away too much of the underlying biology (if they are constructed to be tractable) and that they predict more than is in principle predictable. Their real value (from the standpoint of adaptability) is as a tool for thinking about the stability of functional organizations. The conscious incompleteness of the formalism may dissatisfy those who have hopes for more powerful tools. But the tradeoff between completeness of description and generality of conclusion appears to be fundamental in biology. This does not mean that more completely descriptive tools cannot eventually be developed which give more complete answers to the questions posed. But it does mean that one cannot expect to develop these tools by attempting to fulfill traditional expectations. By posing more modest questions than are naturally posed with other tools, I believe it has been possible to obtain conclusions which are more generally applicable.

It is interesting to compare the point of view of adaptability theory to that of classical biostatistical analysis. Both are basically probabilistic approaches. The difference lies in how the variability of data is viewed. The adaptability theorist views variability as having functional significance, whereas the statistician seeks to extract, with a stated degree of confidence, a prototype correlation which the variability is presumed to mask. There are fundamental relationships in living systems and there are situations in which it has been useful to view variable observations as error which obscures these relationships. But I shall argue that for living matter the variability of data is at least as fundamental as any prototype relationships which could

be extracted from it and that in the typical situation a more fruitful
hypothesis can be constructed about the variability than can be extracted
from it. This functional view of variability is already present in models of
evolutionary processes having their origin in statistics. But by and large
variability of data is still viewed as a nuisance by most experimental and
field biologists. According to adaptability theory this nuisance phenomenon
is especially pronounced in biological materials because of its great importance for life. There is an interesting analogy to the situation in physics.
Originally the variability of data was viewed as an extraneous nuisance. But
now it is known that at least some uncertainty in measurement is due to
quantum fluctuations and that such fluctuations, rather than being a
nuisance, are responsible for the forces which hold our universe together.
The variabilities of biological systems are essential to their integrity in
different but equally significant ways. In both physics and biology there
must therefore be a point at which the paradigm of a prototype reality
masked by error becomes inappropriate. This intrinsic importance of variability was recognized much earlier in the history of biology than in the
history of physics, certainly not later than the appearance of the
Darwin–Wallace theory of evolution. The problem is that this recognition
has not extended to as many areas of biology as it should. It seems to me
that an enormous amount of useful biological data is every day being
ignored or discarded because of the great desire to extract prototype
relationships from it and because of the absence of a suitable adaptability
theoretical framework for interpreting it.

This book shares a profound public debt to the many individuals
associated with the ideas which have contributed to it. The best place to
acknowledge these debts—insofar as it is possible—is in the text itself. But
the debt would be inadequately acknowledged if I did not point out a
fundamental inaccuracy in the historical picture which I have so minimally
outlined. It would be a mistake to imagine that these different lines of
thought developed in isolation. I believe that a strong case can be made that
they have intersected at the most pivotal junctures. Certainly cross-correlations between them have been the subject of deep studies. The problem is
that they are isolated insofar as they have become attached to institutional
structures. The imagery which they engender—life as a physical process, as
a machine, as an expression of stable dynamical forms, as an evolution
process, as an irreducible organization—are often antithetical as well. The
lineage which can be identified as theoretical biology has in this respect
played a special role. As a discipline basically without institutional support
it has provided the necessary but all too narrow conduit of interchange and
has served to maintain the thread of a tradition which at potentially pivotal
times has played the pivotal role. I conjecture that careful evaluation of the
historical evidence would show this claim to have merit.

Thinking back on the individuals with whom I have worked or have had the benefit of discussion I am struck by the extent to which they reflect these different lineages of thought.

The physiologist E. S. Castle sponsored my initial work on adaptability when I was a senior undergraduate in the Harvard Biology Department. I recall that the framework was an independent research course which I called Models and Analogs in Biology. I wrote a primitive version of the book during the summer of 1964, just before moving to the Biophysics Program at Stanford. H. H. Pattee was one of the individuals to whom I showed this manuscript and I am greatly indebted to him for numerous invaluable discussions starting at that time on the compatibility of physics and biology. I was also stimulated by discussions about morphology and organization with A. K. Christensen and by discussion of the automaton paradigm with Michael Arbib. At Stanford I emphasized computational modeling of evolutionary processes. This work is not in evidence in this book, but it served as a laboratory for testing and developing a number of ideas which play an important role in adaptability theory. I returned to the problem of adaptability in a concentrated way during two postdoctoral periods spent at the Center for Theoretical Studies at the University of Miami. I thank Behram Kursunoglu for sponsoring these fellowships and for encouraging me to give a seminar series on Biological Organization in 1969. The specific form of the theory grew out of this series. The formalism itself developed largely while I was a postdoctoral scholar in the Mathematics Department at the University of California at Berkeley. I am indebted to Hans Bremermann for many extremely valuable discussions on mathematical biology during this period and subsequently.

The first segment of the book was written in 1973 while I was a faculty member at the Institute for Information Science at the University of Tübingen and during subsequent visits to Tübingen. I thank Werner Güttinger for encouraging me to give a course on adaptability theory and for discussions of stability theory. I am very specifically indebted on many points over many years to Mario Dal Cin (first at Miami and later in Tübingen) and to Otto Rössler, located in the Theoretical Chemistry Institute at Tübingen. Dal Cin, Rössler, and I ran an informal seminar on adaptation in 1974 which served very effectively to sharpen the problems and to compare algorithmic, statistical, and dynamical approaches.

Segments were written while I was a faculty member in the Department of Biology at the City College of New York and while a member of the Department of Computer and Communication Sciences at the University of Michigan. I acknowledge discussions with members of both faculties, including discussion of a number of interesting statistical problems at CCNY. For perceptive suggestions I thank students who attended my course on adaptability theory at Michigan as well as the interdepartmental community which attended our theoretical biology tea. I acknowledge discussions with John Holland on the process of adaptation.

Major portions of the book as well as major additions to earlier chapters have been completed since joining the Computer Science and Biological Sciences Departments at Wayne State University. I thank colleagues in both departments for valuable discussion and acknowledge the unusually innovative milieu in the Computer Science Department and its Intelligent Systems Laboratory. I acknowledge discussions with M. A. Rahimi on technology and adaptability, collaboration with Roberto Kampfner on evolutionary adaptation from an algorithmic point of view, and discussions with students involved with my course on natural information processing.

For discussions of adaptability theory which have been especially valuable I thank Robert Rosen, Bernard Patten, Harold Hastings, and R. M. Williams. Chapter 10 was written in the summer of 1979, while I was a visiting scholar at Cavendish Laboratory, Cambridge University. I acknowledge intensely stimulating conversations with B. D. Josephson during this visit and at other times on the connection between physics and life phenomena and on the deeper problems connected with adaptive intelligence.

I did major work on Chapter 12 during an interacademy exchange visit to the USSR and major work on Chapter 13 during an interacademy exchange visit to East Germany. I thank the U.S. National Academy of Sciences for sponsoring both visits. I acknowledge collaboration with Efim Liberman on the cyclic nucleotide system of intraneuronal information processing as well as extensive exchange of ideas about the underlying mechanisms of biological computing while at his laboratory at the Institute for Problems of Information Transmission, Moscow. I had the pleasure of many perceptive discussions with Michael Volkenstein on information and evolution. I completed small but crucial pieces during shorter visits to the Institute for Cybernetics in Baku, the Institute for Biophysics in Tashkent, the Institute for High Molecular Compounds in Leningrad, and the Institute for Biological and Chemical Physics in Tallinn. I acknowledge discussion on physics and evolution with Werner Ebeling of the Humboldt University and also discussions of the dynamics and stability of agroecosystems with members of the Institute for Cybernetics and Information Processes in East Berlin. I did bits and pieces at the Institute for Biophysics and the Carl Ludwig Institute for Physiology at the University of Leipzig, at the Institute for Microbiology and Experimental Therapeutics in Jena, and during a visit to the Information Science Institutes at the Dresden Technical University. But I acknowledge these institutes and the many individuals with whom I had the pleasure of interacting less for the writing done at the time than for an important contribution to work done after returning to Detroit.

In the text I acknowledge discussions or articles which may have been stimulating and which draw attention to related work. I have given numerous talks on adaptability theory since 1969 and I am afraid there is no way of crediting all the sharp questions which have sent me home to clarify this or that point. For reading and commenting on the manuscript I would like to thank M. Dal Cin, H. Hastings, R. Kampfner, K. Kirby, R. Rada, O. Rössler, F. E. Yates, and B. Ziegler.

My deepest acknowledgment is to my wife, Deborah, who deciphered and critically examined each page, who drafted the diagrams, and who has been my companion in the laboratory as elsewhere.

I have learned a great deal as a result of working on this book in a number of different lands and in a number of different disciplinary frameworks. There are intense problems bearing on the world and everywhere there is pressure to achieve laudable goals in agriculture, industry, and medicine. This is true in both the capitalist and socialist countries. It occurred to me that the single-mindedness with which these goals are being pursued is so great that our treasury of potentialities—our adaptability—is being cultivated much less than it ought to be. Science is unifying and, properly viewed, our problems could be unifying as well.

Michael Conrad

Detroit, Michigan

Contents

Important Symbols

Page numbers refer to page of first occurrence.

ω — Transition scheme of biota, p. 54

ω^* — Transition scheme of environment, p. 54

$\hat{\omega}$ — Transition scheme of biota as determined in most uncertain allowable environment (stressed transition scheme), p. 56

$\hat{\omega}^*$ — Transition scheme of most uncertain allowable environment, p. 56

$\underline{\hat{\omega}}$ — Stressed transition scheme in information transfer picture, that is, defined over fine (selectively equivalent) states, p. 176; cf. also p. 63

$\hat{\omega}_{ij}$ — Stressed transition scheme of compartment i at level j, p. 96

ω_{h0}^* — Transition scheme of region h of the environment, p. 93

$\hat{\hat{\omega}}_{ij}$ — Stressed partial transition scheme of compartment i at level j, that is, with state of compartment c_{ij} specified in terms of its subcompartments at the next lower level, p. 96

$H(\hat{\omega})$ — Potential behavioral uncertainty of biota, p. 56

$H(\hat{\omega}|\hat{\omega}^*)$ — Potential behavioral uncertainty of biota given behavior of environment (anticipation entropy), p. 56

$H(\hat{\omega}^*|\hat{\omega})$ — Potential behavioral uncertainty of environment given behavior of biota (indifference), p. 56

$H(\omega^*)$ — Actual uncertainty of environment, p. 54

$H(\hat{\omega}_{ij})$ — Potential behavioral uncertainty of compartment c_{ij}, p. 162; cf. also p. 93

$H_e(\hat{\omega}_{ij})$ Effective entropy of c_{ij} (modifiability of c_{ij} + dependence terms), p. 96; cf. also p. 94

$H_e(\hat{\omega}_{ij}|\Pi\hat{\omega}_{h0}^*)$ Effective anticipation entropy of c_{ij} ($\Pi\hat{\omega}_{h0}^* = \omega^*$), p. 96

β^u State u of biota, p. 51

ε^v State v of environment, p. 51

E Energy, p. 12

S Entropy, p. 13

ε Efficiency, p. 144

1

The Ecosystem Process

By an ecosystem I mean a living system which is capable of sustaining itself indefinitely and without help from any other such system. Clearly no single molecule has this property, nor does any single cell or organism, or even any isolated aggregate of organisms, for this aggregate could not sustain itself without a proper environment. Moreover, the aggregate itself must in general be much more complicated than any single organism, or at least any single present-day organism, for in general it takes many interacting organisms to maintain a suitable environment.

In this chapter I would like to describe the minimum components and processes required for a "piece of nature" to have this completeness property. The best way to do this is start off with some actual piece of nature itself, or at least keep some simple, natural system in mind. The most pleasant, perhaps, is a pond, or even better, pond water in a flask.

1.1. POND WATER IN A FLASK

Our flask and its most obvious components are illustrated in Figure 1.1. The first point about the flask is that it is an open system, with energy (usually sunlight) coming from the outside. The second point is that it consists of a living part and a nonliving part. The living part is called the *biota* (or biotic community) and the nonliving part the *environment*.

The environment includes all the components in the system which are not located within the boundaries of the biota. Naturally the environment can assume various states—for example, various temperatures, salinities, pH values, states of mechanical agitation, and so forth. These states may change in a predictable way—for example, with the natural periodicity of the year or day. They may also change in an unpredictable way. In this case we will say that the environment is uncertain. It is of fundamental importance that the environment has geometry (since the biota and environment are of course distributed in real space).

1

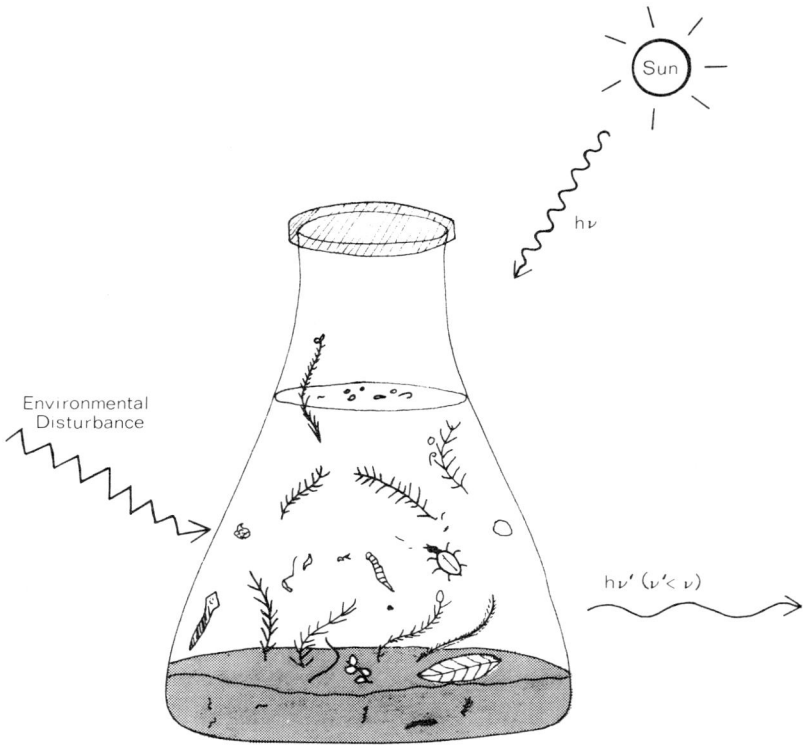

FIGURE 1.1. Flask ecosystem. Illustrated are phases and interfaces of the environment (solid, liquid, vapor), flow of energy through the system (from light to heat), and components of the biotic community. Parameters of the environment (temperature, mechanical forces, chemical concentrations) are determined partly by the outer environment (outside the flask) and partly by its internal dynamics. Important is the fact that the behavior of the environment is in general uncertain, so that the community is subject to unexpected disturbances. Such flask ecosystems exhibit periods of change and periods of stability (i.e., successional processes), but so far as is known they can sustain themselves indefinitely.

The question naturally arises as to what we mean by the boundaries of the biota. It is very hard to answer this question in a precise way, or at least to make a simple, general statement with no exceptions. However, the biota consists of individual organisms which are capable of reproducing themselves either individually or in pairs. Each of these organisms is surrounded by a boundary (for example, a membrane) which separates it from its environment. This boundary is a real physical object, indeed a thermodynamic necessity, since the processes of self-reproduction involve the creation of new order out of the environment and are therefore entropy-reducing. This means that each self-reproducing system must export heat to its environment. The boundary is the geometry which distinguishes the heat bath from the regions which export heat to it.

The flask is of course a physical system and must therefore obey the first and second laws of thermodynamics, but with the provision that it is an open system. Thus, we expect that the entropy of the whole world (including the space around the flask itself) will increase, despite the local entropy-decreasing processes associated with self-reproduction. Also, we expect that the amount of matter in the system will be conserved in the sense that the number of each kind of atom in the system will stay the same, aside from those leaving by evaporation and those entering from the external environment. Thus it is clear that we must expect a flow of energy through the system, with the input of relatively high-quality energy from the outside and the export of heat. This is necessary to support the self-reproduction process. Also, we expect that this self-reproduction process will be concomitant to a flow of matter in the system, and more particularly, to a cycle of matter.

The flow of energy and matter conforms to certain ubiquitous organizational principles (Figure 1.2). Ordinarily the flask must contain certain

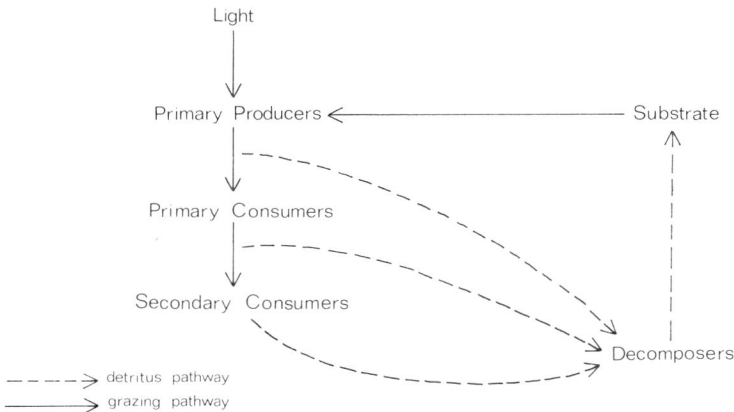

FIGURE 1.2. Schematic matter cycles. Actually the cycle is embodied in an intricate web of matter flow among the various species in the community and not all species fit uniquely into one of the categories (e.g., primary consumers, decomposers). Also, the detritus pathway itself has a complex structure, with different species mediating different steps of decomposition and with intermediate-sized "primary decomposers" linking the two pathways.

components (called primary producers) which convert incoming light to chemical energy. These are the photosynthetic organisms, including the pigmented plants and microorganisms. The chemical energy stored by the primary producers is in general used by other organisms (primary and secondary consumers), forming a web of matter and energy flow. However, this part of the web, which is sometimes called the *grazing* pathway, is not complete. This is because the primary producers would soon run out of chemical elements to synthesize into low-entropy, high-energy forms if there were not some pathway for cycling materials back to their original form. The organisms which do this are called decomposers and the pathways of decomposition the *detritus* pathway. The detritus pathway is an essential, indeed inevitable, part of any ecosystem.

The necessity for cycling arises from the fact that organisms not only are born but also die or decompose, in short, that they have a life cycle. Thus they constantly have either to reproduce or at least to repair themselves, which inevitably involves a flow of matter. The existence of such a life cycle obviously means that the number of organisms increases if the birth rate is greater than the death rate, and conversely that it decreases if the death rate is greater than the birth rate.

The biota has another major characteristic. This is its hierarchical (or, more precisely, compartmental) structure (Figure 1.3). The most important unit of organization is the organism (since it is the birth and death rates of organisms which are important from the standpoint of changes in the composition of the biota). Above the organism level the most obvious units of organization are the population (organisms of a given type or reproductive associations), demes (or spatially restricted subpopulations), social organizations (symbioses, societies), and trophic levels (primary producer, primary or secondary consumer, detritus pathway organism). Below the organism level the most important unit is the cell (since this is the minimal unit of self-reproduction), subcellular organelles (e.g., membranes, mitochondria, the genome), and finally atoms or molecules, which are either bound in or flow through these higher level units.

It is convenient to divide the hierarchy below the organism level into two general categories: genome and phenome. The phenome includes everything within the boundaries of the organism, excluding the genome. The genome, of course, is what principally enables the organism to transmit traits to the offspring. The organism itself may be a single cell or a multicell system, all of whose cells originate from a single cell (and which therefore have the same genome) and whose offspring start from a single cell derived from it. By a population I mean a group of organisms which either have very similar genomes (because they derive from one individual) or which exchange genetic material. This is the same as a species, except that we are thinking only of those members of the species which are within the flask.

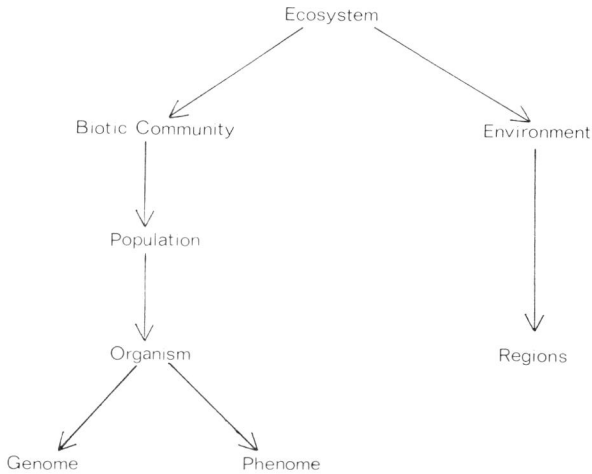

(a)

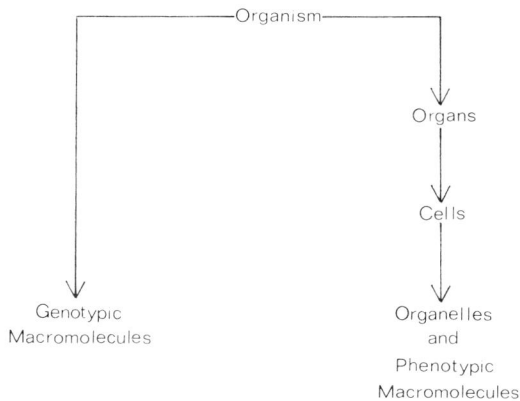

(b)

FIGURE 1.3. Level structure of the ecosystem. Illustrated in (a) is a simplified and schematic decomposition of the environment into regions of space and a decomposition of the community into populations, organisms, and suborganism levels of organization. More levels could be included in a more detailed description (e.g., social organization, families, trophic associations or units, organ systems). Some sublevels of the organism are pictured in (b), with the right-hand side (organs, cells, ...) representing sublevels of the phenome in (a). The organ level should actually be thought of as including extracellular compartments and similarly the organelle level as including small molecules as well as proteins and other macromolecules. The phenotype and genotype are linked through genetically directed protein synthesis. There are many units at each level and a branching, compartmental structure would therefore give a more accurate impression (see Figure 1.4). In some cases levels collapse into one another, e.g., organism and cell may be identical.

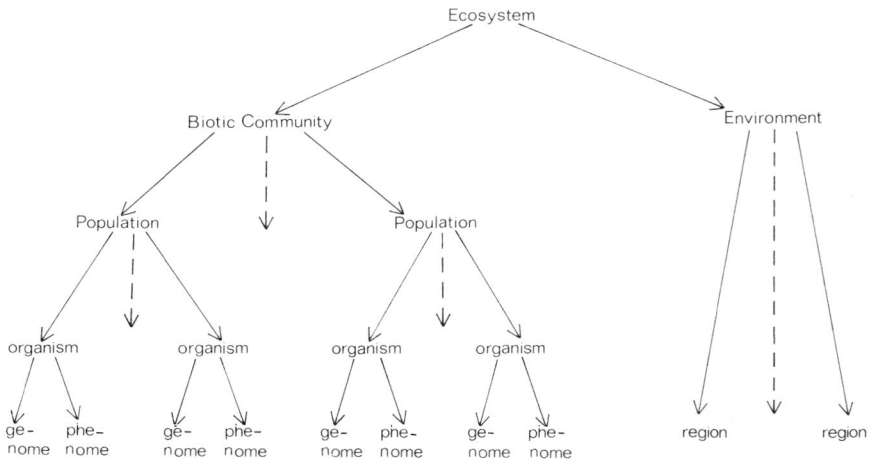

FIGURE 1.4. Schematic compartmental structure of an ecosystem. Levels of organization correspond to the levels in Figure 1.3(a) and regions to local environments. Actually the biotic part of the ecosystem consists of many biotic communities and the global environment of many local environments.

The organization is compartmental as well as hierarchical because what we really have are boxes nested within boxes (Figure 1.4). Each such box is defined by a boundary. The boundary may be a boundary to the flow of energy and matter, as in the case of the cell or organism. Or it may be a boundary to information flow, as in the case of gene flow within the members of a species, in which case it is much less concrete.

In general it is very difficult to give simple, precise definitions of these boundaries—nature rarely sees fit not to try out some important exceptions. Nevertheless, the compartmental aspect of the biota is an outstanding and obvious feature of its organization.

1.2. THE UNCERTAIN ECOSYSTEM

So far we have talked about energy and matter flow, self-reproduction, boundaries, and compartmental structure. All these components and processes are permeated by a single feature which is so obvious that it is sometimes forgotten.

This is the ubiquitous variability of biological materials. We have already described the environment as more or less uncertain. But this uncertainty is even more striking in the case of the biota. No two organisms are quite the same, no two cells quite the same. The differences may be due to the mutability of the genome, or they may be due to random processes in development, or they may be due to different influences from the environment.

The uncertainty may manifest itself at each level of organization. At the level of the genome we speak of gene pool diversity; at the level of the cell or organism we speak of developmental, physiological, and behavioral plasticity. At the level of the population we speak of fluctuation in numbers and culturability (if the increase in numbers occurs in the presence of a favorable environment). At the level of the ecosystem as a whole there is variability also, in this case variability which at least potentially manifests itself in the routability of matter and energy flow.

Does this ubiquitous variability have any significance? The language which biologists use to describe it practically answers the question. Gene pool diversity; developmental, physiological, and behavioral plasticity; and routability of matter and energy flow are all ways of dissipating the unexpected disturbances of the environment or the unavoidable disturbances arising from processes within the system.

I will call this ability to cope with the unexpected disturbances of the environment *adaptability*, and the ability to cope with internal noise *reliability*. In order to function and persist a biological system must clearly be both adaptable and reliable. Another way of saying this is: the variability of biological matter is the *sine qua non* for its ability to persist despite internal noise and external perturbation.

1.3. BALANCE

Another major fact about our flask is that it will keep changing (by definition) until it reaches a stable state. If it reaches such a state we will say that it is a balanced ecosystem.

The general requirement for balance is either that the activities of the system have no net effect on the state of the environment or that they restore this state periodically. More particularly, this means that the atomic and molecular components withdrawn from the environment for self-reproduction and repair must be cycled back and that the ratio of births to deaths, for each type of organism, must be one. The problem of balance is thus the problem of ensuring that there are enough species, appropriately interconnected, to cycle materials in such a way that the environment is restored and of ensuring that the properties of the species and their interlinkages are such that the number of organisms in each species is regulated. Furthermore, the system must be able to do this despite periodic or unexpected changes in the environment, for example, in the input of energy or in the conditions affecting the use of energy.

The problem of balance is thus not only a problem of regulating the environment and the number of organisms (cycling and birth control), but also a problem of reliability (maintaining this regulation despite noise) and adaptability (maintaining it despite unexpected disturbances from the environment). In general, each different type of environment—still water, running water, oceanic, desert, tropical, arctic, and so forth—imposes different limiting factors and therefore stresses the organisms in a different way. Thus the particular properties which enable the system to balance under these different circumstances may be quite different; but in each case the system will have to be both reliable and adaptable.

The sequential change of the ecosystem to a balanced state (or climax community) is called ecological succession. If new species arise during this sequential change it is called evolution. Thus ecological succession is a model for evolution, except that in succession the origin or modification of species (as opposed to their invasion from some other area) does not play a role.

1.4. THE THEORY OF EVOLUTION

The fundamental properties and processes which we have described so far either arise or develop in the course of historical evolution. The fundamental conceptual scheme of biology is the theory of how this happens. This is the theory of evolution by variation and natural selection.

According to the theory of evolution by variation and natural selection the historical development of living matter is based on four fundamental properties:

1. *Self-reproduction.* Evolutionary systems contain units which are capable of producing offspring whose traits are correlated to their own.

2. *Variability.* Errors in self-reproduction give rise to new properties or new combinations of properties, that is, parent and offspring are not in general interchangeable.

3. *Strong principle of inheritance.* Modifications in the parent induced by the environment are not communicated to the offspring.

4. *Differential growth rates.* Whether or not a given unit reproduces depends on its environment and on its properties, that is, the relative growth rates of different populations are modified by the environment.

The evolutionary system can perhaps be defined as a system with interesting behavior, one capable of undergoing significant, sometimes even dramatic changes in behavior, but at the same time having an indefinite half-life. In short, an evolutionary system is an ecosystem and evolution is the ecosystem process.

Properties (1)–(4) just mean that the ecosystem contains organisms (or parts of organisms) which are capable of reproducing themselves with variations; but which ones actually reproduce is determined by the environment. The effect of the environment, which is usually called natural selection, ultimately derives from its finiteness, that is, from the fact that the inevitable scarcity of factors essential for self-reproduction inevitably distorts the relative rates of reproduction. The properties which arise because of this distortion (that is, because of differential growth rates of populations) are sometimes called *adaptations*.

By a reversal of language we sometimes say that adaptations enable biological systems to reproduce preferentially. Moreover, we sometimes think (or deceive ourselves into thinking) that we can characterize the reason for this. This characterization (to the extent that it is feasible) is a functional description of the adaptation.

The above reversal of language is in a certain sense circular. First we say that differential growth rates give rise to adaptations and then we say that these adaptations give rise to the differential growth rates. This circularity, however, reflects a real circularity in the phenomena; indeed, none other than the life circle of self-reproduction. It is this circularity which ensures that the offspring of organisms most likely to reproduce inherit just those properties which again make successful reproduction most likely, and which therefore confers a significance on these properties which has no counterpart in the nonliving world. This (functional) significance is sometimes expressed, figuratively, by saying that the properties of the organism fit to the properties of the environment. In this terminology the organisms which actually reproduce are called "fit."

The above circularity has suggested to many authors that the theory of evolution is itself circular, or, as it is usually put, tautological. This is not so. The theory of evolution is just the theory described in (1)–(4), namely the theory that the historical development of living matter arises from the environment's classification of mutable, self-reproducing systems into those which actually reproduce and those which do not. However, it does not predict, or claim to predict, which organisms will actually reproduce. The theory is thus a theory about the mechanisms of historical development, not a theory about the actual course of evolution itself. It answers the question, how do the properties of biological matter arise, but not the question, which properties will arise? In order to answer this question we need a separate theory, one which makes some statements about which types of biological organization are likely to persist and which are not. In short, we must make a theory of fitness.

This theory of fitness does not have to answer every question about the functional value of this or that adaptation. In fact, this is quite impossible because particular adaptations are so particular and specific. However, this is not true for adaptability. This is because adaptability is adaptation to the uncertainty of the environment, therefore adaptation to something which can in fact be characterized in a very general way, independent of the specific features of different environments. Moreover, adaptability and also reliability (or adaptations for coping with noise) are just the adaptations which are crucial from the standpoint of determining the directions of evolution, for not only are they what determines whether some particular biological organization has the "right to persist," but they also encompass the genetic search strategies which determine the probability with which these different organizations come into existence.

My purpose in this chapter has been to provide a definite image of the global structure of the systems with which we will be dealing. I shall, in the following chapters, classify and characterize the types of variability which can be exhibited by such systems, starting with the fundamental physical nature of these variabilities and proceeding on to describe their functional significance for adaptability at progressively higher levels of organization.

2

The Laws of Dissipation

We begin at the most general level, with the basic principles of thermodynamics. It is the appropriate place to begin, even though the problem of adaptability is not a problem for thermodynamic technique. To cope with a disturbance means to be capable of forgetting it, and to be capable of forgetting it means in some sense to be capable of dissipating it away. What I want to do in this chapter is to develop the thermodynamic undercoat of this thought. In the next chapter I continue the discussion, but in the special context of biological and ecological systems.

There is one other purpose in introducing thermodynamic ideas at this point. The flow of energy is a crucial aspect of ecosystem dynamics and mechanisms of adaptability associated with disturbances to the flow of energy are therefore of critical importance. Thus the material in this chapter will also be useful when we are in a position to deal with this problem (in Chapters 7 and 12).

2.1. ENERGY AND ENTROPY TRANSFORMATIONS IN OPEN SYSTEMS

The planet's entire ecosystem is essentially closed—which means that it is open to energy (primarily light from the sun), but closed to chemical matter. The flask ecosystem is completely closed so long as it is tightly sealed. If there is an opening for gas exchange (i.e., if we are dealing with part of a global ecosystem), then we have an open system. To describe the energy processes of the ecosystem it is therefore necessary to write down and describe the basic laws of thermodynamics as they apply to open systems.

According to the first law of thermodynamics energy is conserved. For open systems this may be written (following essentially the notation of

Prigogine, 1961):

$$\frac{dE}{dt} = \frac{d_eQ}{dt} + \frac{d_eW}{dt} + \frac{d_eR}{dt} + \sum_{i=1}^{n} e_i \frac{d_e n_i}{dt} \qquad (2.1)$$

where E is the energy of the system, d_eQ is the heat flowing across its boundaries, d_eW is the work performed at the boundaries, d_eR is the light energy entering the system, e_i is the molar energy of component i, $d_e n_i$ is the change in the number of moles of component i due to flow across the boundaries (altogether there are n components), and t is the time. In the case of constant temperature and pressure the molar free energy is given by

$$e = (\partial E / \partial n_i)_{Tpn_i'} \qquad (2.2)$$

where p is pressure, T is temperature, and n_i' are the mole numbers of all components other than n_i. Actually the assumption of constant T and p is reasonable in most situations of interest and sufficient for our purposes. We also assume, for simplicity, that $d_eW = -p\,dV$, where V is the volume.

The energy, E, is a function of the state of the system as specified by relevant state variables, e.g., temperature, pressure, and mole numbers. Thus the change in energy is represented by the exact differential, dE/dt. The meaning of the terms on the right-hand side of the equation is quite different, however, for these all represent flows across the boundaries of the system and therefore cannot be expressed as the differentials of any function of the state of the system. In other words, d_eQ/dt is really only a symbol for the heat exchange with the environment during an infinitesimal amount of time, d_eW/dt a symbol for the work performed during an infinitesimal amount of time, and so on. Such quantities are sometimes called inexact differentials, for they cannot be integrated to give a state variable of the system. For example, work is not a state variable, for work performed on the system might be stored in potential energy coordinates, converted to light or chemical potential energy, or converted to heat. Moreover, the particular conversions are a "function" of the path which the system follows between its initial and final state and not the states themselves.

According to the second law of thermodynamics the possible interconversions are subject to certain restrictions. Gradients (asymmetries) of temperature, concentration, and mechanical forces (such as pressure) tend to disappear in isolated systems, but the reverse never occurs. All these tendencies can be summed up in one statement: systems in nature undergo spontaneous changes only to states of greater macroscopic symmetry, where entropy serves as the measure of symmetry.

The entropy change itself can be divided into a reversible part (associated with flows across the boundaries) and an irreversible part (associated with flows within the system). Thus we may write

$$T\frac{dS}{dt} = T\frac{d_e S}{dt} + \phi \qquad (2.3)$$

where S is the entropy, $d_e S$ is the entropy change due to flow across the boundaries, and $\phi \geqslant 0$ is the product of the irreversible contribution to the entropy production per unit time and the temperature. We call ϕ the dissipation and say that the process is reversible only if $\phi = 0$.

In closed systems the reversible contribution to the entropy change is due solely to the flow of heat. In open systems it is associated with the flow of matter as well. Thus, for open systems, we write

$$T\frac{d_e S}{dt} = \frac{d_e Q}{dt} + T\sum_{i=1}^{n} s_i \frac{d_e n_i}{dt} \qquad (2.4)$$

where $s_i = (dS/dn_i)_{Tpn_i'}$ is the specific entropy of component i (and the prime indicates all components other than n_i). The specific entropy is roughly the entropy per mole.

Combining the first and second laws,

$$\frac{dE}{dt} - T\frac{dS}{dt} = \frac{d_e W}{dt} + \frac{d_e R}{dt} + \sum_{i=1}^{n} e_i \frac{d_e n_i}{dt} - T\sum_{i=1}^{n} s_i \frac{d_e n_i}{dt} - \phi \qquad (2.5)$$

where again the terms on the left are exact differentials and the terms on the right are inexact. Recalling the definition of the Helmholtz free energy ($F = E - TS$), we may rewrite equation (2.5):

$$\frac{dF}{dt} = \frac{d_e W}{dt} + \frac{d_e R}{dt} + \sum_{i=1}^{n} e_i \frac{d_e n_i}{dt} - T\sum_{i=1}^{n} s_i \frac{d_e n_i}{dt} - \phi \qquad (2.6)$$

where we assume that the temperature is constant. Thus it is clear that the free energy inevitably decreases unless ϕ (which is always positive in any real process) is compensated by work performed on the system, absorption of light or high-grade chemical energy, or the export of high-entropy chemicals. In biological systems, of course, the important process is the absorption and dissipation of light and high-grade chemical energy, with the concomitant export of heat and low-grade chemical energy.

If the system is to maintain a steady state, all state variables must by definition be constant. Hence the condition for the steady state is $dF/dt = 0$

(or $dE/dt = dS/dt = 0$). In this case

$$\phi = \frac{d_e W}{dt} + \frac{d_e R}{dt} + \sum_{i=1}^{n} e_i \frac{d_e n_i}{dt} - T \sum_{i=1}^{n} s_i \frac{d_e n_i}{dt} \qquad (2.7)$$

which just means that the dissipation must be exactly compensated by the flow of matter and energy across the system's boundaries.

We might also consider what happens in the case of a closed system, e.g., the world ecosystem or a sealed-off flask system. In this case $d_e n_i/dt = 0$, so our equation reduces to

$$\frac{dF}{dt} = \frac{d_e W}{dt} + \frac{d_e R}{dt} - \phi \qquad (2.8)$$

which means that in the steady state the dissipation and work performed must be compensated by the radiation input. Under these conditions the dissipation can be interpreted as the heat produced within the system. This can be seen by combining equations (2.3) and (2.4):

$$T\frac{dS}{dt} = \frac{d_e Q}{dt} + T \sum_{i=1}^{n} s_i \frac{d_e n_i}{dt} + \phi \qquad (2.9)$$

For closed systems at the steady state

$$\phi = - d_e Q/dt \qquad (2.10)$$

from which it follows that the dissipation must be exactly compensated by the heat exported to the environment.

2.2. THE IMPORTANCE OF DISSIPATION

Living systems certainly possess some definite energy, and, with only mild assumptions, are characterizable by some entropy as well. This is true whether we are dealing with a cell, an organism, a population, or an entire ecosystem. Such systems are heterogeneous and extremely changeable. For example, they may be in a stable, complex motion rather than a simple steady state. If the temporal or spatial rate of change becomes very fast it may become difficult to describe them in terms of normal state variables. Nevertheless, the transition between states is often so fast that a thermodynamic description of the initial and final states is sufficient. Also it is usually possible to give a thermodynamic description of sufficiently macroscopic local parts of the system. Even if these assumptions become suspect, we can still expect the relationship between energy and entropy which emerges from the formalism to retain in-principle validity and therefore its usefulness in the arguments to be developed.

The key concept is dissipation. This is the criterion, indeed the measure, of irreversibilty, in other words, the measure of the system's inability to run backwards. This inability is fundamental for life and is the basis of adaptability, for a system which could run backwards would have no possibility for holding on to whatever unusual characteristics distinguish it from its environment (e.g., high energy content, low entropy, details of organization which form the basis of biological function). But even more important for adaptability, it would have no possibility for forgetting the insults visited upon it by this environment.

Just imagine how intrinsically incoherent the concept of a reversible man would be. Or, more simply, imagine a mechanical system such as an undamped harmonic oscillator and suppose that we perturb the system by imparting to it an extra bit of energy. The system will certainly never forget this perturbation, for its energy (kinetic plus potential) is constant. If a motion picture of such a system were run backwards it would make absolutely no difference to the viewer, excluding the occurrence of the perturbation. In other words, it has no history of its own.

Now imagine that we add some friction to the oscillator. The end state will be the same (no motion) regardless of the initial conditions or any subsequent perturbations. Indeed, in this case all would be forgotten and a reversed motion picture would certainly look both peculiar and surprising, for it now would show an infinite number of possible diverging past histories.

The frictional oscillator (for example, a damped pendulum) is clearly dissipative, for the rubbing of parts is concomitant to a rise in temperature and subsequent equilibrating flow of heat to the environment, with a net increase in the entropy of the world. An even simpler example is the compression and subsequent free expansion of a gas. The final state of the gas has complete amnesia for the particular characteristics of the compression. In each case the automatic tendency for macroscopic symmetry to increase gives a preferred direction to the behavior of natural systems in time. It is a preference which biological systems must combat to maintain their high order and complexity, or, more precisely, which they must utilize to drive the local production of such order and complexity. But from our point of view this preference plays an even more important role, for it is also the preference which ultimately dissipates the effects of the unexpected.

We can restate the above remarks in terms of our thermodynamic equations. Suppose that we split the dissipation into two parts

$$\phi = \phi_{\text{big}} + \phi_{\text{small}} \qquad (2.11)$$

where ϕ_{big}, the large contribution to the dissipation, is associated with the basic processes of metabolism, growth, maintenance, irritability, movement, and ultimately reproduction, and would be so associated even in a system adapted to an environment free of disturbance; whereas ϕ_{small}, the small contribution to the dissipation, is associated with the fading away of the effects of disturbance. Substituting this splitting into equation (2.5),

$$\frac{dE}{dt} - T\frac{dS}{dt} = \frac{d_eW}{dt} + \frac{d_eR}{dt} + \sum_{i=1}^{n} e_i\frac{d_en_i}{dt} - T\sum_{i=1}^{n} s_i\frac{d_en_i}{dt} - \phi_{big} - \phi_{small}$$

(2.12)

Clearly the main thermodynamic requirement for maintaining a steady state is

$$\phi_{big} \approx \frac{d_eW}{dt} + \frac{d_eR}{dt} + \sum_{i=1}^{n} e_i\frac{d_en_i}{dt} - T\sum_{i=1}^{n} s_i\frac{d_en_i}{dt}$$

(2.13)

But the ability of the system to draw energy across its boundaries and use it to maintain an improbable and complex steady state is an unlikely ability, one that must be protected from environmental disturbance by the higher-order ability associated with ϕ_{small}. Thus while ϕ_{small} might well be insignificant in the purely energetic sense, it is extremely significant in the functional sense. The above splitting of the dissipation is of course entirely notational, since processes which contribute to the one may also contribute to the other.

2.3. STATISTICAL SIGNIFICANCE OF DISSIPATION

The preference for increasing macroscopic symmetry reflects a deeper, microscopic symmetry. We say that a thermodynamic system is symmetrical to the extent that an observer would have difficulty detecting the interchange of small parts of the system. The parts can be made smaller and smaller, until we are finally permuting the basic elements of the system. The most probable macroscopic state is clearly the one which is most symmetrical from this microscopic point of view, i.e., amenable to the greatest number of rearrangements equivalent from the macroscopic point of view.

For example, in the case of our freely expanding gas, it is clear that there are many more microscopic configurations associated with the expanded state than with the contracted state, hence many more transitions which lead to the latter from the former than conversely. From the macroscopic point of view it appears as if some force impels the system to fall to its final state of maximum symmetry. Indeed, the free energy (usually the Gibbs free energy $G = E + PV - TS$) is called the thermodynamic potential. But of course the fall is strictly a matter of the probabilities favoring the macroscopic state whose parameters are homogeneous for all macroscopic purposes.

The link between macroscopic and microscopic in ordinary thermodynamic systems makes it possible to set up microscopic models for the entropy. The most famous and indeed virtually the only model seriously considered is the Boltzmann entropy,

$$S = -k \sum_{i=1}^{n} f_i \log f_i \qquad (2.14)$$

where f_i is the probability of microscopic configuration i, $k = 1.38 \times 10^{-16}$ ergs/degree is Boltzmann's constant, and n is the number of possible states. If we allow that each configuration is equally probable we can rewrite this as

$$S = k \log \omega \qquad (2.15)$$

where $\omega = 1/f_i$ is the number of possible configurations. This is the usual form of the equation.

Equations (2.14) and (2.15) are measures of the spread of microscopic states. Roughly speaking, if this spread increases, the macroscopic state with which they are associated becomes more symmetrical in the sense described above, or, to say the same thing in the other direction, as the macroscopic entropy increases the ensemble of associated microscopic states also increases. This is one reason why (2.14) and (2.15) provide good microscopic models of macroscopic entropy. Another is that, like macroscopic entropy, they are additive (since logarithms are additive).

The statistical definition of entropy makes it possible to describe dissipation in statistical terms. To make this explicit consider an isolated system, i.e., a system in which the entropy change is due solely to dissipation. In this case

$$\phi = kT \frac{d}{dt} \log \omega = \frac{kT}{\omega} \frac{d\omega}{dt} \qquad (2.16)$$

which is just a combination of equations (2.3) and (2.15) (assuming T constant and ω a function of t). Thus the dissipation is in general associated with an increasing number of possible microstates, although it could also be associated with an equalizing of the probabilities among these states. This is true whether we are dealing with ϕ_{big} or ϕ_{small}. Also note that dissipation is inversely proportional to the number of microstates. For ϕ_{big} this is of no particular importance—since ϕ_{big} is determined by equation (2.13) in any case. It is important for ϕ_{small}, however, for it means that by maintaining in some form a large ensemble of microstates, biological systems can reduce the dissipation concomitant to forgetting perturbation.

The basic property of ϕ is that it is positive in any real process; thus $d\omega/dt$ should be also. The difficulty is that the fundamental equations of motion are time-reversible. If we have a certain initial uncertainty about the detailed state of the system, the uncertainty at any later time will be precisely the same. In other words, the entropy is a conserved quantity, which it should not be.

2.4. BREAKING THE CONSERVATION LAW

To break the conservation of entropy it is necessary somehow to pass from a reversible to an irreversible description. The only way to do this is to discard (in a justifiable way) certain details about the system's behavior.

The first consideration comes from the way in which the statistical entropy is calculated. The procedure is ordinarily to specify the macroscopic parameters of the system and then figure out how many microscopic configurations are compatible with these parameters. In this case the f_i are *a priori* probabilities. Moreover, they are usually postulated to be equal [cf. equation (15)]. This is important, for it is the basis of the fact (discussed in the previous section) that the macroscopically most symmetric state is the one most probable from the microscopic point of view. The reason is that a macroscopically asymmetrical system is consistent with only those microscopic states which themselves have some asymmetry property, by which we mean that they have some basic physical property not invariant under as many interchanges of the particles as possible. Since these states are necessarily in the minority, the increase in macroscopic entropy can be regarded as a direct consequence of the assumption of equal *a priori* probabilities, for if the system starts out in one of the unlikely microstates it will inevitably visit and remain practically forever in the set of likely ones.

The postulate of equal probabilities seems to be quite well justified in practice for the type of systems ordinarily dealt with in thermodynamics, viz. systems without peculiar constraints. Actually, great efforts have gone into the attempt to justify the postulate on a more fundamental basis.

Roughly speaking, the idea is that the system should visit every accessible point in its phase space and that the process of making such visits so transforms the space into itself that the probabilities become homogeneous. This is the so-called ergodic hypothesis. It does not seem to be rigorously true, at least in three dimensions; nevertheless it can probably be regarded as a reasonable explanation for the approximate validity of the assumption of equal *a priori* probabilities under conditions of no unusual constraint.

This still does not answer the initial question, for even if the evenness of the probabilities ensures that the system inevitably falls into the largest equivalence class of microstates, this still does not imply that there is any increase in the uncertainty as to which one of the states is occupied at any given time. One idea, first developed by Boltzmann, is that any initial ignorance about the positions and momenta of the particles in the system is spread by collisions among these particles, so that one would eventually lose all knowledge of the details other than what could be inferred from the macroscopic state. Thus the information which is discarded is information which on the one hand is of no macroscopic significance and on the other is lost because of the impossibility of following the microscopic dynamics.

There is a subtle dichotomy here, one that has been discussed by many authors. Is the loss of information (hence the increase in entropy) an objective or subjective phenomenon? Certainly the second law of thermodynamics describes an objective phenomenon, viz. increase in macroscopic symmetry. From the microscopic point of view, however, it sometimes seems that this just corresponds to our loss of detailed information and that another observer, who did know the detailed state, would see no increase in entropy. Where we see disorder he sees order.

It is perhaps easiest to appreciate this difficulty in a special case. Suppose that instead of starting with one of the unlikely microstates we start with one of the likely ones. We would still lose information about the details, but the loss would not in this case be associated with any increase in the macroscopic entropy, from which it follows that this quantity is really quite indifferent to our knowledge of the system.

The source of the difficulty is that I have ignored an essential point, namely that the acquisition of information about the system is a physical process with physical consequences. An observer able to measure the ensemble of microscopic states directly must be able to somehow perceive the microscopic structure of the system. In order to make such observations, however, he must add energy to the system, thereby increasing its entropy. Thus, if we start with one of the likely states and attempt to determine which one in detail we will certainly increase the entropy of the world and therefore produce a situation in which there are even more likely states.

Roughly this is because the amount of energy imparted to the system must be at least enough for the observer to distinguish his signals from thermal noise.

The connection between entropy and information was first discussed by Maxwell, who realized that a device capable of perceiving the microstates of a system directly could select high-energy particles and therefore violate the second law of thermodynamics (by generating a temperature gradient). Much later Szilard showed that such a device (demon of Maxwell) would itself have to suffer a more than compensating entropy increase. From the present standpoint what is important is that as the number of possible microstates increases the number of measurements required to specify which state the system is actually in increases, so that the actual entropy increase concomitant to the acquisition of information will be greater as the system falls into more and more likely sets of states.

2.5. FURTHER REMARKS ON THE ORIGIN OF IRREVERSIBILITY

So far I have not said what I mean by *information*. Actually it is a fairly easy concept to define when bound to the uncertainty as to which microstate of its ensemble a system is in. In this case we can write

$$I_b = S_{\text{initial}} - S_{\text{final}} \tag{2.17}$$

where S_{initial} and S_{final} serve as the measures of initial and final uncertainty and I_b is the bound information, so called because it represents the reduction in the uncertainty as to which of its possible microstates the system is in (cf. Brillouin, 1962). What Boltzmann showed, in his famous H-theorem, is that if S_{initial} is zero, I_b always tends to decrease (where he used $H = kI_b$). What Szilard showed was that if I_b is positive (e.g., if the final state has zero entropy) this must be paid for by an overall entropy increase, i.e., that the laboratory must provide both a source of high-grade energy and a sink for low-grade energy.

Later (cf. Section 3.5) we will see that information can be defined over quite different sets of possibilities, with concomitant loss of the physical role so far attributed to it. Moreover, the well-defined use indicated above does not apply so readily when discussing the organization of biological systems, for in this case it is difficult to specify a natural initial set of possibilities.

Another point which I have not mentioned is the role of measurement *per se*. Needless to say, measurements must be performed on the system to get information. The measurement process, however, has the fundamental peculiarity that it is a distinct process, not derivable from the equations of motion presumed to govern the measuring system, at least not derivable without added assumptions. Moreover, it is a probabilistic, inherently irreversible process, always concomitant to a bona fide increase in the number of microstates and therefore to an increase in entropy from the purely microscopic point of view. In other words, the very attempt to determine the initial conditions of a system results in an increase in its entropy and therefore loss of information.

The reason for this is that what is governed by the reversible equations of motion is no longer a classical state but what might be thought of (in terms of the rough instrument of ordinary language) as a set of alternative states. To make a measurement means to specify some subset of these alternatives. Thus the measurement process (or, what is equivalent, any attempt to actually obtain a classical picture of the system) inevitably involves the reduction of these alternatives, which is clearly an inherently irreversible and therefore distinct process. Sometimes the reduction is called an acausal jump since it is probability-generating. To put matters into the usual terminology, call the set of alternative states a wave function. The entropy of a system is defined over its possible wave functions, i.e., it is a measure of the uncertainty as to which of its sets of alternative possible states the system is in. In this case we say that the system is a mixture of such sets. Just as in the classical case, the entropy of such a mixture is conserved; however, if a measurement is performed, one of the alternatives must be projected out, which means that the entropy of the mixture must increase. According to Szilard's analysis the very minimum of this increase is $k \ln 2$, where k enters because the entropy expended to acquire information must be above thermal noise and $\ln 2$ enters because we cannot accept more than a 50% chance of error. In short, measurement produces a bona fide increase in the number of microscopic states.

This relation between classical and microscopic descriptions is the substance of the famous complementarity principle of Bohr. Reality consists of a superposition of alternatives, but descriptions of reality useful for action cannot consist of such a superposition, hence the necessity for starting and ending with reduced descriptions. But this must not be taken to mean that nature's tendency to increase in macroscopic entropy is a result of measurement. This is still connected with the fact that the majority of microstates are associated with increased macroscopic entropy. What it does imply, however, is that any attempt to capitalize on the fundamental reversibility of microscopic laws is doomed to be annulled by the fundamental irreversibility of measurement, or by this irreversibility in combination with the purely thermal considerations of the previous section.

The relation among measurement, information, and entropy has been most thoroughly described by von Neumann (1955).* In addition to his formal analysis, von Neumann argued that the process of measurement introduces an essential element of consciousness into quantum mechanics. The argument, I believe, is faulty, but extremely instructive. Von Neumann imagined a measuring system, such as an individual biological system, and supposed that it could be symbolically represented by a pure case wave function. The problem arises when this system interacts as a measuring device with another system in a pure state, for the whole system will always remain in a pure state. However, measurement always converts a pure case into a mixture, or a mixture into a larger mixture. Von Neumann argued that this difficulty can only be circumvented by the intervention of consciousness, since otherwise there would be an infinite regress of measuring systems trying to register measurements which earlier systems tried to register, none of which would lead to an increase in entropy, therefore none of which would allow a measurement. Some famous paradoxes are based on this argument, such as the paradox of Wigner's friend (Wigner, 1967). An object and an observer in a box are described by pure state wave functions. The joint wave function should remain pure so far as an outside observer is concerned, even if the individual on the inside makes a measurement on the object, provided that the outside observer does not himself make a measurement. But if the inside observer is really making a measurement it is reducing the wave function of the object; therefore the entropy of the box should increase. The argument suggests that the act of measurement in the box introduces a new interaction into physics which amends the linearity of the equation of motion, otherwise one is left with the solipsistic conclusion that only the outside friend can make a bona fide observation. Wigner argued, in effect on the basis of von Neumann's regress argument, that this new interaction must involve what appears to be the distinctive feature of the measurement process, namely, the consciousness of the internal observer.

This interpretation is a possible one. But it does not follow from the argument. An alternative conclusion is that the concept of a pure case wave function or even an entropy-conserving mixture as providing a description of a system which can make measurements is inherently inconsistent. The

*See also London and Bauer (1939) and Jauch (1968) for alternative presentations of this analysis. Two discussions of interest from the standpoint of biology include that of Bohm (1951), who observed the analogy between the irreversibility necessary in the measuring instrument and the irreversibility inherent in biological systems, and that of Pattee (1973), who points to the analogy between the constraints occurring in measuring devices and those occurring in biological systems.

problem is the failure to recognize that the assumption of an entropy-conserving description is inconsistent with the most basic requirements which must be fulfilled in order for a system to be an evolutionary system. All measuring systems are either biological systems, as in Wigner's paradox, or systems which interface with biological systems. Any individual biological system which can perform a measurement is competing for mass and energy with other individuals in a larger system. Any complete description of that individual which could be predictive requires as a precondition a complete description of the whole system. On these grounds alone individual biological systems are inherently dissipative; therefore they can never be represented in terms of a pure state or mixture which can be manipulated according to the usual rules (Conrad, 1969). As a consequence the individual system need not be entropy-conserving and in fact cannot be entropy-conserving; otherwise it could never come into existence or go out of existence. Paradoxes such as that of von Neumann's regress and Wigner's friend arise from an in-principle unacceptable representation of individual biological systems in terms of a wave function which can really only be properly defined for a very much larger system. Any description of the entropy increase accompanying measurement must refer to this larger system, not to a regress of individuals within it which can only be terminated by escaping any individual. On the basis of this consideration it is at least a possibility that the interaction responsible for the entropy increase in measurement is connected to the constraints of the type which make evolution possible.

But the dissolution of the regress argument does not resolve the problem of measurement, nor does it explain the irreversibility concomitant to the reduction of the wave function in the process of measurement, nor does it explain the conscious experience which accompanies at least some measurements. It actually complicates the situation by referring the process of reducing the wave function to the whole ecosystem. Unfortunately the problem cannot be dismissed. It has critical significance for the fundamental conceptual scheme of biology. If statistical descriptions are only useful fictions for dealing with highly complex deterministic processes, it follows that the theory of evolution by (statistical) variation and natural selection must be no more than a useful fiction as well. This is the ergodic interpretation of the theory of evolution. According to this interpretation a more complete description of the evolution process exists in principle. Our inability to find this description is due to the impracticality of describing a macroscopic system in microscopic terms or to the inherently partial character of any description which refers to a single individual. A variant of the ergodic interpretation is that it is in principle impossible to set up and compute a more complete, microscopic model of evolution due to the physical limits of computation. In this interpretation the theory of evolution assumes a status which might be called quasifundamental.

It is interesting to consider alternative interpretations which involve measurement processes, excluding interpretations based on ergodic models of measurement.* One possibility is that the statistical processes of evolution assume a fundamental significance only because any attempt to obtain information which would allow a more detailed description is doomed to result in an even more statistical description due to measurement. Under these conditions nonfictional probabilities would be present, hence the requirement necessary for evolution theory to be fundamental would necessarily be met, but peculiarly without measurement processes entering into the dynamics of evolution itself. But it would then have to be admitted that an unsatisfying element of subjectivity would accrue to the theory. The observer would be playing the same probability-creating role as in measurement of nonliving systems, but would necessarily be placed outside of any living system since we are now dealing with living systems and have assumed that no measurement processes occur within the system observed. This possibility returns us to von Neumann's interpretation of measurement as registration in consciousness, though without being forced to it by any compelling argument. The connection between any measurement which we can positively identify and our conscious experience of it is a logical necessity, but its placement in an observer external to any physically defined system is not.

A more interesting possibility is that measurement processes do occur within evolutionary systems and that at some level variation and selection are a measurement process. As a consequence, probability arises in a fundamental way and the theory of evolution assumes a fundamental status in principle. But the internal occurrence of measurement processes would imply that evolutionary systems are not governed by the reversible (so-called time evolution) equations of quantum mechanics, even assuming a description that refers to the complete system. According to this interpretation the duality of the strongly causal law of motion and the less than strongly causal law of measurement implies a duality of processes in nature, with

---

*Such ergodic models involve the coupling of the microsystem to a macroscopic measuring device. The reduction of the wave function is accompanied by the multiplication of each of its components by a random phase factor which has the effect of removing the interference terms. The acausal probability process corresponds to this randomization of phase. Ergodic models of phase randomization in measurement would have the same consequences for the status of the theory of evolution as purely classical ergodic models, and in fact would require the measurement process to be interpreted as a useful but fundamentally fictional approximation as well. For example, the entropy increase inherently associated with the openness of local biological systems is compatible with their registering a measurement, but the entropy increase of the whole ecosystem of which the local systems are a part would in an ergodic interpretation be due to the asymmetry of the initial conditions rather than to measurement.

measurement processes intervening in the time development in a prominent way only in living systems. A variant of this interpretation is that the general law of time development exhibits some reduction of the wave function, and therefore has a probability-generating feature, but that for some reason this is not seen in systems which have been sufficiently simplified to become objects of a predictive law. The tenability of this variant interpretation is at the moment unsubstantiated, but it may correlate with the proposal of Prigogine (1978) of a microscopic dynamics which displays irreversible features. According to this variant the overreliance on idealization has led to artifacts, such as microscopic reversibility and strong causality, which have fictionalized the conceptual scheme of physics, implying that it is the conceptual scheme of biology which is fundamentally sound and that of physics which is fundamentally faulty.*

These alternatives have a bearing not only on the relation between the fundamental conceptual scheme of biology, namely evolution theory, and that of physics, but also on the relation between both of these and the interpretation of psychological primitives. Josephson (1974) has made the subtly interesting observation that with the advent of quantum mechanics a psychological element enters into physical theory, not just because of the intuitive connection between subjective experience and the projection process, but because the construction of projection operators and the decision to apply them, now factors contributing to the time development of the system, cannot be derived from the theory. In this respect they appear to involve the notion of choice, just as the construction and use of measuring equipment appear to involve the notion of choice. In a discussion of evolution a tempting observation is that choice, a primitive of psychology, is closely associated with variation and selection, primitives of evolutionary theory. The problem of whether choice is fictional or fundamental is the same as the problem of whether variation and selection are fictional or fundamental.

The problem of measurement has been and remains one of the issues most pondered by scientists and philosophers in this century. The astonishing point is not that one or another interpretation is compelling, but that careful analysis shows that the most primitive concepts of physics, biology, and psychology are deeply linked and connected to the antinomies of reversibility–irreversibility and objectivity–subjectivity. It is interesting to consider the extent to which it is reasonable to take these interpretive questions into the structure of a scientific theory and the extent to which it may reasonably be expected that the clashing conceptual schemes of biology

---

*In this respect it is pertinent to note that for the purposes of generating probabilities each measurement-type process can be accompanied by an arbitrarily small entropy increase, since in this case there is no reason to exclude an error rate greater than 50%.

and physics can ever be reconciled. It seems to me that reconciling these antinomies and more broadly reconciling biology and physics is possible only to the extent that the reconciliation does not lead to the solution of fundamental philosophical antinomies, such as the antinomy between the freedom of the mathematician to create physical theories and the necessity which such theories impose on the mathematician and on the life process in general. The consistency of scientific theories, whether physical or biological, would be suspect if they were not open as regards answers to such inherently arguable questions. Explanting an apt formulation of Bohr (1961), the logical ordering of the sciences should allow the scientist to avoid truth which is too deep, a criterion which has merit independent of Bohr's framework of complementary descriptions.

This is not a clearcut criterion since it is unquestionably difficult to disentangle scientifically answerable and scientifically unanswerable questions. But on the basis of this admittedly imperfect criterion it appears that the present conceptual situation of physics and biology is unsatisfactory. Either the conceptual scheme of one or the other is a fiction, in which case the logical structure of science enforces inherently arguable interpretations, or there is no possibility of a contradiction-free logical structure, in which case the scientist is permitted to deduce any conclusion. On these grounds it should not be excluded that developments could in the future lead to a common nonfictional basis for both physics and biology which would do justice to the explanatory modes of both sciences in principle. The process of measurement is a candidate for playing a role in this respect since it bears on the main point of conflict, namely the objective reality of probability generation, while at the same time it is open with respect to the arguable interpretive questions. As an illustration, the objective reality of probability generation does not allow the conclusion that free will is a reality, but the nonexistence of probability generation would guarantee that it is an illusion. The important point is that when the assumptions of the two schemes bear on the same feature of nature they should not be in contradiction. Elucidating a common nonfictional basis for physics and biology does not mean demonstrating that the assumptions of the one are sufficient to entail the assumptions of the other. Such a demonstration is impossible due to the limitations of computation and to the limitations of experiment. Such a strong reduction would itself violate the principle of philosophical neutrality by giving permission to the scientist to deduce inherently arguable conclusions, for example about the nonpublic aspects of experience, and as such would be epistemologically as untenable as the present-day absence of even a weak reduction of biology to physics sufficient to demonstrate consistency (or, as I expect, expansion of physics to biology).

These questions are subtle and tempting, but I think that for the immediate purposes the best strategy is to put them aside, while remembering that they have been put aside. The view will simply be that the theories of thermal physics require certain steps to accompany the passage of a detail from significance to insignificance, and that this is so regardless of the problems concomitant to justifying each of these steps.

2.6. FORGETTING PERTURBATION

One type of detail is the type which arises through perturbation. Thus the ability of a physical system to forget details includes the ability to forget perturbations. The essential steps are

1. The perturbed system is, by definition, driven into an atypical macroscopic state.
2. If this atypical state is compatible with a relatively small set of microscopic states of the system, if the original macroscopic state is compatible with a much larger set of microscopic states, and if the latter states are accessible from the former, we say that the system is stable to the perturbation, for in time it will surely visit and remain practically forever in the original set.
3. If the atypical perturbed state is compatible with a larger number of microscopic states than the original, normal state, it is still possible to return to the latter from the former by dissipating high-grade energy into the environment (i.e., utilizing the same type of mechanisms required for the development and maintenance of the state in the first place). In this case the system is stable to the perturbation because it is driven back to the original set by the fall of the environment from a less to a more likely set.

As soon as the system reaches the original set [either by a passive process, as in (2), or by an active process, as in (3)], we can say that it has forgotten the perturbation, for no macroscopic trace is left at this point. Moreover, it would be completely useless to try to reconstruct the past by trying to specify which state of the likely set of states the system might actually be occupying on the basis of our knowledge of the previous set of less likely states (of either the system or environment). There is no choice about throwing out information about these details, for we are really only throwing out information whose collection would result in an objective increase in entropy and therefore loss in information.

2.7. IGNORING PERTURBATION

There is another process which plays a role in the resistance to perturbation, one which is connected to the ultimate quantum mechanical nature of the system. The essential idea is that energy transfers in nature occur in discrete lumps. Thus if the energy levels of the system of interest are not compatible with the energy of some environmental event, this event will not perturb the system. In effect, the system ignores the potentially perturbing event.

It is amusing and worthwhile to try to capture the reason for this. The root idea in quantum mechanics is that angular momentum (or action) comes in discrete units, or quanta. Another root idea is that energy is a product of a quantum of action and frequency ($E = h\nu$) and that momentum is a product of a quantum of action and wave number ($p = h/\lambda$, where λ is wavelength). Thus as soon as we impose some boundary conditions on the system we have a situation in which only certain frequencies and wave numbers are possible. It follows that all energy transfers must occur in acceptable lumps.

I will call such a restriction on energy transfer a selection rule (although this is ordinarily not used in such a general sense). The important point is that the system has only a finite number of possible states, so that not every perturbation can change the state. The simplest and most important case is a system subject to thermal perturbation. The random fluctuations in the environment bump it from one level to another, but the number of possible new states formed by these bumps is limited (which is another way of saying that specific heat is finite) and the readiness with which the system is bumped depends on the closeness of the levels. If the levels are relatively far apart, many fluctuations will have no effect. There are extreme cases of this in which the differences are greater than thermal energies, effectively isolating the system from its thermal environment. As an example, imagine that the energy levels are crowded into two groups with a big gap between them. Such a system would be able to classify its inputs into two groups, and quite reliably, since random variations in each input would generally only bump the system around within one group or the other. The highly reliable semiconductor switching elements in computers are organized in this way. Other more dramatic examples are superconductivity and superfluidity. These phenomena are based on the pairing of electrons (or other particles governed by the exclusion principle) at low temperatures. The persistent motion results from the condensation of the pairs into a coherent, low-energy state. Many authors have speculated on a possible role of such coherent phenomena in life processes (e.g., London, 1961; Schrödinger, 1944; Pattee, 1968), although none of these speculations have so far been confirmed.

The protective aspect of quantization does not imply that major environmental disturbances are ignorable—this is clearly out of the question. But it appears plausible that quantization plays a crucial role in connection with reliability of function in the face of small (thermal) perturbation. Later we will see that reliability has an important connection to adaptability.

2.8. REDUCING PERTURBATION AND THE SIGNIFICANCE OF QUANTUM VARIABILITY

The thermal fluctuations are determined by temperature, and therefore belong to the givens which biological systems must either cope with or utilize. The connection between quantum fluctuations and the stability of atoms and molecules is more intimate. It is most interesting to consider the principles involved, since they are of major importance for the maintenance of biological structure, for the storage of information, and for the origin of biological variability.

According to the uncertainty principle there is an intrinsic relationship between the statistical stability of a variable and the statistical instability of its conjugate variable. Thus, according to the position–momentum uncertainty relation $\sigma p \sigma x \sim h$, where σp is the standard deviation of momentum measurements and σx is the standard deviation of spatial measurements. One important condition for molecular stability is therefore electronic delocalization. Since delocalization means that the variability which would result from position measurements (σx) would be increased, the variability which would result from momentum measurements (σp) would be smaller. This in turn means that the variability of the kinetic energy ($p^2/2m$) is reduced under conditions of delocalization; therefore the chances of an upward fluctuation of kinetic energy sufficient to decompose the system are reduced. This is the basis of chemical resonance, that is, of the stability exhibited by those molecular structures which allow for many alternative electronic configurations. A more important example is the covalent bond. In this case the delocalization is based on the fact that electrons with opposite spins can share the same space. The stabilizing power of the consequent "electronic positional variability" is obviously connected to the fact that the information accumulated by biological systems in the course of evolution is locked into covalent structures which are unlikely to form or break in the absence of a suitable catalyst. The catalyst presumably opens up a new pathway of energy exchange or enhances destabilizing fluctuations by imposing constraints which reduce σx.

A second important condition for stability is "energy variability." According to the time–energy uncertainty principle, $\sigma E \sigma t \sim h$, where σE is the standard deviation of the energy measurements on the system and for the present purposes σt can be interpreted as the standard deviation of the times at which these measurements are made. During short intervals of time it is possible for the energy fluctuations to be large. It is therefore possible for two particles to exchange momentum during this short interval of time in a way which would ordinarily violate the conservation of energy. For example, an electron and a proton may exchange momentum through the emission and absorption of a transiently existing quantum of light, or two electrons might exchange a transiently existing quantum of sound which allows them to pair at low temperatures. It is such exchanges of virtual particles which are the basis of the attractive and repulsive forces which determine the structure of matter.

The situation is that at the microscopic level the stability and instability of matter are fundamentally based on violations of the macroscopic thermo-dynamic principles laid out at the beginning of this chapter. Quantum variability of energy—that is, short-lived breakdowns in energy conserva-tion—give rise to the attractive forces which stabilize the structure of matter. Quantum variability of momentum—that is, breakdowns in momentum conservation under conditions of reduced positional variabil-ity—is one basis for the change in this structure. Another basis is the conversion of heat into work in the form of displacement forces acting on molecules, in violation of the second law of thermodynamics (Figure 2.1).

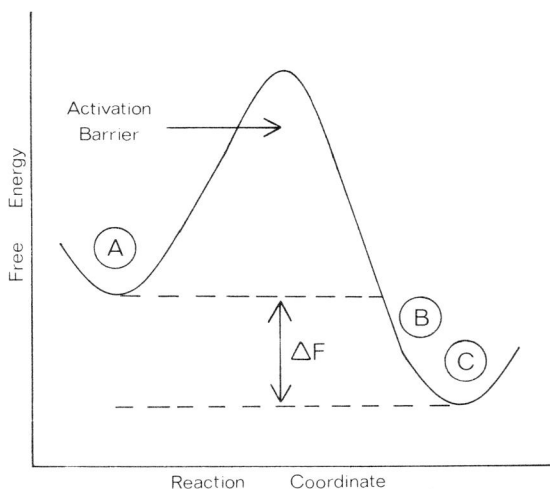

FIGURE 2.1. Free energy diagram for reaction $A \rightleftharpoons B + C + \Delta F$. Fluctuations lift the reactants over the activation barrier, but ΔF, not the height of the barrier, determines the final equilibrium.

But of course violation here only determines the rate at which macroscopic symmetry increases. It can lead to no macroscopic violation of thermodynamic laws.

The stability of macroscopic biological structures and their variability ultimately depend on these processes. It is now possible to identify the relationship between such biological variability and the fluctuation phenomena considered. This variability is not the same as the fluctuation phenomena, such as thermal and quantum fluctuations, nor is it solely due to these phenomena. All jumps between stable states, such as the jumps of the kind which occur in chemical processes, are driven by fluctuations. The chemical changes which occur, say, in the process of correct gene transcription are just as much driven by fluctuation as those which occur in gene mutation. The physical fluctuation phenomena lead to mutation and other forms of biological variability when they can lead to a variety of macroscopic outcomes or to macroscopic outcomes which depend on an unobserved external situation. Mutation is thus a consequence of both fluctuations in the microscopic circumstances of a gene and of a greater or lesser degree of sensitivity of the outcome to the differences in these circumstances. If the fluctuations are true accidents, then the mutations have a truly accidental aspect. But whether or not they are true accidents returns us to the difficult issue of the place of measurement in nature.

REFERENCES

Bohm, D. (1951) *Quantum Mechanics*. Prentice-Hall, Englewood Cliffs, New Jersey.

Bohr, N. (1961) "Discussion with Einstein on Epistemological Problems in Atomic Physics," reprinted in *Atomic Physics and Human Knowledge* (cf. p. 66). Science Editions, New York.

Brillouin, L. (1962) *Science and Information Theory*. Academic Press, New York.

Conrad, M. (1969) "Some Problems Associated with the Physical Description of Evolution Processes," pp. 17–27 in *Computer Experiments on the Evolution of Co-adaptation in a Primitive Ecosystem*, Ph.D. Thesis, Biophysics Program, Stanford University, Stanford, California.

Jauch, J. M. (1968) *Foundations of Quantum Mechanics*. Addison-Wesley, Menlo Park, California.

Josephson, B. D. (1974) "The Artificial Intelligence/Psychology Approach to the Study of the Brain and Nervous System," pp. 370–377 in *Physics and Mathematics of the Nervous System*, ed. by M. Conrad, W. Güttinger, and M. Dal Cin. Springer-Verlag, Heidelberg.

London, F. (1961) *Superfluids*, Vol. I. Dover Publications, New York.

London, F., and E. Bauer (1939) *La théorie de l'observation en mécanique quantique*. Hermann, Paris.

Pattee, H. H. (1968) "The Physical Basis of Coding and Reliability in Biological Evolution," pp. 67–93 in *Prolegomena to Theoretical Biology*, ed. by C. H. Waddington. University of Edinburgh Press, Edinburgh.

Pattee, H. H. (1973) "Physical Problems of Decision-Making Constraints," pp. 217–225 in *The*

Physical Principles of Neuronal and Organismic Behavior, ed. by M. Conrad and M. Magar. Gordon and Breach, New York and London.

Prigogine, I. (1961) *Thermodynamics of Irreversible Processes.* Wiley, New York.

Prigogine, I. (1978) "Time, Structure, and Fluctuations," *Science 201* (4358), 777–785.

Schrödinger, E. (1944) *What Is Life?* Cambridge University Press, Cambridge, England.

von Neumann, J. (1955) *Mathematical Foundations of Quantum Mechanics.* Princeton University Press, Princeton, New Jersey.

Wigner, E. P. (1967) "The Problem of Measurement," pp. 153–171, and "Remarks on the Mind–Body Question," pp. 171–184, in *Symmetries and Reflections.* Indiana University Press, Bloomington, Indiana.

3

The Dissipative Ecosystem

The pond ecosystem of Chapter 1 is clearly more complex than the damped oscillator or the freely expanding gas, or even an open system of chemical reactions. These examples are tailored to simplicity in comparison to even the minimal biological system. Yet we would hardly be committing heresy to suppose that a system whose architecture is as intricate as the pond ecosystem obeys the same general laws of dissipation as these more poverty-stricken relatives. On the other hand, it would certainly be heresy, or at least obtuse, to suppose that nothing of interest distinguishes the dissipation of energy in these most complex and simple of situations.

We already know something very general about these special, distinguishing features. They must be just those features which make the pathways of dissipation interesting in the sense described in Chapter 1, viz. in the sense that they mediate self-reproduction and are therefore ultimately both the basis and the product of evolution.

I want to pursue the significance of this point. Even the simplest systems have many possible pathways of dissipation. For example, there are many possible pathways for a gas to spread through a box, or, more generally, for a system to run to equilibrium. Even more generally, there are always many different possible steady states for a given energy input and output, even states of the same energy and entropy. In each case the actual pathway is a matter of rate. Thus, if we could add something to the system which would make one way of dissipating energy faster than the others (e.g., one way of running to equilibrium, one way of maintaining a steady state), this would become the dominant way. Thus, what ultimately distinguishes life from nonlife is that the rates of processes are a selected subset of the possible rates, selected so that the system has the gift of evolution.

I begin by describing the agents which do the selecting in biological systems (viz. enzymes), turn to the naked logic of cellular self-reproduction (basically selective cross-catalysis), and then dress this logic with those extra features relevant to the dissipation of disturbance. The role of energy and entropy in creating structure is important since this is relevant to the self-reorganization of systems subsequent to disturbance as well as to their initial self-organization. The role of boundary structures is important since it has implications for the openness of biological systems and therefore for the characterization of their state. It is also necessary to consider critically the various concepts of organization and information in biology, and in particular to distinguish these concepts from the behavioral uncertainty associated with state-to-state behavior. These various distinctions, e.g., between macroscopic and microscopic uncertainty of state transitions and between these and the homogeneity or inhomogeneity of the states, play a fundamental role in the development of the theory.

3.1. SELECTIVE DISSIPATION AND SELF-REPRODUCTION

The first question is, what does the selecting? The immediate answer is the enzyme, by which we here mean any entity which speeds certain processes and not others. In general these entities assume the form of protein molecules, and indeed they are conventionally defined as catalytic molecules which selectively increase the rates of certain biochemical processes. However, nucleic acids and other biochemical structures can also selectively increase the rates of processes without themselves being consumed.

The first and fundamental process which these enzymes have to speed up is their own production. The trick, derobed of all its exquisite detail, is essentially cross-catalysis of nucleic acids and proteins. In modern-day cells the nucleic acids specialize as an information store, with the sequence of bases in DNA coding for the sequence of amino acids in proteins. The protein molecules fold up into some three-dimensional shape on the basis of the weak interactions among these amino acids. The specificity (or selectivity) of the proteins is determined by their shape. The two *sine qua non* protein functions are transcription of the nucleic acid into daughter nucleic acid and translation to protein, except that the latter is mediated by a system of nucleic acid adaptors (or transfer RNAs). The basic flow of information and the role of folding are illustrated in Figure 3.1.

There are of course many types of proteins produced, each speeding processes subserving either the basic function of self-reproduction or one of the subsidiary functions. The evolution process is a direct consequence of this reproduction process along with the distortion of the rates of reproduction by environmental factors (more commonly called natural selection). Thus the ultimate answer to the question of what selects the enzymes and

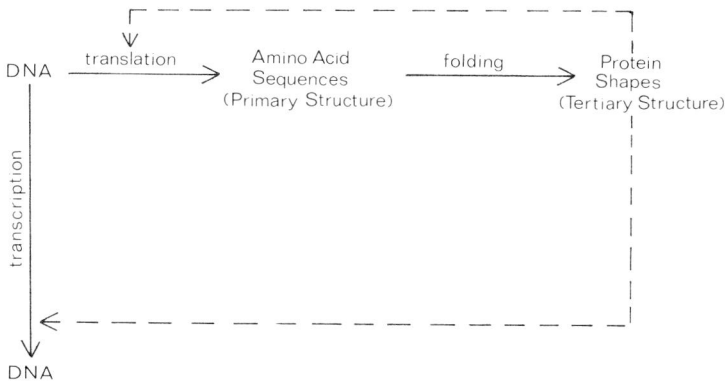

FIGURE 3.1. Flow diagram of selective cross-catalysis underlying biological self-reproduction. Broken arrows point to steps controlled most notably by protein enzymes and, in the case of translation, by transfer RNAs (not shown). DNA also satisfies the definition of a catalyst in transcription and translation.

therefore the rates is in some sense the rates themselves, but indirectly through the probabilistic process of variation and selection.

There is an interesting analogy and a more interesting disanalogy between the molecular biological mechanism of self-reproduction and designs for self-reproducing machines of the type originally proposed by von Neumann (1966; cf. also Arbib, 1969). Von Neumann's design consisted of a description (or blueprint), a constructor, a replicator, and a control device which decides when the parent should split into parent and offspring. The constructor is universal if it can interpret the blueprint and build any structure described by it. If the blueprint describes the constructor, replicator, and control device, the system is *a fortiori* self-reproducing. Von Neumann argued that the replicator is necessary, for if the constructor were responsible for constructing the blueprint, another blueprint would be required, and so forth, leading to a regress. That is, the functions of construction and replication are logically distinct. The blueprint is clearly analogous to DNA; the constructor, to the protein and nucleic acid machinery responsible for reading DNA; and the replicator, to the polymerase which transcribes it. In effect the blueprint and constructor together catalyze the production of more construction and replication machinery, while the replicator catalyzes the production of more blueprint. At a deeper level, however, the analogy breaks down. In the machine the blueprint is a rule which controls the operations of the constructor. In the cell the DNA provides a structural description of the amino acid sequences of the proteins, although it must of course code for control over the reading of genes as well. The processes of the cell are therefore influenced by an energy-dependent folding of the structures encoded in the blueprint rather than prescribed by this blueprint in a completely formal way. In Chapter 10 it will be shown that this energy dependence confers a gradualism on the function changes exhibited by biological systems in response to mutation which makes them especially suitable for evolution by variation and selection, although at the expense of introducing an element of intractibility which makes them much less suitable than machines for prescriptive design.

Our description of self-reproduction is very minimal. At least we should add that the process must be orchestrated in space and time and therefore the spatial matrix is of critical importance. It must also be surrounded by a boundary, or outer membrane. This is a thermodynamic necessity for any self-organizing process. In fact membranes also subserve a selective function (by controlling what chemicals enter or leave the cell) and therefore assume a fundamental rate-determining role as well.

3.2. SELF-ASSEMBLY AND SELF-REORGANIZATION

To some extent the spatial matrix of the reproduction process is a precondition for the process and must pre-exist, in the sense that it is not specifically encoded in the sequence of bases in DNA. However, some of the spatial structure is itself attributable to the complementary shape of proteins, along with the specific stickiness inherent in various short-range interactions (e.g., basically dipole interactions such as van der Waals forces, hydrogen bonding, charge fluctuation interactions, and hydrophobic interactions). According to the so-called principle of self-assembly these enable certain proteins (and other macromolecules) to assemble spontaneously into larger, quite particular spatial aggregates.

One way to think about the self-assembly process is in terms of the competition of energy and entropy. Consider the formula

$$F = E - TS \tag{3.1}$$

Since S tends to increase, the stable situation is one in which the free energy is a maximum. However, at very low temperatures the potential energy dominates, so that the system may develop quite complex structure in the fashion of crystal structure formation. Biological systems are not at low temperature, but the effect is still possible because the very-short-range van der Waals interactions among large molecules that fit to one another with high closeness are in general additive and can add up to be greater than thermal energies. In this circumstance the energy can defeat entropy and it is possible for a specific structure to form (or re-form subsequent to perturbation) without its necessarily being a relatively high-entropy form. Furthermore, the formation process is dissipative, for all spontaneous irreversible processes are dissipative.

We can now ask a hindsightful but relevant question. Why are macro-molecules such as nucleic acids and proteins the fundamental mediators of self-reproduction and evolution? The answer is that the components of a truly self-organizing system must be big enough either to stick together because of their specific shapes (self-assembly) or to stick in an appropriate and transient way to the substrate to be catalyzed, but at the same time they must be small enough to diffuse and find either each other or the substrate.

The size of molecules is thus an essential ingredient in self-assembly and self-organization generally. It is *a fortiori* also an essential ingredient of self-repair (in the sense of self-reorganization). For example, suppose that the structure of a biological system is perturbed by some environmental insult. If the components of the system have the self-assembly capability they will automatically dissipate this insult away, provided it is not too large. From the standpoint of adaptability this is the main significance of the self-assembly principle.

3.3. DISSIPATIVE PATTERNS AND DISSIPATIVE REPATTERNING

By exporting heat a system clearly becomes more ordered, at the expense of the environment becoming less ordered. By taking in high-grade energy and exporting heat it can remain ordered or continue to become more ordered despite dissipation. This is evident from equation (2.6). It is less obvious that in the presence of suitable constraints the dissipation of energy can be used to create dynamic order involving the coherent motion of molecules. In the presence of such constraints patterns which are genuinely asymmetrical in space or time may be the stable ones. Such stable asymmetric patterns are sometimes called dissipative structures since they are created by the dissipation of energy rather than by its minimization (Glansdorff and Prigogine, 1971; cf. also Ebeling, 1976). In order for an inhomogeneous spatial distribution to develop in a chemical system the production and decay of substances must interfere constructively and destructively at different locations. The suitable constraints in this case are the presence of at least two substances which diffuse at different rates and whose production is coupled nonlinearly, such as by autocatalytic and cross-catalytic interactions. A number of interesting dynamical phenomena are possible, such as stable clocklike behavior and bifurcation to different inhomogeneous concentration patterns in response to continuous changes in external conditions.

The key feature of dissipative structures is that the symmetrical pattern is unstable and asymmetrical patterns are the stable ones. They thus exhibit a spontaneous tendency to symmetry breaking, in superficial contradiction to the second law of thermodynamics. As a consequence processes of this type are thought to play an important role in morphogenesis, where a pattern which at least appears to have relatively high symmetry at the beginning undergoes a stable development into patterns which are dramatically asymmetric. Here morphogenesis could involve the control of differentiation in embryogenesis (Turing, 1952), but it could also involve the development of form at other levels of organization. In this respect dissipation is less dependent on spatial scale than is energy as an organizing principle at biological temperatures. However, order arising from minimization of energy is the basis for the initial constraints, such as the catalytic interactions, which allow for the formation of dissipative patterns. It is also responsible for the all-important boundary across which heat must be exported. Without the prior existence of such a boundary there would be no possibility for the development of dynamical order through dissipation. Both types of ordering principles contribute to the organization of biological systems, but as a matter of experimental fact the most crucial features of organization, such as the genetic memory, the spatial geometry necessary for cellular reproduction, and neuronal memory, are based on the energy principle. Those living organisms which can be brought close to the absolute zero of temperature and can be brought back to normal temperatures without undue tissue damage retain all this organization, yet at the absolute zero of temperature all motion ceases aside from fluctuations due to the uncertainty principle.

From the standpoint of adaptability the main significance of order through dissipation is the same as that of the self-assembly principle. If a stable dynamical pattern is perturbed by some environmental insult the system will spontaneously return to this pattern, provided that the insult is not so large that it is driven into the basin of attraction of another pattern. Any system which structures itself through dissipation is automatically a system which can restructure itself through dissipation. As a consequence the mechanisms of structure formation through dissipation are necessarily mechanisms of adaptability as well.

3.4. PATTERNS OF ACTIVITY

The boundary has an even more fundamental significance. Each self-reproducing unit is a boundaried collection of molecules. The peculiarity is that there is no definite set of molecules, for there is an incessant through-flow of substances between the inside and the environment. This is important, for it means that when we think of the cell or any fundamental unit of life above the cellular level we must think of a bona fide process, or, more picturesquely, of a pattern of activity abstracted from a set of particles. Even the structures, including the boundary itself, are aspects of the process, only more slowly changing aspects.

This shift in point of view reflects a really fundamental difference between physics and biology. Suppose, for example, that we wanted to describe such a pattern in terms of the basic laws of physics. We would have to begin with a classical description of the system in terms of the positions and momenta of the particles and then turn this into a quantum mechanical description by replacing these with their operator equivalents. The problem is that we could never say in advance how many molecules would contribute to the pattern without including all of the atoms and molecules in the rest of the system (i.e., all of the molecules in the complete ecosystem). This is because the self-reproducing nature of such patterns along with the finiteness of the number of atoms and energy input inevitably produce the competitive process of evolution, so that it is in general impossible to say when a particular pattern or type of pattern is likely to disappear without reference to the whole system. In other words, we can write an equation for a molecule in terms of fundamental physics; in principle we might be able to write an equation for an ecosystem in terms of fundamental physics; but we can't write an equation for anything in between.

The importance of this feature has already been attested to (in Section 2.5) by paradoxes of measurement, such as that of Wigner's friend, which result from the failure to recognize it. The problem is that recognizing it means recognizing a situation which is intractable from the purely physical point of view. Even if we could write and solve our ecosystem equation we would still have a very hard time specifying what we mean by the state of one of the patterns in terms of the solution. However, the starting assumption of a pattern of activity already implies that there must be some set of properties in terms of which it is recognizable. This is confirmed by biological practice, for the biologist daily distinguishes and deals with various possible conditions of cells, organisms, and populations, and does so without reference to the whole ecosystem. But the properties in terms of which these conditions are specified can no longer be the fundamental observables of physics; rather they must be new properties, suitable to the unit of organization.

Thus we have two descriptions, one emerging from the practice of physics and inherently global and the other emerging from biological practice and essentially local. To the extent that both of these are valid there must be some principle of correspondence. This is: each state (or significantly different condition) of a unit of biological organization is an equivalence class of microscopic physical states of the entire ecosystem.

Thermodynamics itself provides the first famous example of such a situation. Each macroscopic state (as specified by temperature, pressure, volume, mole numbers) is an equivalence class of microscopic states, and one which is computable by a definite procedure. The idea is much more complicated in biology, however, for what properties are relevant depends on the functions performed by the system. In the case of a population there are clearly certain natural properties, namely number and relative location of the organisms. In the case of the organisms themselves the properties are those discussed in the textbooks of physiology, but it is clear that it would be hard to guess these on an *a priori* basis.

Some biologists have on occasion characterized the properties associated with biological systems at each new level of organization as emergent properties. But states described in terms of these properties are more usefully thought of as equivalence classes of microscopic physical states, with the properties being various computed (e.g., average) values of these equivalence classes. Certainly, one is obliged to admit that it would be difficult and probably not feasible to discover such properties on the basis of this idea (cf. Rosen, 1969). However, this does not at all detract from its potential value as a correspondence principle linking the state of a biological system as a distinguishable pattern of activity to its microscopic physical basis.

3.5. INFORMATION UNBOUND

The multiplicity of possible pathways of dissipation (and therefore the possible alternate forms of what we have called the patterns of activity) is the basis of a second and much more important information concept. So far we have considered only information which is bound in the sense that it is given by the entropy taken over the set of possible microstates of the system. Now we can also consider measures of entropy (and therefore information) taken over the set of possible pathways of dissipation.

For example, imagine an enzymatically controlled set of chemical reactions in the steady state (as in Figure 3.2). There are practically an infinite number of possible steady states, for the choice of enzymes determines what chemical forms are present and for all practical purposes there are an indefinite number of such choices. Thus an entropy defined over the possible steady states would be very large, assuming each had an *a priori* equal probability of occurring. The actual thermodynamic entropy

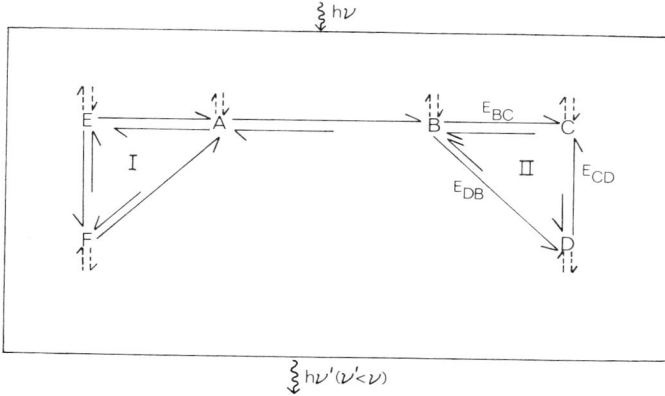

FIGURE 3.2. Enzymatic selection of pathways of dissipation in a chemical system. E_{IJ} are enzymes speeding the interconversion of substances I and J. Broken lines represent the virtually infinite number of possible reactions and reaction cycles not shown explicitly. In the present case dissipation of relatively high-grade energy ($h\nu$) into heat ($h\nu'$) is mediated by cycle II. Different enzymes would select cycle I or one of the other possible pathways of dissipation (after Conrad, 1972, by permission).

of any one of these steady states, however, is quite definite and limited, for the number of compatible microstates is definitely limited.

Does this mean that our new concept of information swamps the thermodynamic entropy? Definitely not! In fact it is ordinarily quite small in comparison to the thermodynamic entropy. This is because we certainly cannot make any assumption of equal *a priori* probabilities. The number of potential steady states (or, more generally, pathways of dissipation) is very large, but the probability of autonomously jumping from one state to another is not usually large. After all, each steady state is a distinct macroscopic state and therefore what we are talking about is an ensemble of macroscopic states.

This purely thermodynamic insignificance, however, does not mean that entropies defined over possible pathways of dissipation are functionally insignificant. Suppose that our system is such that some external influence is capable of pushing it into any one of a large number of its possible modes of dissipation. If we are unaware of this external influence (say a perturbation), the probabilities for the different pathways of dissipation will depend on the probabilities of the different types of influences. In this case the entropy defined over the set of possible pathways could become very large, yet it would make virtually no contribution to the thermodynamic entropy.

The reason is that there is an external source of energy, perhaps very small, which switches the system from one pathway to another.

Suppose further that the pathways of dissipation resulting from the external perturbation eventually decay back into the original pathway. Then we say (as in Section 2.6) that the system is stable to the perturbation. How stable it is depends on whether there are enough modes of dissipation with this property that all the likely perturbations are forgotten through the decay process. The decay process is, of course, itself dissipative. We know this because any forgetting of the past must be associated with dissipation. However, the amount of dissipation mediating the stability of the system may be quite small in comparison to the amount of energy dissipated by any one of the patterns. In fact, this dissipation, pulling the various patterns back to the "attractor" pattern, is just our ϕ_{small} from Section 2.2.

Information taken as the entropy of some abstract set of possibilities is sometimes called "free information," to distinguish it from information defined over the set of possible microstates. The latter, bound information, is basically the same as the thermodynamic entropy. This is clearly not the case for the free information. What we propose, however, is that any form of free information is ultimately defined over some set of possible pathways of dissipation. These pathways of dissipation are generally recognizable in terms of some macroscopic properties, which properties therefore define the macrostate of the system. In biological systems the macrostate is generally best thought of as a pattern of activity, so that what we are dealing with is a set of possible dissipative patterns. If the dissipation is very slow this degenerates to the equilibrium case where we deal with the set of possible equilibrium states.

What about something as abstract as a set of symbols? These must have some physical embodiment (such as the ink marks on this page or the nerve impulse running from eye to brain) whose creation at one time required the performance of work and which is subject to erosion, however slow (in the case of print) or fast (in the case of the impulse). Moreover, the resistance of such forms to erosion or more malicious perturbation depends on the dissipation of energy, for their maintenance or reconstruction requires the further performance of work. In this case, however, the arbitrariness of the form is inconsistent with its being an attractor for the perturbed forms, either on the basis of energy (self-assembly) or thermodynamic entropy (increase in the ensemble of microscopic states). Systems with this arbitrariness property must therefore be repaired by some external system. Since these must also be repaired, we must eventually reach a system that repairs itself and which therefore must capitalize on the self-organizing properties of macromolecules. Indeed, the logical limit of this regress is self-reproduction (which brings us back inevitably into the circle of biological systems and evolution). This is the most powerful and universal mechanism of self-repair, for in producing the child nearly all the sins of the father are forgotten at once.

3.6. INFORMATION AND ORGANIZATION

Unfortunately there is yet another concept of information, viz. the entropy reduction in creating the form itself. This is not the same as the number of macrostates or the ensemble of microstates, but rather has to do with an initial set of microstates from which the present set is selected out. In other words, our initial entropy is an entropy calculated over a relevant initial set of microstates before the particular macrostate is prepared and the information is taken as the difference between this initial entropy and the final entropy calculated over the final set of microstates.

This organizational concept of entropy and information is much less definite than our former concepts, for there is in general no arbitrary way of specifying a relevant initial set of states. What was the initial set of possible states of matter which was contracted to form the typewriter on which I am now writing? What were the initial set of states from which our present world ecosystem was selected? There is really no satisfactory or nonarbitrary way of answering these questions. Indeed, another, entirely macroscopic, way of seeing this is to ask how much entropy change would be required to produce the given form in a reversible process. Clearly one would have to pick a standard state and it is not obvious what a sensible choice would be.*

Organization is important in biology. The ability of biological organizations to forget perturbation and the mechanisms which underlie this ability are certainly not independent of the nature of the organizations. We will eventually have much to say about this, but fortunately we can avoid using any concept of organizational entropy in the above sense. The reason is that biological systems come in units and subunits (populations, organisms, cells), so that it is possible to describe organization in terms of the diversity of different types of units. For now, however, it is only necessary to keep as clear as possible the distinction between organization and our other, nonarbitrary concepts of entropy and information.

---

*According to the third law of thermodynamics the entropy of the system at the absolute zero of temperature can always be taken equal to a constant, e.g., zero. We could envisage determining the entropy change in a reversible process which takes a system of atoms and molecules from this standard state to the actual state, but this is clearly not feasible for living systems.

3.7. THE CHESSBOARD ANALOGY

It is perhaps worth restating some of these distinctions in terms of an analogy which might be called the chessboard analogy. The thermodynamic entropy is the ensemble of microstates corresponding to any particular position of the pieces on the board. There are, of course, a very large number of board positions. This large number does not contribute to the thermodynamic entropy, however, for each piece is a macroscopic object which remains in place practically forever if undisturbed. Nevertheless, the large number of possible board positions is the key to the game and we could, if we wanted to, measure the entropy over this set of possibilities by constructing a set of *a posteriori* probabilities on the basis of a very large number of actual games. The probabilities are necessarily *a posteriori* (or at least contingent on external factors) for in each case it requires an external push on one of the pieces to change the board position.

The entropy calculated over the board positions is essentially an index of the diversity of the actually occurring board positions. We could also consider the probability of any given board position given the preceding board position and calculate an entropy over these probabilities. In this case we have an entropy measure on the possible sequences of states, i.e., a measure on the spread on the possible behaviors of the system. Each one of the different sequences involves the performance of work, but the way in which the work is applied is different for each distinct sequence.

So far we have tacitly assumed that our entropies are calculated over functionally distinct board positions, i.e., for the purpose of the game we do not consider different positions of the pieces within the squares. However, we could consider such functionally equivalent but distinct states and calculate an entropy over these, in terms of either states or sequences of states. The size of the ensemble of such states is actually important, for as they become larger it is less likely that a piece will accidentally be pushed into the wrong position. Furthermore, if a piece gets too close to one of the edges of its proper square the player can always put it back in the middle. This, of course, requires a computation on the part of a player—either a human or a computer would have to calculate that the piece is out of position and set it back into its calculated proper position. This type of error correction plays an important role in reliability in general, so that in fact we cannot ignore these functionally equivalent but distinct states, even though they are irrelevant to the game itself.

The entropies discussed so far are all distinct from the information (or entropy change) associated with the organization of the system, i.e., with the physical form or structure of a typical configuration of the pieces (which in this case is almost entirely due to the form of the pieces themselves). This is

TABLE 3.1
Classification of Entropy Measures

Entropy concept	Characteristic	Chessboard analogy
Thermodynamic entropy	Measures ensemble of microscopic states	Ensemble of possible microstates compatible with macroscopic state of pieces and board
Macroscopic diversity (functionally distinct)	Measures ensemble of functionally distinct macroscopic states or (more generally) pathways of dissipation	Ensemble of possible board positions distinct from the standpoint of the rules of the game
Macroscopic diversity (functionally equivalent)	Measures ensemble of functionally equivalent macroscopic states or (more generally) pathways of dissipation	Ensemble of possible board configurations equivalent from the standpoint of the rules of the game
Behavioral uncertainties	Measure the ensemble of behavioral sequences (i.e., measure macroscopic diversity conditioned on the past and possibly other, external factors)	Ensemble of possible board configurations given previous position
Entropy of organization	Measures difference between size of initial set of microstates and size of final set, or, alternatively, equals thermodynamic entropy change concomitant to irreversible preparation of the system starting from standard state (in general an impractical and arbitrary concept)	Thermodynamic entropy change concomitant to formation of pieces and board in a reversible process
Structural diversity (diversity of components)	Measures number of types of components in the system and the evenness of their relative occurrence	Measures indicating number of different types of chess pieces (pawns, knights, kings...) and their relative frequency (one king, two knights...)

the information concept which we are a bit suspicious of, for it is difficult to determine an initial set of states from which the selection is to be made.

Another index of organization is the diversity of pieces, i.e., the number of different types and their relative frequency. As an index of this diversity we can take the entropy calculated by considering the probability of drawing at random a piece of a given type (e.g., pawn, rook, or queen). This index clearly increases as the number of types of components increases and their relative frequencies become more equal. Later we will have more to say about the relationship among diversity, organization, and the other entropies.

The chessboard analogy is useful in picturing the different conceptions of entropy: thermodynamic entropy, diversity of functionally distinct macroscopic states, diversity of distinct but functionally equivalent macroscopic states, entropy of organization, and diversity of components (cf. Table 3.1). The chessboard is not a biological system and we should also point out where the analogy breaks down. First, the chessboard is a very special case, since it is a static structure, not requiring constant throughflow of matter and energy. Thus ϕ_{big} is not at all big, we are not dealing with a dissipative pattern, and we need not think of the macroscopic states in terms of such patterns. Also, the chessboard is a completely macroscopic system, with no crucial functional role for individual molecules. As soon as individual molecules have a key role, as they do in biological systems, the possibility for self-organization and self-assembly comes into play, whereas a chessboard must be organized (or reorganized after shaking) with the help of an external computing and effector system, and at most could be made more reliable by making the moves more difficult (increasing the size of the board squares, hooking pieces to the squares).

3.8. THE FORGETFUL ECOSYSTEM

Now we turn back to the ecosystem picture developed in Chapter 1 (i.e., the pond water system in a flask) and reexamine it in the light of the energy, entropy, and information concepts described in this chapter.

Ecosystems consist of a biotic community and physical environment with some external source of energy (e.g., the sun). The community consists of subunits at various levels of organization, e.g., cells, organisms, populations. Each organism (or, at minimum, cell) is a boundaried pattern of activity with a potentiality for self-reproduction (with variation). The maintenance, growth, or reproduction of this pattern of activity requires the import (across the boundary) of relatively high-grade energy (light, high-grade chemical potential energy) and the export of heat and low-grade chemical energy to the environment. The patterns are thus bona fide dissipative patterns. Moreover, the flux of chemical matter through the boundaries, and the critical dependence of this flux on activities of other units means that each macrostate of the unit corresponds to a set of microstates of the entire ecosystem. In general, however, each unit is so integrated as a unit of function that we can describe it in terms of certain relevant properties, e.g., anatomical, physiological, and behavioral properties in the case of cells or organisms, numerical and structural properties in the case of populations. In the case of biological systems the dissipation associated with growth, reproduction, and repairing the inevitable effects of dissipation are the main components of dissipation. Furthermore, there is here no contradiction, for all local, entropy-defying processes within the boundaried region are driven by the unlocking of much larger, global entropy increases.

The statistical spread of microstates compatible with the properties of the pattern serves as the measure of its thermodynamic entropy. However, the pattern itself might be more or less organized, in the sense that we could imagine a highly heterogeneous form (i.e., one requiring a large amount of selection to produce it) with a large number of compatible microstates or a simple form (one not requiring very much selection to produce it) with the same number. The difficulty is that there is no feasible way of specifying the initial set of microstates from which the selection is made. However, the compartmental nature of biological systems (cells within organisms within populations) makes it possible to index the heterogeneity of organization in terms of the diversity of components.

Next we have the sets of possible macroscopic organizations of the system, i.e., sets of possible physiological and morphological states, sets of possible population numbers and relative positions of organisms, sets of possible patterns of energy transfer among populations, and so forth. The entropy now is the measure of the uncertainty or spread of macroscopic organizations actually exhibited by cell, organism, population, or community. If the entropy is defined over the states themselves, without any reference to prior history, what we really have is a measure of the diversity of possible states, i.e., a measure of the number and relative frequency of the different macroscopic organizations which the system can assume. More important, however, is the uncertainty of the entire sequence of behavior, for the state at one instant of time might be more or less uncertain once the state at the previous instant of time is specified. There are many such conditional entropies (or uncertainties). The simplest is the uncertainty of the state of the biota given the state of the ecosystem at the previous instant of time. The next simplest is the same, except that we also admit to knowing the contemporaneous state of the environment (in which case we have a measure of the biota's ability to anticipate the environment). We could even go further and define uncertainties of the above type for individual populations, organisms, or even cells. Finally we can distinguish between the uncertainties measured on the functionally distinct states of the community or any of its subunits and those measured on the macroscopically distinct, but functionally equivalent states.

The significance of these conditional uncertainties is that they give a bona fide measure of the behavioral indeterminacy of the system, as opposed to a measure of the diversity of states. For example, an organism might assume quite a diversity of states in the course of its life cycle, but this is quite distinct from its behavioral uncertainty.

Now suppose that our community is perturbed by the environment. This situation is inevitable, for the environment is in general unpredictable and bound to assume states which are both unanticipated by and disturbing to the community or one of its subsystems.

There are two possibilities. The first is that the disturbance is completely forgotten. In this case any alteration which it produces in the community is completely dissipated away. The time scales, however, may be short or relatively long. If the disturbance initially only pushes the system around within its set of microscopic states, the time scale of the dissipation process is very short. If it is absorbed in the set of macroscopically distinct but functionally equivalent states, the dissipation process is longer, but not too long, for either entropy maximization, energy minimization, or repair by an auxiliary device pulls the system back into whatever state it would have reached anyway. Finally, the disturbance may drive the community or some of its subsystems into functionally distinct states, for example, by modifying population size, inducing different modes of organism development or behavior, or pushing organisms into different physiological states. In the first case the time scale of forgetting involves generations, in the second at least a single generation, and in the last perhaps only a small part of a single generation.

The second possibility is that the disturbance is not completely forgotten. In this case we say that evolutionary change occurs and that adaptability takes the special and important form of the development of adaptations. The process is still dissipative, however, because the biota must generate a sufficiently large ensemble of states (by genetic variation) for the disturbance to make a suitable selection. In general, the permanence of the change is based on changes in gene sequence or on changes in the genetic composition of the community due to changes in the abundance of different types of organisms. One must be a bit cautious, however, for permanent changes are also propagated by language and culture.

In current discussions of evolution the importance of ϕ_{big} is often emphasized, not only in maintaining so-called dissipative structures, but also in driving local, entropy-decreasing processes. Actually, from the standpoint of the dynamics of evolution it is ϕ_{small} which is really crucial, for this is the concomitant of adaptability.

REFERENCES

Arbib, M. (1969) "Self-reproducing Automata: Some Implications for Theoretical Biology," pp. 204–226 in *Towards a Theoretical Biology*, Vol. 2, ed. by C. H. Waddington. Edinburgh University Press, Edinburgh.

Conrad, M. (1972) "Information Processing in Molecular Systems," *Currents in Modern Biology* [now *BioSystems*] 5, 1–14.

Ebeling, W. (1976) *Strukturbildung bei Irreversiblen Prozessen*. B. G. Teubner, Leipzig.

Glansdorff, P., and I. Prigogine (1971) *Thermodynamic Theory of Structure, Stability and Fluctuations*. Wiley, New York.

Rosen, R. (1969) "Hierarchical Organization in Automata Theoretic Models of the Central Nervous System," pp. 21–35 in *Information Processing in the Nervous System*, ed. by K. N. Leibovic. Springer, Heidelberg.

Turing, A. M. (1952) "The Chemical Basis of Morphogenesis," *Philos. Trans. R. Soc. London Ser. B 237*, 5–72.

von Neumann (1966) *The Theory of Self-Reproducing Automata*, ed. by A. W. Burks. University of Illinois Press, Urbana.

4

Statistical Aspect of Biological Organization

Now I want to put the general biological picture described so far into formal, precise language. The essential idea is that the ecosystem (indeed any biological system) can be described in terms of sets of states and laws (perhaps unknown and generally probabilistic) governing the state-to-state transitions. In the case of the ecosystem the sets of states are the states of the biota and the states of the environment. For example, in the case of the flask each biota state is a collection of properties sufficient for characterizing everything in the flask within certain boundaries, while each environment state is the collection of properties sufficient for characterizing everything within the flask outside these boundaries. Also, we assume that the state-to-state transition probabilities are similar in similarly prepared flasks. However, we will consider the nature of the states and transition probabilities much more carefully later on.

4.1. BEHAVIORAL DESCRIPTION

For simplicity assume a finite set of states (which is physically realistic) and a discrete time scale (which is unrealistic, but good enough for present purposes, mathematically simpler, and accurate if the points on the scale correspond to the times at which observations are made). Our system is then described by a quadruple

$$\langle \mathbf{B}, \mathbf{E}, \Omega, \mathbf{T} \rangle \qquad (4.1)$$

where $\mathbf{B}$ is the set of biota states ($\beta^i \in \mathbf{B}$), $\mathbf{E}$ is the set of environment states ($\varepsilon^j \in \mathbf{E}$), Ω is the transition scheme of the ecosystem, and $\mathbf{T}$ is the set of times. Note that $\beta^i, \beta^j, \beta^k, \ldots$ indicate individual states of the biota and that $\varepsilon^i, \varepsilon^j, \varepsilon^k, \ldots$ indicate individual states of the environment. The transition scheme is

$$\Omega = \{ p[\beta^u(t+\tau), \varepsilon^v(t+\tau)|\beta^r(t), \varepsilon^s(t)]|u, r \in I, v, s \in J \} \quad (4.2)$$

where t is time, τ is a definite time interval, I is the index set of the biota, J is the index set of the environment, and no assumption about stationarity is made.*

The transition scheme is just the set of probabilities which determine the state of the biota and environment at time $t + \tau$ given their states at time t. In order to specify the scheme completely it is necessary to specify the initial probabilities of these states, i.e., the set

$$\{ p[\beta^r(t), \varepsilon^s(t)]\} \quad (4.3)$$

The initial probabilities play the role of initial conditions and the transition scheme plays the role of the equation of motion. The probabilities at time t may of course be determined by initial probabilities at some earlier point in time by continued application of the scheme. Also note that the state of the whole ecosystem is completely specified by specifying the states of the biota and the environment.

The above is a purely behavioral description of the ecosystem. As such it is virtually unimpeachable, for, admitting our minor simplifying assumptions, it says nothing at all about the world. Indeed, we may properly regard it as a kinematics rather than a dynamics. To make a more powerfully predictive theory it is necessary to say something about the probabilities themselves. This is possible, and in a quite general way, if we recruit the fundamental principles of biology to impose restrictions on these probabilities. This is precisely our program, but in order to develop it in a useful way it is first necessary to define statistical measures suitable for characterizing sets of probabilities.

4.2. STATISTICAL MEASURES

What we want, of course, is a measure of uncertainty, for it is the uncertainty of the environment which is significant from the standpoint of adaptability. The most convenient tools for this purpose are the entropy

---

*Strictly speaking, the transition scheme assigns the probabilities to the states, i.e., Ω is given by $\mathbf{B} \times \mathbf{E} \times \mathbf{B} \times \mathbf{E} \times \mathbf{T} \to \{ p(\beta^u, \varepsilon^v) \}$.

measures of information theory [Shannon and Weaver (1962); also Ashby (1956), who first used such measures to study the relation between "variety" and regulation]. Historically these measures derive from statistical thermodynamics, as statistical correlates of classical thermodynamic entropy. But the properties which make them serve this function so well also make them (and their generalizations) serve as ideal measures of statistical spread.

The entropy of a set of probabilities is

$$H[p(1),...,p(n)] = - \sum_{i=1}^{n} p(i)\log p(i) \qquad (4.4)$$

where H is entropy and $p(i)$ is the probability of event i. The entropy of a set of conditional probabilities is

$$H[p(1|j),...,p(n|j)] = - \sum_{i=1}^{n} p(i|j)\log p(i|j) \qquad (4.5)$$

where $p(i|j)$ is the probability of event i given event j. The conditional entropy is usually defined by weighting this quantity by the probabilities of the different possible given events:

$$H[p(1|1),...,p(i|j),...,p(n|m); p(1),...,p(m)]$$

$$= - \sum_{j=1}^{m} \sum_{i=1}^{n} p(j)p(i|j)\log p(i|j) \qquad (4.6)$$

The above definitions carry over with only slight generalization to transition schemes. Thus, the entropy of the ecosystem transition scheme is given by

$$H(\Omega) = - \sum p[\beta^r(t), \varepsilon^s(t)] \, p[\beta^u(t+\tau), \varepsilon^v(t+\tau)|\beta^r(t), \varepsilon^s(t)]$$

$$\times \log p[\beta^u(t+\tau), \varepsilon^v(t+\tau)|\beta^r(t), \varepsilon^s(t)] \qquad (4.7)$$

where the sum is taken over all $r, u \in I$ and $s, v \in J$ and for convenience H is written as $H(\Omega)$ rather than as a function of all its arguments. Notice that the entropy of the scheme is weighted by the probability of the initial ecosystem state. This is necessary if it is to serve as a measure which characterizes the whole scheme (Khinchin, 1957). Also, notice that in the present case we are dealing with a joint probability. Thus it is possible to define a marginal entropy for the behavior of the biota:

$$H(\omega) = -\sum p[\beta^r(t), \varepsilon^s(t)] \, p[\beta^u(t+\tau)|\beta^r(t)\varepsilon^s(t)]$$

$$\times \log p[\beta^u(t+\tau)|\beta^r(t), \varepsilon^s(t)] \tag{4.8}$$

where

$$\omega = \{ p[\beta^u(t+\tau)|\beta^r(t), \varepsilon^s(t)]\} \tag{4.9}$$

and the sum runs over u, r, and s. Likewise we can define a marginal entropy for the behavior of the environment:

$$H(\omega^*) = -\sum p[\beta^r(t), \varepsilon^s(t)] \, p[\varepsilon^v(t+\tau)|\beta^r(t), \varepsilon^s(t)]$$

$$\times \log p[\varepsilon^v(t+\tau)|\beta^r(t), \varepsilon^s(t)] \tag{4.10}$$

where

$$\omega^* = \{ p[\varepsilon^v(t+\tau)|\beta^r(t), \varepsilon^s(t)]\} \tag{4.11}$$

and the sum runs over v, r, and s. The first of these marginal entropies represents the average uncertainty in the behavior of the biota given the initial state of the biota and the initial state of the environment. The second represents the average uncertainty in the behavior of the environment, again given the initial states of the biota and environment.

There are two important conditional entropies for the scheme. First, suppose the state of the ecosystem at time t and the state of the environment at time $t + \tau$ are known. The conditional entropy of the complete scheme given the environment transition is

$$H(\omega|\omega^*) = -\sum p[\beta^r(t), \varepsilon^s(t), \varepsilon^v(t+\tau)]$$

$$\times p[\beta^u(t+\tau)|\beta^r(t), \varepsilon^s(t), \varepsilon^v(t+\tau)]$$

$$\times \log p[\beta^u(t+\tau)|\beta^r(t), \varepsilon^s(t), \varepsilon^v(t+\tau)] \tag{4.12}$$

where the sum is taken over all $r, u \in I$ and $s, v \in J$ and for convenience we again write H as $H(\omega|\omega^*)$ rather than as a function of all its arguments. Equation (4.12) is reasonable since we are now dealing with the average uncertainty in the biota transition given the initial state of the biota, the initial state of the environment, and the environment transition.

Similarly, it is possible to define an average uncertainty for the environment transition given the initial state of the biota, the initial state of the environment, and the biota transition. This is the conditional entropy of the complete scheme given the biota transition:

$$H(\omega^*|\omega) = -\sum p\left[\beta^r(t), \varepsilon^s(t), \beta^u(t+\tau)\right]$$

$$\times p\left[\varepsilon^v(t+\tau)|\beta^r(t), \varepsilon^s(t), \beta^u(t+\tau)\right]$$

$$\times \log p\left[\varepsilon^v(t+\tau)|\beta^r(t), \varepsilon^s(t), \beta^u(t+\tau)\right] \qquad (4.13)$$

where we use the same notation as above and the sum is again taken over all r, u, s, and v.

4.3. FUNDAMENTAL IDENTITY

The entropies defined above are connected by the following fundamental identity:

$$H(\omega) - H(\omega|\omega^*) \equiv H(\omega^*) - H(\omega^*|\omega) \qquad (4.14)$$

The left-hand side is the information which the behavior of the environment provides about the contemporaneous behavior of the biota (since it is the difference between the uncertainty in the behavior of the biota given the initial environment state and the uncertainty in its behavior given this initial state along with the environment transition). Similarly, the right-hand side is the information which the behavior of the biota provides about the contemporaneous behavior of the environment. Thus the identity simply asserts that the information which the behavior of the environment provides about the behavior of the biota is identically equal to the information which the behavior of the biota provides about the behavior of the environment.

To prove the identity it is only necessary to substitute the original definitions. The proof is given in an addendum to this chapter, in part because we are dealing with a generalized case and in part because it is useful to gain some experience with the indices.

4.4. FUNDAMENTAL INEQUALITY

The equation introduced in the previous section is useful as a starting point, but of course it could not possibly answer our question about the relationship between the statistical properties of the biota and the statistical properties of the environment. This is because it is an identity and therefore expresses a connection which is true regardless of the certainty or uncertainty of the environment, indeed, true whether or not we are dealing with a living or a nonliving system.

What we must do, to turn this connection into a useful statement about the real world, is characterize the statistical properties of the biota in a way which is independent of the statistical properties of its particular environment. This is possible if we imagine stressing the biota (subject to some condition) by varying the statistical character of the environment. The statistical properties of the stressed biota are its inherent (or potential) statistical properties.

The simplest way of stressing the biota is to find the most uncertain environment in which it is capable of remaining alive indefinitely. This is the extreme measure of the adaptability of the biota since it is the extreme measure of its capacity for coping with environmental uncertainty. More precisely, the adaptability of the biota is given by

$$H(\hat{\omega}^*) \equiv \max_{\omega^*} \left[H(\omega^*) \text{ such that } A \right] \tag{4.15}$$

where $H(\hat{\omega}^*)$ is a maximum over all possible transition schemes subject to condition A. This condition is: the half-life of the biota is not decreased at all.

The adaptability of the biota may also be defined, in completely equivalent fashion, as

$$H(\hat{\omega}^*) \equiv \max_{\omega^*} \left[H(\omega) - H(\omega|\omega^*) + H(\omega^*|\omega) \text{ such that } A \right] \tag{4.16}$$

where the condition A is the same as above. This may be abbreviated as

$$H(\hat{\omega}^*) \equiv H(\hat{\omega}) - H(\hat{\omega}|\hat{\omega}^*) + H(\hat{\omega}^*|\hat{\omega}) \tag{4.17}$$

where $\hat{\omega}$ is the biota transition scheme determined when $\omega^* = \hat{\omega}^*$. Thus $H(\hat{\omega}^*)$ is the entropy of the most uncertain environment which does not inevitably cause a catastrophic change in the biota, $H(\hat{\omega})$ is the potential uncertainty in the behavior of the biota, $H(\hat{\omega}|\hat{\omega}^*)$ is the potential uncertainty in the behavior of the biota given the behavior of the environment, and $H(\hat{\omega}^*|\hat{\omega})$ is the potential uncertainty in the behavior of the environment given the behavior of the biota.

Equation (4.17) implies the following fundamental inequality:

$$H(\hat{\omega}) - H(\hat{\omega}|\hat{\omega}^*) + H(\hat{\omega}^*|\hat{\omega}) \geqslant H(\omega^*) \tag{4.18}$$

This follows directly from the definition, for the biota could not remain in an allowable state (or even alive) if its adaptability were exceeded by the uncertainty of the environment.

Notice that the adaptability has three components:

1. *Behavioral uncertainty.* $H(\hat{\omega})$ represents the potential behavioral uncertainty because it is a measure of the ensemble of possible modes of behavior of the biota.
2. *Ability to anticipate.* $H(\hat{\omega}|\hat{\omega}^*)$ is the potential tolerance for decorrelation of the biota from the environment, i.e., the biota's *inability* to anticipate the state of the environment.
3. *Indifference.* $H(\hat{\omega}^*|\hat{\omega})$ is the potential tolerance to decorrelation of the environment from the biota, i.e., the biota's potential for insensitivity or *indifference* to the environment. This indifference may be either selective, in which case the system avoids harmful features of the environment, or nonselective, in which case it misses out on useful features. Nonselective indifference is error.*

The actual magnitudes of these components are not determined by equation (4.18), for a single equation can of course not fix four variables. In fact the actual values are related, but in a way which depends on considerations of reliability.

The imposition of condition A is quite strong, especially in the case of a complete ecosystem. However, we could weaken it to depend on the degree of damage to the system, for example, by excluding the pathological states, or by adding the restriction that the ecosystem is not inevitably pushed into a new qualitative mode of behavior (such as an alternative climax). For now, however, the above definition is the simplest and is sufficient for a first analysis of adaptability.

It is clear that direct measurements of adaptability are in general quite impractical (regardless of our choice of condition A). The problem of the theory of biological adaptability is not to make such direct measurements but rather to establish connections between adaptability and more conveniently measurable biological properties, or to compare measurable biological properties of systems which are prepared in such a way that their adaptability (in relation to some standard) is known (cf. Section 6.11).

---

*This distinction may be expressed notationally by the splitting

$$H(\hat{\omega}^*|\hat{\omega}) = H(\hat{\omega}^*|\hat{\omega})_s + H(\hat{\omega}^*|\hat{\omega})_n$$

using s and n to indicate the selective and nonselective components of indifference, respectively. The distinction is here treated as a conceptual one, valid in principle. In practice the problem of determining whether indifference is selective or nonselective is as difficult as determining whether or not it is advantageous from the standpoint of evolution.

4.5. REGULAR CAPACITY

The expression

$$H(\hat{\omega}^*) - H(\hat{\omega}|\hat{\omega}^*) \qquad (4.19)$$

equals the information which the environment provides about the biota under the statistically most unfavorable conditions. We may also define inherent statistical properties for the biota based on the maximization of information transfer. More precisely,

$$\max(\omega^*)[H(\omega) - H(\omega|\omega^*)] \equiv H(\tilde{\omega}) - H(\tilde{\omega}|\tilde{\omega}^*) \qquad (4.20)$$

where the maximization is taken over all possible transition schemes, and $\tilde{\omega}^*$ and $\tilde{\omega}$ are the transition schemes which give the maximum. This will be called the regular capacity since it is an alternative definition of adaptability which has the same structure as the capacity ordinarily defined in information theory (but I wish to emphasize that symbolic information processes have not so far been considered).

The capacity may also be defined, in equivalent fashion, as

$$\max(\omega^*)[H(\omega^*) - H(\omega^*|\omega)] \equiv H(\tilde{\omega}^*) - H(\tilde{\omega}^*|\tilde{\omega}) \qquad (4.21)$$

Thus we can write

$$H(\tilde{\omega}) - H(\tilde{\omega}|\tilde{\omega}^*) + H(\tilde{\omega}^*|\tilde{\omega}) \equiv H(\tilde{\omega}^*) \qquad (4.22)$$

which is, of course, just a special instance of our fundamental identity, equation (4.14). There are three possible cases:

$$H(\tilde{\omega}) - H(\tilde{\omega}|\tilde{\omega}^*) + H(\tilde{\omega}^*|\omega) \begin{cases} < H(\omega^*) \\ = H(\omega^*) \\ > H(\omega^*) \end{cases} \qquad (4.23)$$

where

$$H(\omega^*) \leqslant H(\hat{\omega}^*) \qquad (4.24)$$

(since the biota could not otherwise persist indefinitely). In the second of these cases (=), the system achieves the optimal potential ability to use information about the environment for adaptability, as originally defined, but may not utilize the selective component of indifference as effectively as possible. In the first case (<) the system pays for its adaptability in terms of a relatively high decorrelation and error. In the third case (>) the system maintains unused potentialities in one or more components of the adaptability.

4.6. TIME SCALES AND INFORMATION FLOW

Recall that the term on the left-hand side of our fundamental identity, equation (4.14), is the information which the behavior of the environment provides about the contemporaneous behavior of the biota, while the term on the right-hand side represents the information which the biota provides about the contemporaneous behavior of the environment. These interpretations carry over to the "stressed" equation (4.18), except that in this case we are dealing with the potential information which the biota provides about the environment (or conversely) under the most adverse allowable conditions. They also carry over to the capacity equation (4.23), except that in this case the potential information is by definition a maximum.

Needless to say these remarks are significant only if

$$H(\omega) - H(\omega|\omega^*) > 0 \qquad (4.25a)$$

$$H(\omega^*) - H(\omega^*|\omega) > 0 \qquad (4.25b)$$

This in turn is possible [for example, in the case of (4.25b)] only if either of the following conditions hold:

$$p[\beta^r(t), \varepsilon^s(t)] \neq p[\beta^r(t), \varepsilon^s(t), \beta^u(t+\tau)] \qquad \text{(all } r, s, u)$$

$$(4.26a)$$

$$p[\varepsilon^v(t+\tau)|\beta^r(t), \varepsilon^s(t)] \neq p[\varepsilon^v(t+\tau)|\beta^r(t), \varepsilon^s(t), \beta^u(t+\tau)]$$

$$\text{(all } r, s, u, v) \quad (4.26b)$$

This follows directly from definitions (4.10) and (4.13), since these imply that $H(\omega^*) - H(\omega^*|\omega) = 0$ only if both conditions fail. Actually the failure of (4.26a) implies the failure of (4.26b), as proved at the end of this section. Right now, however, I wish to consider the significance of these conditions.

Condition (4.26a) may be rewritten as

$$p[\beta^u(t+\tau)|\beta^r(t), \varepsilon^s(t)] \neq 1 \qquad (4.27)$$

Thus the first condition just means that the future state of the biota must not be completely determined by its present state and the state of the environment (for otherwise the amount of information which the biota transition could provide would certainly be zero).

The second condition (4.26b) is a bit more subtle. Clearly this condition implies that the state of the biota at time $t + \tau$ affects or is affected by the state of the environment at time $t + \tau$. But, of course, this violates causality, for events in the one could not be instantaneously communicated to the other. Naturally, the biota may be able to anticipate (more or less) the behavior of the environment. But knowledge of $\beta^u(t + \tau)$ is still superfluous in this case once $\beta^r(t)$ and $\varepsilon^s(t)$ are given, for the degree of correlation could not exceed the correlation between the state of the ecosystem at time t and the state of the biota at time $t + \tau$.

The above difficulty arises from the fact that we have allowed observations at only certain discrete intervals of time (for simplicity, intervals of equal length) and have regarded all the observable states on an equal footing. But, of course, the ecosystem is alive and functioning between these observations. This, in fact, is the solution to our problem, for the information which the biota provides about the contemporaneous behavior of the environment on the longer time scale equals the information which flows from the environment to the system on the shorter (in reality continuous) time scale. This is realistic because changes in the environment often take place in a series of correlated steps, the first of which have no significant effect on the system. For example, clouds and wind may precede the onset of rain. The slight change in light intensity or net force on an organism may have hardly any significant physiological effect, but it does signal the onset of a situation which is likely to be of considerable importance. Alternatively, the change in the environment may be quite sudden and long-lasting, but nevertheless require a considerable amount of time to have a significant effect on the biota. For example, the sudden onset of rain may also serve as a signal as long as the system reacts quickly enough. In either case the states of both the biota and the environment at the end of the time interval are correlated to the state of the environment during the time interval. But since we have not included these intermediate states, it will appear that the end state of the biota provides information about the (contemporaneous) end state of the environment.*

We could, of course, also consider the opposite direction, that is, equation (4.25a). In this case the end state of the environment provides information about the contemporaneous end state of the biota. In either

---

*Anticipation may also involve adjustment to favorable changes, as in seed germination. Built into the trigger mechanism of germination is a process (adjusted in the course of evolution) which in effect calculates whether present moisture or other relevant factors are sufficient to indicate a sufficient long-term supply of water, assuming a situation in which this is a critical problem for the growing plant. The seed state is thus dryness-insensitive, but nevertheless not so insensitive that present moisture cannot serve as a signal. Once the seed germinates, it clearly switches into a dryness-sensitive mode of behavior, one in which degree of moisture can no longer be regarded as a signal, which is why seed germination is so often such a peculiarly stubborn process.

case, however, the actual direction of information flow is from environment to biota (since it is the biota, not the environment, which responds to signals and therefore creates the correlation).

PROOF THAT THE FAILURE OF (4.26a) IMPLIES THE FAILURE OF (4.26b). First suppose that condition (4.26a) fails but that condition (4.26b) holds. According to (4.27) this failure implies

$$p[\beta^u(t+\tau)|\beta^r(t),\varepsilon^s(t)]=1 \tag{4.28}$$

Now consider the identity .

$$p[\beta^r(t),\varepsilon^s(t),\beta^u(t+\tau)]\,p[\varepsilon^v(t+\tau)|\beta^r(t),\varepsilon^s(t),\beta^u(t+\tau)]$$

$$\equiv p[\beta^r(t),\varepsilon^s(t)]$$

$$\times p[\varepsilon^v(t+\tau)|\beta^r(t),\varepsilon^s(t)]\,p[\beta^u(t+\tau)|\varepsilon^v(t+\tau),\beta^r(t),\varepsilon^s(t)]$$

$$\tag{4.29}$$

If (4.26a) fails this may be replaced by

$$p[\varepsilon^v(t+\tau)|\beta^r(t),\varepsilon^s(t),\beta^u(t+\tau)]$$

$$\equiv p[\varepsilon^v(t+\tau)|\beta^r(t),\varepsilon^s(t)]$$

$$\times p[\beta^u(t+\tau)|\varepsilon^v(t+\tau),\beta^r(t),\varepsilon^s(t)] \tag{4.30}$$

This implies that (4.26b) holds only if $p[\beta^u(t+\tau)|\varepsilon^v(t+\tau),\beta^r(t),\varepsilon^s(t)]$ is not equal to one or zero. But this contradicts (4.28).

Now suppose that condition (4.26b) fails. In this case (4.29) may be replaced by

$$p[\beta^r(t),\varepsilon^s(t),\beta^u(t+\tau)]$$

$$\equiv p[\beta^r(t),\varepsilon^s(t)]\,p[\beta^u(t+\tau)|\varepsilon^v(t+\tau),\beta^r(t),\varepsilon^s(t)] \tag{4.31}$$

which is entirely consistent with condition (4.26a).

The connection between (4.26a) and (4.26b) means that the behavior of the biota provides information about the contemporaneous behavior of the environment only if the former is not completely determined by the prior state of the ecosystem. But even in this case there are two possible sources of information. The first is the correlation between the state of the biota and environment arising from events occurring during the time interval. The second arises directly from the time development of the ecosystem itself. This is inevitable because we begin, for example, by specifying $\beta^r(t)$, $\varepsilon^s(t)$, and $\beta^u(t+\tau)$. Thus specifying $\varepsilon^v(t+\tau)$ certainly adds to our information, provided that this state is not completely determined by the others.

Note that if the time development of the ecosystem is completely determined there can be no causality violation, but that if there is no causality violation the time development of the ecosystem can still be indeterminate. Thus our formalism is actually consistent with causality in the sense that it is possible for the equations to be nontrivial in the absence of any instantaneous transfer of influence from environment to biota. However, this leaves open the really important question of why the state of the biota would ever assume a state appropriate to the state of the environment. To deal with the mechanism of correlation it is necessary to expand the formalism in such a way that condition (4.26b) assumes a form consistent with causality.

4.7. INFORMATION TRANSFER PICTURE

The apparent inability of our equations to deal with the mechanism (as opposed to the consequences) of correlating is possible because we have thrown out certain information, or, more precisely, because the processes of the biota are such that certain details of its behavior, important for information transfer, are dissipated away.

In order to include this dissipation it is necessary to allow a more detailed description of the ecosystem, one which admits events occurring between the intervals of observation, and which also admits that certain states of the biota differ from one another only in the sense that they are transmitting different signals (Conrad, 1977). Suppose that each biota state is really a family of essentially similar states:

$$\beta^i = \{\beta^{i1}, \ldots, \beta^{in}\} \tag{4.32}$$

where each of the β^{ir} is macroscopically distinct but identical as regards each organism's probability for reproducing. The transition scheme is now

$$\underline{\Omega} = \{\, p\big[\beta^{ui}(t+\tau), \varepsilon^v(t+\tau), \beta^{gm}(t+\tau_B), \varepsilon^h(t+\tau_B),$$

$$\beta^{ek}(t+\tau_E), \varepsilon^f(t+\tau_E)|\beta^{rj}(t), \varepsilon^s(t)\big]\} \tag{4.33a}$$

or equivalently

$$\underline{\Omega} = \big\{ p\big[\beta^{ui}(t+\tau), \varepsilon^v(t+\tau)|\beta^{rj}(t), \varepsilon^s(t), \beta^{ek}(t+\tau_E), \varepsilon^f(t+\tau_E),$$

$$\beta^{gm}(t+\tau_B), \varepsilon^h(t+\tau_B)\big]\, p\big[\beta^{ek}(t+\tau_E), \varepsilon^f(t+\tau_E), \beta^{gm}(t+\tau_B),$$

$$\varepsilon^h(t+\tau_B)|\beta^{rj}(t), \varepsilon^s(t)\big]\big\} \qquad (4.33b)$$

where superscripts u, r, e, g run over index set I as usual and the added superscripts i, j, k, m run over the new index set K, and in general $\tau > \tau_B > \tau_E$. τ_E is the correlation time of the environment, i.e., $\tau - \tau_E$ is the maximum time the system has to adjust to $\varepsilon^v(t+\tau)$, or, alternatively, τ_E is the smallest time interval for which it is possible to eliminate $\varepsilon^v(t+\tau)$ from (4.33) without significantly changing the probabilities. τ_B is the correlation time of the environment, i.e., the minimum amount of time required for the biota to assume a state appropriate to the environment at time $t+\tau_E$ [or $t+\tau$, since $\varepsilon^v(t+\tau)$ and $\varepsilon^f(t+\tau_E)$ are highly correlated].

Equation (4.33a) [or equivalently (4.33b)] is just the analog of our original transition law, equation (4.2), but extended to include intermediate states so that the behavior of the ecosystem is now described by a higher-order Markov process. We can also write analogs of the marginal transition laws, viz.

$$\underline{\omega}^* = \big\{ p\big[\varepsilon^v(t+\tau), \varepsilon^f(t+\tau_E), \varepsilon^h(t+\tau_B)|\beta^{rj}(t), \varepsilon^s(t)\big]\big\} \qquad (4.34a)$$

$$\underline{\omega} = \big\{ p\big[\beta^{ui}(t+\tau), \beta^{ek}(t+\tau_E), \beta^{gm}(t+\tau_B)|\beta^{rj}(t), \varepsilon^s(t)\big]\big\}$$

$$(4.34b)$$

Thus the analog of the condition for information transfer is

$$p\big[\varepsilon^v(t+\tau), \varepsilon^h(t+\tau_B), \varepsilon^f(t+\tau_E)|\beta^{rj}(t), \varepsilon^s(t)\big]$$

$$\neq p\big[\varepsilon^v(t+\tau), \varepsilon^h(t+\tau_B), \varepsilon^f(t+\tau_E)|\beta^{ui}(t+\tau), \beta^{ek}(t+\tau_E),$$

$$\beta^{gm}(t+\tau_B), \beta^{rj}(t), \varepsilon^s(t)\big] \qquad (4.35)$$

This does not necessarily violate causality since, e.g., $\beta^{ui}(t+\tau)$ may provide some information about $\varepsilon^f(t+\tau_E)$. However, what is important is that it provide some information about $\varepsilon^v(t+\tau)$, and of course this feature does violate causality.

Now suppose that

1. $\varepsilon^v(t + \tau)$ and $\varepsilon^h(t + \tau_B)$ are strongly correlated to $\varepsilon^f(t + \tau_E)$.
2. $\beta^{gm}(t + \tau_B)$ is strongly correlated to $\varepsilon^f(t + \tau_E)$.
3. $\beta^{ui}(t + \tau)$ is strongly correlated to $\beta^{gm}(t + \tau_B)$.
4. $\beta^{ek}(t + \tau_E)$ belongs to β^r.
5. $\beta^{gm}(t + \tau_B)$ belongs to β^u.

Conditions (1)–(3) are restatements of our comments about correlation times and just mean that the environment state at the earlier times serves as a message to the biota about the environment state at the later time. Condition (4) means that the biota is in a state which is essentially similar to its initial state when it receives this message. Condition (5) means that its state is essentially similar to either the initial or final state, but our description would have to be still more refined to include these.

Conditions (1)–(4) imply that the first probability in equation (4.33b)

$$p\big[\beta^{ui}(t + \tau), \varepsilon^v(t + \tau)|\beta^{ek}(t + \tau_E), \varepsilon^f(t + \tau_E),$$

$$\beta^{gm}(t + \tau_B), \varepsilon^h(t + \tau_B), \beta^{rj}(t), \varepsilon^s(t)\big] \qquad (4.36)$$

is essentially equal to either zero or one (depending on whether the correlations are positive or negative). This means that it is possible (for the purpose of calculating entropies) to approximate Ω by

$$\big\{ p\big[\beta^{gm}(t + \tau_B), \varepsilon^f(t + \tau_E)|\beta^{rj}(t), \varepsilon^s(t)\big]\big\} \qquad (4.37)$$

since high correlation means that the final states (at time $t + \tau$) are irrelevant to behavioral uncertainty in any case. Expression (4.37) is the same as the second probability in equation (4.33b), except that $\beta^{ek}(t + \tau_E)$ and $\varepsilon^h(t + \tau_B)$ are thrown out [since they are redundant according to conditions (1) and (4)]. It describes the behavioral uncertainty generated by the ecosystem in time, as opposed to expression (4.36), which describes the information transfer underlying the correlation between biota and environment. The redundant states, $\beta^{ek}(t + \tau_E)$ and $\varepsilon^h(t + \tau)$, can also be discarded from expression (4.36) and in fact the states at time t can be discarded as well.

Using (4.37) the condition for information transfer becomes

$$p\big[\varepsilon^f(t + \tau_E)|\beta^{rj}(t), \varepsilon^s(t)\big] \neq p\big[\varepsilon^f(t + \tau_E)|\beta^{rj}(t), \varepsilon^s(t), \beta^{gm}(t + \tau_B)\big]$$

$$(4.38)$$

This allows $\beta^{gm}(t + \tau_B)$ to provide information about $\varepsilon^f(t + \tau_E)$. This is also possible in equation (4.35), but it is now the significant point since condition (1) makes it possible to replace equation (4.38) by

$$p\left[\varepsilon^v(t + \tau)|\beta^{rj}(t), \varepsilon^s(t)\right] \neq p\left[\varepsilon^v(t)|\beta^{rj}(t), \varepsilon^s(t), \beta^{gm}(t + \tau_B)\right]$$

$$(4.39)$$

This definitely does not violate causality and moreover allows anticipation of the environment. Furthermore, we may pass back to the original condition (5) to replace $\beta^{gm}(t + \tau_B)$ with $\beta^u(t + \tau)$. Thus this condition simply expresses the ability of biological systems to react to the environment in such a way that they utilize temporal correlations in the environment. The expression of this ability appears as a violation of causality because we have ignored the preliminary changes in the environment and also the reaction of the biological systems to these changes. The resulting description is useful, however, because the possible reactions are restricted to the members of a family of essentially similar states. Indeed, this is necessary, for if the states which mediate the reactions have significantly different effects on reproduction probabilities the environment states which precede them could not be regarded as signals.

Notice that in general $\beta^{gm}(t + \tau_B)$ is not the most suitable biota state relative to $\varepsilon^f(t + \tau_E)$. In other words, the correlation process described above does not avoid error. What it does, however, is reduce the functional importance of error to a minimum.

In one sense we are back where we started—to the point where the accumulation of information in the purest sense only arises from the time development of the ecosystem [condition (4.26a)]. Why then insist on our information transfer condition? The reason is that the correlation process is without doubt a real process, indeed a *sine qua non* of life. Thus the details which we discard are not subjective or arbitrary. Quite to the contrary, they are just the details discarded by biological systems themselves. However, like all forms of irreversibility, this loss of information is hard to discern if examined in too much detail.

4.8. FURTHER REMARKS ON INFORMATION TRANSFER

So far we have considered only two intermediate instants of time, τ_E and τ_B. However, with only straightforward generalization we could consider any number of instants, $\tau_1, \ldots, \tau_n$, where $\tau_n = \tau$. The results are exactly the same, except that we can now describe systems which operate on the basis of a number of different correlation times (which can be taken to be the minimum of τ_B minus the minimum of τ_E). For example, suppose that the environment is described (as it should be) in terms of a number of significant variables—light, temperature, salinity, chemical concentrations, and so forth—each of which may change at different rates and produce errors which are more or less significant for the organism. The importance of correlation time and the relative strengths of the various correlations [(1–5) of the previous section] may assume different strengths for these different aspects of the environment, including significant strengths over time periods exceeding the arbitrary time interval, τ. In this situation the relaxation time of environmental change is long relative to τ and it is possible for τ_B to exceed τ by quite a bit [in which case τ must be replaced by $n\tau$ in conditions (1–5)]. Essentially the advantages of rapid correlation and response are traded for more economical but slower mechanisms of correlation and (frequently) more permanent forms of response.

Time scales are important since they determine the direction of information flow within the biota and also have a major influence on patterns of adaptability in populations and communities.

4.9. TWO-TIME FORMALISM

Transition schemes (determined from a large ensemble of similarly prepared systems) make it possible to define statistical quantities for the behavioral uncertainties generated by ecological or other, less global, biological systems (Conrad, 1976). An alternative procedure is to use the same large ensemble to determine the probabilities for the occurrence of the various states at pairs of times, t and $t + \tau$. The resulting statistical quantities characterize the behavioral uncertainty exhibited by the system (Conrad, 1972a,b). To develop such measures consider the joint entropy $H[\beta(t + \tau), \varepsilon(t + \tau), \beta(t), \varepsilon(t)]$. Expanding this gives a variant form of the fundamental identity

$$H[\beta(t + \tau), \beta(t)] - H[\beta(t + \tau), \beta(t)|\varepsilon(t + \tau), \varepsilon(t)]$$

$$+ H[\varepsilon(t + \tau), \varepsilon(t)|\beta(t + \tau), \beta(t)] \equiv H[\varepsilon(t + \tau), \varepsilon(t)] \quad (4.40)$$

where

$$H[\beta(t + \tau), \beta(t)] = -\sum p[\beta^u(t + \tau), \beta^v(t)] \log p[\beta^u(t + \tau), \beta^v(t)]$$

$$(\text{all } u, v) \quad (4.41)$$

and

$$H\left[\beta(t+\tau),\beta(t)|\varepsilon(t+\tau),\varepsilon(t)\right]$$

$$= -\sum p\left[\varepsilon^v(t+\tau),\varepsilon^s(t)\right]p\left[\beta^u(t+\tau),\beta^r(t)|\varepsilon^v(t+\tau),\varepsilon^s(t)\right]$$

$$\times\log p\left[\beta^u(t+\tau),\beta^r(t)|\varepsilon^v(t+\tau),\varepsilon^s(t)\right] \qquad \text{(all } u,v,r,s\text{)} \quad (4.42)$$

$H[\varepsilon(t+\tau),\varepsilon(t)]$ and $H[\varepsilon(t+\tau),\varepsilon(t)|\beta(t+\tau),\beta(t)]$ are given by analogous expressions. The identity (4.40) is proved in the addendum, again by simply substituting in the definitions. Notice that all these entropies are functions of τ.

By increasing $H[\varepsilon(t+\tau),\varepsilon(t)]$ to the maximum consistent with system function, it is possible (as with the transition scheme formalism) to determine inherent, maximum allowable values of the entropies and therefore to form a variant of the fundamental inequality:

$$H\left[\hat{\beta}(t+\tau),\hat{\beta}(t)\right] - H\left[\hat{\beta}(t+\tau),\hat{\beta}(t)|\hat{\varepsilon}(t+\tau),\hat{\varepsilon}(t)\right]$$

$$+ H\left[\hat{\varepsilon}(t+\tau),\hat{\varepsilon}(t)|\hat{\beta}(t+\tau),\hat{\beta}(t)\right] \geqslant H\left[\varepsilon(t+\tau),\varepsilon(t)\right] \qquad (4.43)$$

Analogs of the information transfer equation can be developed by starting with a joint entropy taken over finer states and a number of time periods, e.g., by starting with

$$H\left[\underline{\beta}(t+\tau),\varepsilon(t+\tau),\underline{\beta}(t+\tau_B),\varepsilon(t+\tau_B),\right.$$

$$\left.\underline{\beta}(t+\tau_E),\varepsilon(t+\tau_E),\underline{\beta}(t),\varepsilon(t)\right] \qquad (4.44)$$

where $\underline{\beta}$ indicates the arguments are probabilities defined over the $\beta^{ui},\beta^{rj},\dots$. By discarding irrelevant details [conditions (1–5), Section 4.7], it is possible to express this equation in a form which reflects the temporal flow of influence responsible for correlation between system and environment:

$$H\left[\hat{\underline{\beta}}(t+\tau),\hat{\underline{\beta}}(t)\right] - H\left[\hat{\underline{\beta}}(t+\tau_B),\hat{\underline{\beta}}(t)|\hat{\varepsilon}(t),\hat{\varepsilon}(t+\tau_E)\right]$$

$$+ H\left[\hat{\varepsilon}(t+\tau_E),\hat{\varepsilon}(t)|\hat{\underline{\beta}}(t+\tau_B),\hat{\underline{\beta}}(t)\right] \geqslant H\left[\varepsilon(t+\tau),\varepsilon(t)\right] \quad (4.45)$$

where τ_E is the time period concomitant to initial change in (some feature of) the environment and τ_B is the time period concomitant to response.

The difference between the two-time and transition scheme formalisms can be seen by comparing equations (4.41) and (4.8). $H(\hat{\omega})$ describes the behavioral uncertainty generated by the biota *per se*, i.e., the uncertainty in its behavior given probabilities for the prior state of the world. $H[\beta(t + \tau), \beta(t)]$ describes the behavioral uncertainty exhibited by the biota, i.e., the uncertainty in its behavior *per se*, without any reference to the initial state of the environment. Thus some of this uncertainty is due to our ignorance of environmental influence, whereas this is not the case for $H(\omega)$. Similarly some of the uncertainty in the decorrelation term in the two-time formalism is due to ignorance about the influence of the initial biota state and the error contribution is partly due to ignorance about the influence of the initial environment state.

Despite the above interpretive differences, the outer form of the two equations is precisely the same. Thus, if we made the following alternative definitions:

$$\omega \rightarrow \{\, p[\beta^r(t)]\, p[\beta^u(t + \tau)|\beta^r(t)]\} \qquad (u, r \in I) \qquad (4.46)$$

$$\omega^* \rightarrow \{\, p[\varepsilon^s(t)]\, p[\varepsilon^v(t + \tau)|\varepsilon^s(t)]\} \qquad (v, s \in J) \qquad (4.47)$$

and substituted in equation (4.40), the resulting expression would clearly have the same structure as equation (4.14), the only difference being in the underlying interpretation of $H(\omega)$, $H(\omega^*)$, $H(\omega|\omega^*)$, and $H(\omega^*|\omega)$. Thus any statement (based on this structure) which can be made about adaptability in one interpretation corresponds to a statement which could be made about adaptability in the other interpretation. This is shown in the Addendum.

Actually it is possible to decompose the two-time formalism into two parts, one which directly reflects inherent statistical attributes of biological processes and one which reflects in part lack of information about initial conditions. This can be seen by expanding, e.g., the two-time anticipation entropy:

$$H\left[\hat{\beta}(t + \tau), \hat{\beta}(t)|\hat{\varepsilon}(t), \hat{\varepsilon}(t + \tau)\right]$$

$$= \tfrac{1}{2}\{H\left[\hat{\beta}(t)|\hat{\varepsilon}(t), \hat{\varepsilon}(t + \tau), \hat{\beta}(t + \tau)\right] + H\left[\hat{\beta}(t)|\hat{\varepsilon}(t), \hat{\varepsilon}(t + \tau)\right]$$

$$+ H\left[\hat{\beta}(t + \tau)|\hat{\varepsilon}(t), \hat{\varepsilon}(t + \tau), \hat{\beta}(t)\right] + H\left[\hat{\beta}(t + \tau)|\hat{\varepsilon}(t), \hat{\varepsilon}(t + \tau)\right]\}$$

$$(4.48)$$

The first term in the curly brackets is a measure of the uncertainty of the past state of the biota given all information about the present state and the past and present state of the environment. Therefore it represents the degree to which the biota can forget the past, including the degree to which it can forget the effects of past disturbances. If the expansion were expressed in terms of finer states, this selective dissipation would also reflect information processing in the strict sense. The second term (which is always larger, or at least never smaller) reflects the same processes as the first, except that in addition it reflects the absence or lack of knowledge about the initial biota state. Thus the difference between the first and second terms represents the amount of uncertainty which arises because of this lack of knowledge and therefore the difference between the exhibited and generated contribution. The third term represents the uncertainty generated by the system in the forward direction of time, and the fourth term represents the uncertainty exhibited by the system. These reflect noise processes when formulated in terms of finer states.

The advantage of the transition scheme formalism is that it directly reflects the inherent statistical attributes of biological processes (because the influence of prior states is included in the transition schemes). An advantage of the two-time formalism is that it is possible to dissect the statistical processes of the biota into more refined components (e.g., into components associated with loss of information about past states of the biota). This is because the influence of prior states is not represented beforehand, so that these components are not merged into the transition scheme. However, this also makes the two-time formalism more cumbersome than the transition scheme formalism.

Note that the alternative forms of expansion (e.g., in terms of $H[\beta(t + \tau)]$ or $H[\beta(t + \tau), \beta(t), \varepsilon(t)]$) are not particularly useful, in the first instance because the important idea of behavioral uncertainty is lost and in the second because the important distinction between biota and environment uncertainty is lost.

4.10. DIVERSITY OF BEHAVIOR

The entropies and conditional entropies defined so far measure the uncertainty of ecosystem schemes (or behavior), not the diversity of ecosystem states. Thus the schemes (4.9) and (4.11) [or (4.46) and (4.47)] may generate sequences which are extremely diverse, but nevertheless quite determinate. For example, the environment, the biota, or some component of the biota might go through a periodically repeating sequence of states. The diversity of states in this sequence would not appear in the measures defined till now because these measures have been defined on the transition probabilities, not on the number of possible states and their relative frequency of occurrence.

The distinction between uncertainty and diversity of behavior is of fundamental importance. This is an especially critical point because the entropy-type measures which we have used to characterize uncertainty are also excellent indices of diversity. The appropriate definitions in this case are:

$$H[\beta(t)] = -\sum p[\beta^r(t)]\log p[\beta^r(t)] \qquad \text{(all } r \in I) \qquad (4.49)$$

$$H[\varepsilon(t)] = -\sum p[\varepsilon^s(t)]\log p[\varepsilon^s(t)] \qquad \text{(all } s \in J) \qquad (4.50)$$

where $p[\beta^r(t)]$ and $p[\varepsilon^s(t)]$ are the relative proportions of the states β^r and ε^s at time t, $H[\beta(t)]$ is the diversity of the biota states, and $H[\varepsilon(t)]$ is the diversity of the environment states. In both cases these diversities may be thought of as the uncertainties generated by a random process selecting states from either the set of biota or the set of environment trajectories. In other words, if the system were entirely deterministic the randomness would arise exclusively from the sampling process rather than from the dynamics of the system itself.

We can also define conditional diversities, $H[\beta(t)|\varepsilon(t)]$ and $H[\varepsilon(t)|\beta(t)]$. These are defined precisely as in equation (4.5), except that what we are dealing with is a random sampling process of, say, biota states which occur at the same time as some particular environment state, but averaged over all environment states.

The following identity, as usual, holds among the diversities:

$$H[\beta(t)] - H[\beta(t)|\varepsilon(t)] + H[\varepsilon(t)|\beta(t)] \equiv H[\varepsilon(t)] \qquad (4.51)$$

Moreover, as before, we can assume that the potential magnitude of the left-hand side is greater than or equal to the maximum tolerable value of the right-hand side, for otherwise the biota would be driven into nonliving states. However, this relationship does not take temporal correlations into account and therefore does not admit an information transfer picture.

Another way of looking at the diversity equation is as a special case of the two-time equation, in the limit where τ goes to zero. Recalling equation (4.43) and taking this limit,

$$H[\hat{\beta}(t)] - H[\hat{\beta}(t)|\hat{\varepsilon}(t)] + H[\hat{\varepsilon}(t)|\hat{\beta}(t)] \geqslant H[\varepsilon(t)] \qquad (4.52)$$

While this relation is true and puts some restriction on the diversities, it completely obscures the distinction between complex but deterministic sequences in both the biological system (e.g., cell cycle) and environment (e.g., cycle of the planet) with uncertain variations on these sequences. For example, in a completely quiet, nondiverse environment, the error might be low but $H[\hat{\beta}(t)|\hat{\epsilon}(t)]$ might be high because it is unknown in which stage of a cycle process the biota is. This term thus reflects functional attributes of the biota quite distinct from anticipation.

The behavioral diversity must be greater than or equal to the behavioral uncertainty. This is because we have included all of the behavioral variation in our definition of diversity. Thus we can write

$$H(\hat{\omega})/H[\hat{\beta}(t)] = \gamma \qquad (4.53)$$

where $H[\hat{\beta}(t)]$ is the maximum value of $H[\beta(t)]$ and $0 \leqslant \gamma \leqslant 1$. γ is an index of the fraction of the biota's total behavioral *variation* associated with uncertainty generated by its state transitions. If $\gamma = 0$, the behavior of the biota is completely deterministic; if $\gamma = 1$, all the behavioral variation arises from variability generated by the transition scheme.

The concept of behavioral diversity perhaps corresponds in an intuitive sense to complexity of behavior, but the notion of complexity raises so many different questions, under active investigation from the standpoint of automata and computers (Bremermann, 1974), that it is safer to use the word *diversity*.

4.11. THE VARIABILITY OF BIOLOGICAL MATTER

The types of states over which our entropies are defined are listed in Table 4.1 (see also Figure 4.1). These include the adaptively distinct states (the $\beta^u, \beta^r, \ldots$), each associated with different reproduction probabilities for the various organisms in the biota relative to the given environment. They also include states which are adaptively equivalent but macroscopically distinct and potentially informationally distinct (i.e., the $\beta^{ui}, \beta^{rj}, \ldots$). These are essentially the same as regards the organism's immediate chance for reproduction, but potentially quite different as regards their effect on future behavior. Thus they serve to represent messages about the future behavior of the environment. Their effect on the reproduction probabilities depends only on the appropriateness of the behavior to which they give rise.

The entropies themselves describe either the variability of the biota and environment or the diversity of their total variation. There are certain general trends connecting these different measures, but pursuit of these must be deferred to Chapter 9. In general terms the potential behavioral uncertainty of the biota must not be less than the actual behavioral

TABLE 4.1
Layers of Behavioral Description[a]

Type of state	Characteristic
1. Adaptively distinct	Associated with different selective advantage relative to the contemporaneous environment, i.e., subserves different fitting relations between system and environment
2. Informationally distinct	Sets of macroscopically distinct but adaptively equivalent states serving as precursors to different adaptively distinct states; subserves information transfer
3. Informationally equivalent	Macroscopically distinct states serving as precursors to the same adaptively distinct state; subserve reliability
4. Macroscopically equivalent	Microscopically distinct states equivalent as regards all aspects of system function; reflects openness of system and contributes to integrity of system as dissipative pattern; may also subserve reliability

[a]Adaptively and informationally distinct states both contribute to functionally distinct macroscopic diversity, informationally equivalent states contribute to functionally equivalent macroscopic diversity, and macroscopically equivalent states contribute to thermodynamic entropy (cf. Table 3.1).

uncertainty of the environment, but with the provision that the former may be decreased at the expense of greater anticipation (based on information transfer) or error. Roughly speaking, these remarks are also true for the diversities, except that the behavioral diversity of the biota may be extremely high even in a constant environment and (we will see) is influenced by quite different factors.

Actually the adaptively and macroscopically distinct states are not sufficient to encompass all aspects of biological variability. This is because the transfer and processing of signals are subject to error and therefore inevitably raise the problem of reliability. Systems often deal with this by what might be called state multiplication. In this case each of the macroscopically distinct states belongs to an equivalence class of informationally equivalent states with some kind of redundancy structure. The equivalence class itself (to be denoted by $[\beta^u]_c$) is genuinely informationally distinct, but its members are informationally equivalent in the sense that they are correlated to the environment and the future behavior of the biota in the same way. They have the property that state transitions resulting from noise usually lead to other states in the same equivalence class. We return to this point in Chapters 8 and 9.

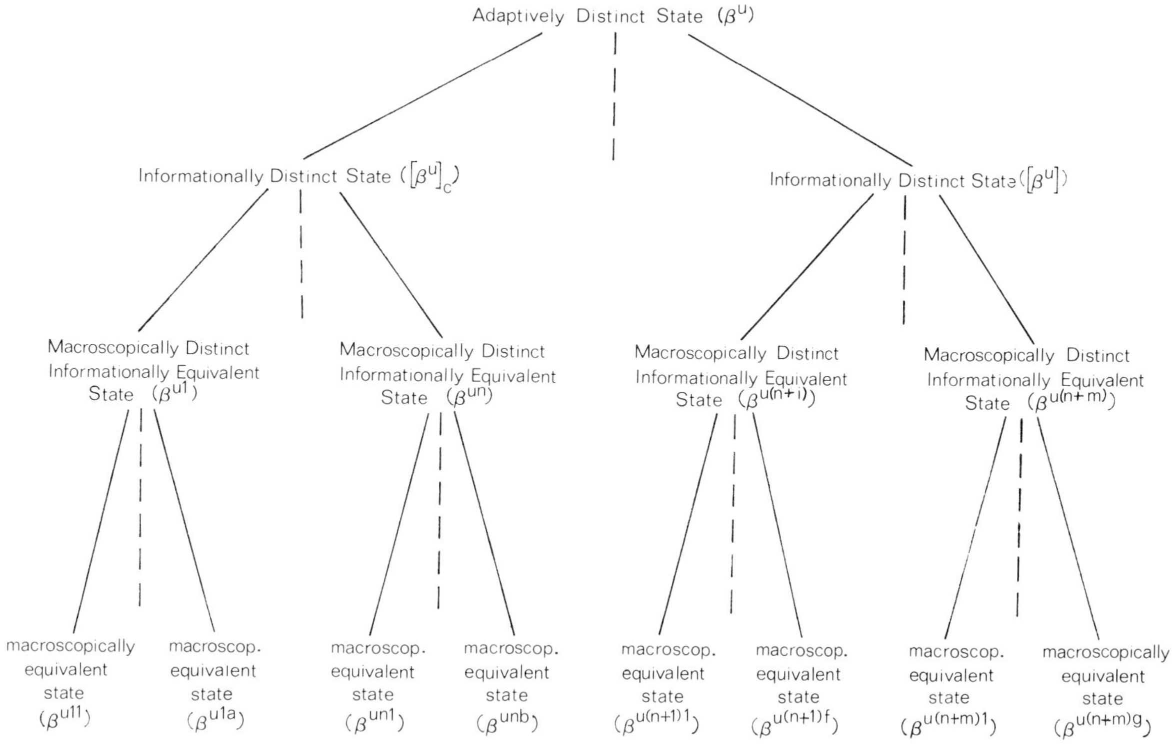

FIGURE 4.1. Layers of biological variability. Each type of state consists of finer states which are essentially equivalent in some respect, e.g., with respect to adaptive value, information transfer, macroscopic distinguishability. States identical as regards information transfer serve the function of reliability by virtue of having some redundancy property. Macroscopically equivalent states reflect the fact that biological systems are open systems, hence most appropriately viewed as patterns of activity. They may also contribute to reliability, but by virtue of the fact that the system tends to assume the macroscopic state which is most probable and therefore which comprises the greatest number of microscopic states.

There is yet another layer of variability. According to classical physics the state of the world at any time is uniquely determined by the initial conditions and laws of motion; in other words, at the finest level of description the world is completely reversible. If we admit a certain degree of ignorance at the very beginning, the classical equations of motion ensure that this degree of ignorance (or lack of information) is conserved. Again this is true for the quantum mechanical equations of motion (but not for the measurement process). These considerations imply that a completely detailed description of an ecosystem should obey conservation laws of the above type (assuming that biological systems obey the laws of physics). But we have not assumed any such restriction in our original transition scheme [equation (4.2)], and indeed have explicitly excluded it in our information transfer scheme [equation (4.26b)], since this necessarily includes the influence of prior states at a number of points in time.

The reconciliation of these fundamentally different types of description again lies in the passage from reversible to irreversible dynamics. In the case of our more detailed description each of the β^{ui} is itself a set of still more refined states, β^{uik}, obeying reversible, Hamiltonian dynamics (see Figure 4.1). These are the states of the ecosystem as they would be defined in physics. The passage to our cruder, irreversible description is possible because certain of these states can be interchanged without having any effect whatsoever on the reproduction probabilities. More formally, $\beta^{uik}(t + \tau)$ is essentially the same as $\beta^{uih}(t + \tau)$ if the transition scheme (4.33) is the same whether β^{ui} is replaced by β^{uik} or β^{uih}.

This type of situation is actually quite familiar. For example, it certainly will not make any difference to the reader which copy of this book he is reading. Nevertheless, there is no doubt that corresponding words in different copies will differ in physical detail. But, of course, the details are dissipated away, or, as it is sometimes put, absorbed into the heat bath. Indeed, this is an inevitable aspect of the correlation process described in the previous section, except that there certain details which are important for relatively short periods of time are also dissipated away.

Actually, defining the physical states of the biota, even in principle, presents some complications because it is an open system in constant and complicated interaction with the environment, implying that each such state is in reality an equivalence class of states of the entire ecosystem (cf. Section 5.3). Also, we might point out that some authors have considered the possibility that what we call the correlation process is essentially a form of measurement. True or false, potentially useful or not, this returns us to the subtle questions of the origin of (selective) irreversibility discussed in the previous chapters.

ADDENDUM: STRUCTURAL CORRESPONDENCE BETWEEN TRANSITION SCHEME AND TWO-TIME FORMALISM

1. PROOF OF FUNDAMENTAL IDENTITY (transition scheme formalism). It is convenient to rearrange the identity [equation 4.14]:

$$H(\omega) + H(\omega^*|\omega) = H(\omega^*) + H(\omega|\omega^*) \tag{4.A1}$$

Substituting definitions (4.8), (4.10), (4.12), and (4.13):

$$-\sum_{u=1}^{n}\sum_{r=1}^{n}\sum_{s=1}^{m} p[\beta^r, \varepsilon^s, \beta^u]\log p[\beta^u|\beta^r, \varepsilon^s]$$

$$-\sum_{u=1}^{n}\sum_{r=1}^{n}\sum_{v=1}^{m}\sum_{s=1}^{m} p[\beta^r, \varepsilon^s, \beta^u, \varepsilon^v]\log[\varepsilon^v|\beta^r, \varepsilon^s, \beta^u]$$

$$= -\sum_{r=1}^{n}\sum_{v=1}^{m}\sum_{s=1}^{m} p[\beta^r, \varepsilon^s, \varepsilon^v]\log p[\varepsilon^v|\beta^r, \varepsilon^s]$$

$$-\sum_{u=1}^{n}\sum_{r=1}^{n}\sum_{v=1}^{m}\sum_{s=1}^{m} p[\beta^r, \varepsilon^s, \beta^u, \varepsilon^v]\log p[\beta^u|\beta^r, \varepsilon^s, \varepsilon^v]$$

$$\tag{4.A2}$$

where we write out the summations explicitly, taking $|I| = n$ and $|J| = m$, consolidate the probabilities, and suppress the dependence on time (which is here immaterial). Now notice that

$$\sum_{v=1}^{m} p[\varepsilon^v|\beta^r, \varepsilon^s, \beta^u] = 1 \tag{4.A3a}$$

for any r, s, u. Similarly

$$\sum_{u=1}^{n} p[\beta^u|\beta^r, \varepsilon^s, \varepsilon^v] = 1 \tag{4.A3b}$$

for any r, s, v. Multiplying the left-hand side of (4.A2) by (4.A3a), the right-hand side by (4.A3b), and combining logarithms,

$$-\sum_{u=1}^{n}\sum_{r=1}^{n}\sum_{v=1}^{m}\sum_{s=1}^{m} p[\beta^r, \varepsilon^s, \beta^u, \varepsilon^v]\{\log p[\beta^u|\beta^r, \varepsilon^s]\, p[\varepsilon^v|\beta^r, \varepsilon^s, \beta^u]\}$$

$$= -\sum_{u=1}^{n}\sum_{r=1}^{n}\sum_{v=1}^{m}\sum_{s=1}^{m} p[\beta^r, \varepsilon^s, \beta^u, \varepsilon^v]$$

$$\times \{\log p[\varepsilon^v|\beta^r, \varepsilon^s]\, p[\beta^u|\beta^r, \varepsilon^s, \varepsilon^v]\}$$

$$\tag{4.A4}$$

But according to the law of multiplication of probabilities,

$$p[\beta^u, \varepsilon^v | \beta^r, \varepsilon^s] = \begin{cases} p[\beta^u | \beta^r, \varepsilon^s] p[\varepsilon^v | \beta^r, \varepsilon^s, \beta^u] \\ p[\varepsilon^v | \beta^r, \varepsilon^s] p[\beta^u | \beta^r, \varepsilon^s, \varepsilon^v] \end{cases} \tag{4.A5}$$

from which it follows that we in fact have an identity. Note that (4.A5) implies that both sides of (4.A4) can be written as

$$- \sum_{u=1}^{n} \sum_{r=1}^{n} \sum_{v=1}^{m} \sum_{s=1}^{m} p[\beta^r, \varepsilon^s] p[\beta^u, \varepsilon^v | \beta^r, \varepsilon^s] \log p[\beta^u, \varepsilon^v | \beta^r, \varepsilon^s]$$

$$\tag{4.A6}$$

where we expand the ou:er set of probabilities. However, this is just the definition of $H(\Omega)$ as given in equation (4.7). Thus,

$$H(\Omega) = \begin{cases} H(\omega) + H(\omega^* | \omega) \\ H(\omega^*) + H(\omega | \omega^*) \end{cases} \tag{4.A7}$$

2. PROOF OF ALTERNATIVE FUNDAMENTAL IDENTITY (two-time formalism). It is again convenient to rearrange the identity [equation (4.40)]:

$$H[\beta(t+\tau), \beta(t)] + H[\varepsilon(t+\tau), \varepsilon(t) | \beta(t+\tau), \beta(t)]$$

$$= H[\varepsilon(t+\tau), \varepsilon(t)] + H[\beta(t+\tau), \beta(t) | \varepsilon(t+\tau), \varepsilon(t)]$$

$$\tag{4.A8}$$

Again substituting definitions [cf. equations (4.41) and (4.42)],

$$- \sum_{u=1}^{n} \sum_{r=1}^{n} p[\beta^u, \beta^r] \log p[\beta^u, \beta^r]$$

$$- \sum_{u=1}^{n} \sum_{r=1}^{n} \sum_{v=1}^{m} \sum_{s=1}^{m} p[\beta^u, \beta^r] p[\varepsilon^v, \varepsilon^s | \beta^u, \beta^r] \log p[\varepsilon^v, \varepsilon^s | \beta^u, \beta^r]$$

$$= - \sum_{v=1}^{m} \sum_{s=1}^{m} p[\varepsilon^v, \varepsilon^s] \log p[\varepsilon^v, \varepsilon^s]$$

$$- \sum_{u=1}^{n} \sum_{r=1}^{n} \sum_{v=1}^{m} \sum_{s=1}^{m} p[\varepsilon^v, \varepsilon^s] p[\beta^u, \beta^r | \varepsilon^v, \varepsilon^s] \log p[\beta^u, \beta^r | \varepsilon^v, \varepsilon^s]$$

$$\tag{4.A9}$$

where summations are again written out explicitly and the notation again simplified by omitting the dependence on time. In this case note that

$$\sum_{v=1}^{m} \sum_{s=1}^{m} p[\varepsilon^v, \varepsilon^s | \beta^r, \beta^u] = 1 \qquad (4.A10a)$$

for any r, u and

$$\sum_{u=1}^{n} \sum_{r=1}^{n} p[\beta^u, \beta^r | \varepsilon^v, \varepsilon^s] = 1 \qquad (4.A10b)$$

for any v, s. Multiplying, as in case (1), the left-hand side of (4.A9) by (4.10a) and the right-hand side by (4.A10b),

$$-\sum_{u=1}^{n} \sum_{r=1}^{n} \sum_{v=1}^{m} \sum_{s=1}^{m} p[\beta^r, \varepsilon^s, \beta^u, \varepsilon^v]\{\log p[\beta^r, \beta^u] p[\varepsilon^v, \varepsilon^s | \beta^u, \beta^r]\}$$

$$= -\sum_{u=1}^{n} \sum_{r=1}^{n} \sum_{v=1}^{m} \sum_{s=1}^{m} p[\beta^r, \varepsilon^s, \beta^u, \varepsilon^v]\{\log p[\varepsilon^v, \varepsilon^s] p[\beta^u, \beta^r | \varepsilon^v, \varepsilon^s]\}$$

$$(4.A11)$$

where again logarithms are combined. The identity follows from noting that

$$p[\beta^r, \varepsilon^s, \beta^u, \varepsilon^v] = \begin{cases} p[\beta^u, \beta^r] p[\varepsilon^v, \varepsilon^s | \beta^u, \beta^r] \\ p[\varepsilon^v, \varepsilon^s] p[\beta^u, \beta^r | \varepsilon^v, \varepsilon^s] \end{cases} \qquad (4.A12)$$

which further implies that both sides of (4.A8) can be written as $H[\beta(t + \tau), \varepsilon(t + \tau), \beta(t), \varepsilon(t)]$, as stated in Section 4.9.

Together (1) and (2) substantiate the statement (made in Section 4.9) that the fundamental identities of the transition scheme and two-time formalism have the same external structure [therefore are expressible in the same notation, using redefinitions (4.46) and (4.47)]. The underlying interpretations are different, however, with the transition scheme formalism counting as biotic uncertainty only that uncertainty which is inherent in the biota (aside from uncertainty in initial conditions) and the two-time formalism making no distinction between inherent biotic uncertainty and biotic uncertainty traceable to environmental influence. The remainder of this book is developed in terms of the transition scheme formalism, unless otherwise stated.

REFERENCES

Ashby, W. R. (1956) *An Introduction to Cybernetics.* Wiley, New York.

Bremermann, H. J. (1974) "Complexity of Automata, Brains and Behavior," pp. 304–331 in *Physics and Mathematics of the Nervous System*, ed. by M. Conrad, W. Güttinger, and M. Dal Cin. Springer, Heidelberg.

Conrad, M. (1972a) "Statistical and Hierarchical Aspects of Biological Organization," pp. 189–220 in *Towards a Theoretical Biology*, Vol. 4, ed. by C. H. Waddington. Edinburgh University Press, Edinburgh.

Conrad, M. (1972b) "Can There Be a Theory of Fitness?" *Int. J. Neurosci. 3*, 125–134.

Conrad, M. (1976) "Biological Adaptability: The Statistical State Model," *Bioscience 26*, 319–324.

Conrad, M. (1977) "Functional Significance of Biological Variability," *Bull. Math. Biol. 39*, 139–156.

Khinchin, A. I. (1957) *Mathematical Foundations of Information Theory.* Dover, New York.

Shannon, C. E., and W. Weaver (1962) *The Mathematical Theory of Communication.* University of Illinois Press, Urbana.

5

Hierarchical Aspect of Biological Organization

The spirit of the previous chapter was completely behavioral. We have talked about states of the ecosystem, the transitions from state to state, and the general relations which must necessarily obtain between the statistical behavior of the biota and the statistical behavior of the environment. So far, however, we have said nothing at all about the nature of these states.

The simplest and certainly the most innocent observation is that the states must be organizations of matter and that each of β^r or ε^s is no more than the name of such an organization.* In principle, these organizations might be described in completely physical terms, for example, in terms of the positions and momenta of all the particles in the system (assuming a classical description). In fact, I have already assumed the existence of a microscopic description in our discussion of the role of dissipation. However, this type of description is much too refined, indeed so much so that it loses all the main features of biological organization. Instead, what we need is a description in which the specification of the state directly expresses the essential structural properties of the ecosystem, just as a biologist would describe them.

5.1. COMPARTMENTAL STRUCTURE OF THE ECOSYSTEM

No property of biological matter should strike the eye with greater force than its hierarchical or, more precisely, compartmental organization. This, of course, just reflects the fact that we describe the ecosystem in terms of biota and environment; that we describe the former in terms of communities, populations, organisms, organs, cells, and organelles (including the

---

*Recall that β is mnemonic for biological system in $\beta^i, \beta^u, \beta^r, \ldots$ and that superscripts $i, u, r, \ldots$ refer to states $i, u, r, \ldots$ of this system. ε is mnemonic for environment.

genome); and that we describe the latter in terms of local environments (or regions of space).

A typical level structure is illustrated in Figure 5.1 (see also Figure 1.3 and the corresponding compartmental structure Figure 1.4). Naturally our description might be more or less detailed. For example, we might include trophic levels, social or family organization, cell types, or even tissues. In particular the cellular phenome could be described in terms of its constituents (e.g., organelles, proteins). Alternatively we could make a simpler (and often more practical) description in which the organism and all its subcompartments (other than the genome) are lumped together in a single *phenome* compartment (Fig. 1.3a). In this case the biotic community consists of populations, populations consist of organisms, and organisms consist of a genome and phenome. In each case the environment is divided into local regions, corresponding to the cubes in a three-dimensional grid extending over the entire space of the system, but with the provision that the state of each of these cubes is determined only by that portion of it which is not occupied by living matter (see Figures 5.2a and 5.2b). In other words, the state of each "local region" subcompartment is determined by everything

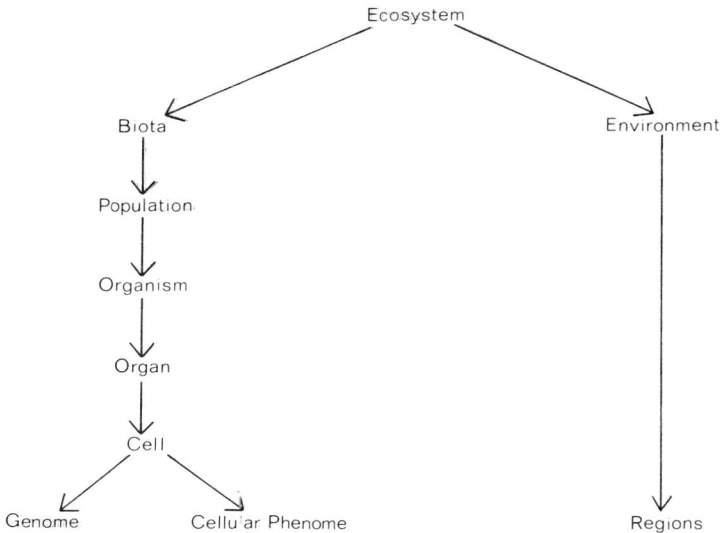

FIGURE 5.1. Typical level structure. A corresponding compartmental structure is illustrated in Figure 5.3. For simplicity the biota is assumed to be identical to the biotic community. In general, however, the biota consists of many biotic communities. More levels could be included, but are unnecessary to illustrate the general principles.

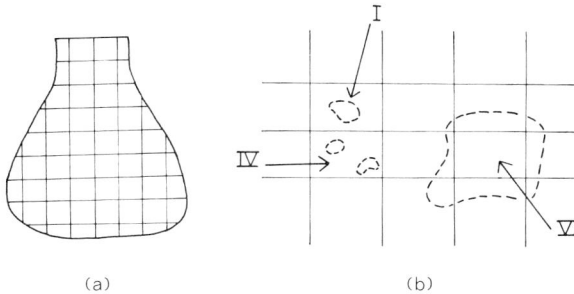

FIGURE 5.2. Decomposition of environment. (a) The space of the ecosystem (e.g., the flask in the case of the flask ecosystem) is coordinatized into small regions of space. The regions are the subcompartments of the environment. In practice more levels of environmental description are often convenient. For example, it may be useful to divide the global environment into local environments associated with each community and these local environments into regions associated with phases or interfaces (e.g., air, water, substrate or air–water interfaces). However, this complication introduces no new principles. (b) The state of each region is determined by that portion of it not occupied by living matter (represented by dotted lines). In the case of regions I and IV this is the area outside the dotted lines and inside the grid square. Region VI, however, is completely occupied by living matter and is therefore regarded as being in a null environmental state. In reality the grid squares are of course grid cubes.

within this subcompartment except the living matter (and is in the null state if it contains only living matter).

So far I have just written down some general compartmental structures and have associated each element of these structures with some biological term. But what about the validity of the compartmental concept itself? Despite its obviousness it might be a merely partial description, or even a misdescription. This is an important question, but one which is better deferred to the end of the next chapter, after we have developed the concept more fully.

Another, related problem is that I have not given definitions for these biological terms. There is truly a difficulty here, one that hinges on the fact that the compartmental organization expresses and is determined by a much deeper functional organization. It is therefore worth addressing a few general remarks to this point, sufficient to fix our terminology relative to standard usage.

First we recall our general conception of the biota as the living part of the ecosystem, i.e., the part surrounded by the outermost boundaries demarcating the inflow of light and high-grade chemical energy and the outflow of heat and low-grade chemical energy (cf. Chapters 1–3). Our conception of the (biotic) community, in our simplified description of the ecosystem, is precisely the same, except that in this case we consider only that portion of the biota associated with some defined region of space. The difficulty of course is that the boundaries may be more or less arbitrary (especially, for example, in a pond), so that one community grades into another. However, what we are thinking of is just the region of space which the biologist chooses to observe. In the laboratory, each community might be confined to a separate flask, so that there is no problem of arbitrariness, at least from the experimental point of view.

By a population I mean exactly what is meant by a species, except restricted to the region of space associated with a particular community. This is perhaps not so helpful, since species concepts are difficulty-strewn and controversial (cf. Mayr, 1957). Roughly, however, we can think of the population as a group of similar organisms in the same area. The similarity may be based on sexual communication (exchange of genetic material, cf. Mayr, 1963), on overall similarity of traits (Sokal and Sneath, 1963), or on the fact that they utilize the same type of resources and are affected by the same environmental factors. The latter type of similarity is based on functional identity and gives rise to what might be called an ecological species concept. To make contact with the other concepts, it is necessary to assume the competitive exclusion principle, according to which no two species (defined on the basis of sexual communication or overall similarity) can coexist in the same space, utilizing the same resources (Gause, 1934; Volterra, 1931).*

By an organism I mean the highest-level compartment which is capable (either individually or in pairs) of integrated self-reproduction. The organism may consist of only a single cell (or minimal units of self-reproduction) or of

---

*The exclusion principle is frequently subject to skeptical scrutiny, but as a practical matter it is fairly robust even if exceptions can in fact be found in nature or constructed theoretically. If different species (defined either on the basis of gene pool structure or on the basis of gross traits ultimately controlled by genes) could in general function equally well in the same environment, selection would not in general be an important factor in evolution. However, insofar as it is *a priori* less likely for systems with distinct differences to be selectively equivalent than selectively inequivalent, selectively inequivalent systems inevitably appear in the course of evolution. As soon as they do so, differential selection begins to act, and act differently at different places and times, thereby amplifying the differences. So in a sense selection inevitably selects selection to be an important factor in evolution. The situation as regards the relation between selection and molecular traits, that is, nucleic acid and protein sequences, is discussed in Chapter 10. There it will be seen that selection can also act to increase its own effectiveness by acting to create systems in which differences at the molecular level are selectively less inequivalent.

many cells, in which case the cells typically derived from a single one contain the same genetic information, although perhaps differently expressed, and are capable of producing offspring as an integrated unit. Thus the organism level marks a fundamental divide as far as the evolution process is concerned.

The cell is the minimal unit of self-reproduction and therefore in a real sense the minimal unit of life. Sometimes it is useful to regard it as consisting of two parts, the genome and the cellular phenome (including the other organelles, the membrane, and so forth). By the genome I mean the sequence of bases in DNA potentially communicable to daughter cells through mitosis or meiosis. (But note that this definition is narrower than is typical in some current practice, where "genome" is used to refer, in a general way, to the entire genetic apparatus.)

Famous is the fact that the sequence of bases in DNA codes for the sequence of amino acids in protein, which in turn folds to assume some three-dimensional shape and, concomitant to that shape, a specific function. These specific functions are the prime determinants of the higher-level structure and behavior of biological matter, and therefore of all the compartments. Thus it is clear that a very complex and circular functional picture lies behind the compartmental concept, so it is not surprising that we would have to invoke broader biological considerations to define the compartments precisely.

There are many side difficulties so far not touched upon. The main one is that the compartments collapse into one another at the extremes. Clearly the cell may be the organism, but the organism may also be a continuum of cytoplasm shared by many nuclei. In other cases the organism may be hard to distinguish from the population; in still others the population may be a hybrid complex of diverse types, or, just the opposite, a set of asexual lines related only through historical origin. But rather than negating the compartmental concept, the difficulty at the extremes is really a confirmation of it, for what is happening at these points is a collapsing together of one or more compartments, or perhaps a coalescing which is sometimes consummated and sometimes broken by a complex life cycle. It is not that the compartmental concept is an illusion, but only that biological matter, with its evolutionary fluidity [or opportunism, as Simpson (1949) has put it] inevitably uses the possibilities of compartmental organization in the most flexible way.

5.2. STATES OF COMPARTMENTS

In Chapter 4 we defined the state of the biota in terms of a set $\{\beta^1, \beta^2, \ldots\}$ of adaptively distinct states. Each such state was further subdivided into finer states, associated with the transmission of different signals (or identical signals in the case of redundancy). Finally, each of these finer states is an equivalence class of detailed states, as defined in physics.

Now what we must do is redescribe each of the β^r in terms of the states of the various subcompartments of the biota and likewise the state of the environment in terms of the states of its various subcompartments (or regions). The state set of each biota compartment can also be described in terms of finer states, i.e., the adaptively distinct states of each compartment can be divided into states associated with the processing of different signals, with each of the latter being an equivalence class of physically defined states. The problem (whether the description is gross or refined) is that compartments are nested within one another, so that it is redundant to specify the state of the biota by specifying the state of each compartment. To avoid this redundancy we must introduce a new concept, viz. the concept of partial states.

The idea of the partial state is a little like the smile on the Cheshire cat, except that we are eventually interested in finding the cat. The way to do this is to define the state of a compartment (such as the biota) in terms of the compartments at the next lower level (in this case the population), but without distinguishing any variations within the boundaries of these compartments. The process is repeated for each of the "black box" compartments, in terms of their compartments at the next lower level, until we finally reach the bottom-level compartments. The states of these are defined as an equivalence class of states in the physical sense, just as was originally the case for the biota as a whole.

Roughly speaking, the partial state of any compartment can be thought of as the configuration of its subcompartments, but excluding any observations within the boundaries of these subcompartments. More precisely (in operational terms) we think of the complete state of a compartment as being determined by some set of possible observations, while its partial state is determined by some subset of these observations, with those observations being thrown out which primarily derive from measurements within the boundaries of the subcompartments. Thus, in general, the partial state of any compartment in terms of its subcompartments is given by specifying (1) the number of subcompartments and (2) the configuration of the subcompartments (e.g., their connectivity, spatial location, or relative spatial location).

The partial states associated with various levels of organization and their interpretation in terms of biological observables are listed in Table 5.1. This table is built by asking and answering the following questions:

1. *How is the biota broken up into populations and what is its configuration in terms of populations?* The division into populations answers the question, which species are present? These species may be

TABLE 5.1
Biological Observables Correlated with Partial States

Compartment	Partial state[a]	Primary observable properties
Biota	$\dot{\beta}^r_{15}$	Species composition; food web structure
Population	$\dot{\beta}^r_{u4}$	Number of organisms; spatial location of organisms
Organism	$\dot{\beta}^r_{u3}$	Number of organs present; relative position of organs
Organ	$\dot{\beta}^r_{u2}$	Number of cells present; contiguity relations among cells
Cell	$\dot{\beta}^r_{u1}$	Relative location of genome and cellular phenome; pattern of gene activation in cell
Cellular phenome	β^r_{u0}	Physiological state of cell[b]
Genome	$\beta^r_{(u-1)0}$	DNA sequence

[a]β^r_{ij} is state r of compartment i at level j. Dot notation indicates partial state.
[b]To be specified in terms of classical physiochemical variables.

viewed as connected by channels over which energy and matter may flow. Thus the configuration of the biota in terms of populations is the food web structure of the biota.

2. *How are the populations broken up into organisms and what are their configurations in terms of these organisms?* The division into organisms answers the question, how many organisms are present? The configuration is the spatial arrangement of these organisms.

3. *How are the organisms broken up into organs and what are their configurations in terms of these organs?* The answer to the first question specifies the number of organs in each organism. The answer to the second specifies the relative locations of these organs. (Note that the number of organs may change in growth, for example in plants, and the relative position in motion, in particular animal motion.)

4. *How are the organs broken up into cells and what are their configurations in terms of these cells?* The first question is, how many cells are present in each organ? The configuration is the topographical relation (or relative spatial position) of the cells, or any other distinguishable relationships (external to the cell membrane) affecting their connectivity. Also, note that extracellular fluids within the boundaries of the organism may be regarded as associated with a compartment at the level of the cell.

5. *How are cells broken up into the genome and cellular phenome and what is the configuration in terms of these subcompartments?* By definition each cell has one genome and one subcellular phenome. The configuration of the cell in terms of genome and cellular phenome is specified by specifying the relative location of genome and phenome (which may change, for example, during the cell cycle) and the particular genes which are turned on or off at any given time (which corresponds to the association between genome and phenome constituents).

6. *How are the genome and cellular phenome broken up into subcompartments and what is their configuration in terms of these subcompartments?* We do not answer this question. Instead, the procedure is to specify the base states of the genome and cellular phenome. The base state of the genome is the sequence of bases in heritable DNA (which of course is actually a configuration of more elementary components). The base state of the cellular phenome is the physiological state of the cell, excluding the state of the genome, its relative position, and the particular genes which are on or off.

Naturally, we could also describe the cellular phenome in terms of levels of organization, so that our base states are specified by subcellular components. Such a detailed description is interesting from the standpoint of analyzing control processes within the cell, but it is not necessary and indeed impractical from the standpoint of analyzing these processes within the ecosystem as a whole. Also, we could eliminate some levels, for example by describing the organism only in terms of genome and phenome. This lower-resolution description is practical and useful for many purposes, but it is insufficient to distinguish certain important ecosystem principles.

There are two points to note. The first is that any choice of compartmental structure is associated with and indeed defines a procedure for "dividing" observations among the compartments. To make the proper associations, however, may require more complete observations, in particular if it is necessary to characterize the type of the system. For example, counting the number of populations is contingent on recognizing these populations, which in turn requires at very least some observations of gene flow and (in the absence of applicability of a sexual, so-called biological species concept) observations of other traits as well. The compartmental structure is thus invisible to a purely top-down or bottom-up identification procedure. The second point is that the partial states, like the complete states, are sets of finer, functionally similar states. Some of these are associated with a horizontal flow of information among compartments at a particular level, others with a vertical flow between levels. In the former case the finer states are configurations of compartments at the next lower level.

For example, the partial state of the biota consists of a finer set of states, corresponding to different configurations of the interspecies communication network. The same consideration extends to each of the other levels, all the way down to the different pattern of signal flow between genome and subcellular phenome. In the case of vertical flow, for example from organ to genome, the finer states are still configurations of compartments at the next lower level, but the signals themselves are associated with configurations at a number of levels (which means that the finer states of these levels are correlated, as they should be).

Finally it is necessary to say what is meant by the environment state. This is the macroscopic state of the environment (temperature, chemical concentrations, lighting, external forces...) specified at each location in the space occupied by living matter. The local environment, as a subcompartment of the total environment defining the biotic community in question, is specified in the same way as are regions, or subcompartments of the local environment. The reason for "resolving" down to regions is that it is desirable to include the topographic features of the environment, for example the various phase boundaries. This is important because the different species and even the different organisms among these species are all exposed to very different environmental conditions. Roughly speaking, the regions correspond to the microenvironment except that what is a microenvironment and what is not depends on the size of the organism. We could, of course, avoid this difficulty by describing the environment in terms of even more levels, but this would introduce no new ecological principles. We will always assume (for simplicity) that the environment can be adequately described by specifying the state of each region.

5.3. REFERENCE STRUCTURES

The compartmental structure changes with time. New types of populations enter the area or come into being; others leave or become extinct. More rapidly, within each population, there is the incessant birth and death of organisms, with a concomitant entry of new organisms into the current structure and deletion of others. At least as rapidly, and generally more so, within each organism there is a production and demise of organs—for example, in plants a constant addition or loss of leaves or flowers. Even more rapidly, within the organ there is an incessant birth and death of the cell, with the accompanying duplication or division of genome and cellular phenome. All these fluxes are the inevitable concomitants of the cycle of life. But this is not the only cycle in the ecosystem. There is the equally inevitable cycle of matter and therefore the even more rapid flux of atoms and molecules through the ecosystem. Indeed, the compartmental structure is most properly thought of as a pattern of activity embedded in this deeper flux.

It is clear that our description of the state of the biota must incorporate this fundamental impermanency of the compartmental structure, for the transition schemes [e.g., equations (4.2) and (4.33)] require us to describe the state not at one particular instant of time but at two or more instants of time. There are a number of ways of handling this problem, but the simplest is to use reference compartmental structures (Conrad, 1975). The idea here is to construct a single reference structure with many extra compartments, but with numerous of the compartments (at any given time) being either in the unborn state or consigned to the graveyard. The advantage is that rather than dealing with numerous different compartmental structures, each reflecting the state of the biota at a particular point in time, we deal with one grand structure which corresponds to each of these particular structures when appropriate compartments are assigned to either the unborn or dead state.

A convenient reference structure is illustrated in Figure 5.3. I adopt the convention that c_{ij} is compartment i at level j. c_{16} is the ecosystem as a whole (but for clarity we also use the name of the compartment in the diagram). In the simplest case the ecosystem consists of one biotic community c_{15}, and its environment, c_{25}. The biotic community contains n potential types of populations, where n is much larger than the possible number of compartments that can exist at one time at any level. Thus the population compartments range from c_{14} to c_{n4}. This procedure is continued all the way down to the cell level, in each case choosing n as the potential number of subcompartments of any given compartment. The cell itself (by convention) has only two compartments, the genome and phenome, with the genome being labeled by odd i and the phenome being labeled by even i. The environment is broken up into m regions, with the convention that these are at the bottom level. Thus $c_{(2n^4+i)0}$ is the region i of the environment.

Adopting the above conventions makes it possible to describe the state of the ecosystem or any of its subsystems in compartmental terms. For the rth complete state of c_{ij} at time t we write $\beta_{ij}^r(t)$. Thus we may write

$$\beta^r(t) \equiv \beta_{15}^r(t) \tag{5.1a}$$

$$\varepsilon^s(t) \equiv \varepsilon_{25}^s(t) \tag{5.1b}$$

where the left-hand sides are our previously used labels for the states of the biota and environment and we continue to use β's for the biota and ε's for the environment (and indeed for all subcompartments of the biota and

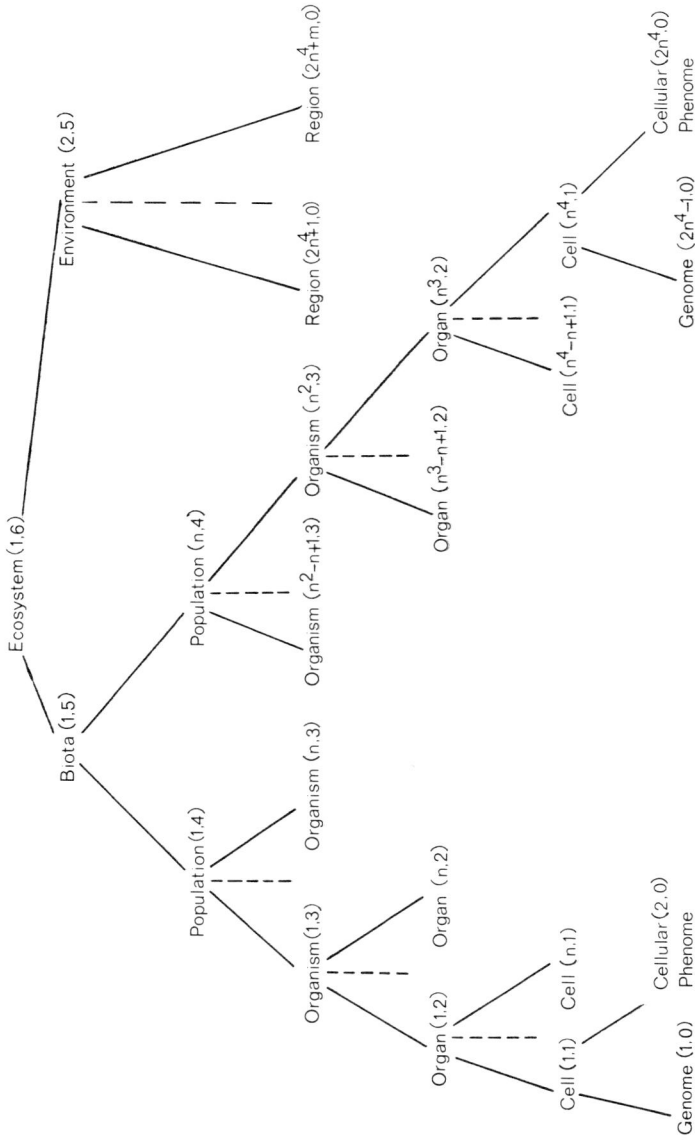

FIGURE 5.3. Reference compartmental structure. $c(i, j)$ is compartment i at level j. In the diagram, however, c is replaced by the name of the compartment, e.g., $c(n,3) =$ organism $(n,3)$. Note that there are n compartments at each level of the biota, except at the bottom level, and that n is much larger than the largest number of possible compartments that exist at any given time. There are m regions of the environment, considered to be at the bottom level.

environment respectively). For the rth partial state of c_{ij} at time t we write $\dot\beta_{ij}^r(t)$. Thus the rth partial state of the biota is given by $\dot\beta_{15}^r(t)$; the rth partial state of the organism compartment c_{i3} by $\dot\beta_{13}^r(t)$, and so forth.

It seems reasonable that the complete state of the biota (or of any compartment) can be expressed as the set of partial states of all its subcompartments, including its own partial state. This means that values of the superscripts can be found such that

$$\dot\beta_{15}^r(t) = \left[\dot\beta_{15}^a(t), \ldots, \dot\beta_{ij}^b(t), \ldots, \dot\beta_{(2n^4)0}^c(t) \right] \tag{5.2}$$

where $0 < i < 2n^4$ and $0 \leqslant j < 5$ and the set is written in the form of a many-tuple (with the ordering a matter of convention). For the environment this many-tuple may be written

$$\varepsilon_{25}^s(t) = \left[\varepsilon_{(2n^4+1)0}^a, \ldots, \varepsilon_{(2n^4+m)0}^b \right] \tag{5.3}$$

since the only states of concern are the complete states of the regions.

MATHEMATICAL NOTE.   Strictly speaking, the complete state of a compartment is the intersection of the partial states of all its subcompartments, including itself, where each partial state is the set of complete states of the compartment such that the subcompartment has a particular property. For example, the biota has a certain set of complete states such that there are k organisms in a particular population, which we now call the partial states of that population (ignoring spatial location). In this case the postulate expressed in equation (5.2) can be rewritten as:

$$\dot\beta_{15}^r(t) = \bigcap_{i,j} \dot\beta_{ij}^u(t) \tag{5.4a}$$

where the set theoretic intersection ($\cap$) runs over index sets determined by the reference structure (starting with $i = 1$ and $j = 5$) and the equation holds for some choice of u's.

Similarly for the environment it is possible to write

$$\varepsilon_{25}^s(t) = \bigcap_h \varepsilon_{h0}^v(t) \tag{5.4b}$$

This notion of state, which is fundamental to the axiomatic formulation of the theory of probability (cf. Kolmogorov, 1956), makes explicit our observation that partial states are properties of the whole system. Thus the postulate is simply that the complete state of any compartment is the same, for all practical purposes, as the intersection of those sets of complete states of that compartment determined by our choice of subcompartmental properties.

I will normally omit explicit reference to the index sets, assuming that they run over the entire compartmental structure. Also, note that the interpretation in terms of set theoretic intersection implies that the commas in equation (5.2) can be interpreted as "and" symbols. The intersection symbol runs over sets of either biotic or environmental states, but I will continue to separate these sets by a comma.

5.4. TRANSITION SCHEMES AGAIN

Now the problem is to combine the structural description of the ecosystem with the behavioral description developed in the previous chapter. In particular, recalling the first version of the ecosystem transition scheme, equation (4.2), and subscripting transition schemes according to the same rules as compartments,

$$\omega_{16} = \left\{ p\left[\beta_{15}^u(t+\tau), \varepsilon_{25}^v(t+\tau) \middle| \beta_{15}^r(t), \varepsilon_{25}^s(t) \right] \middle| u, r \in I, v, s \in J \right\}$$

$$(5.5)$$

where $\omega_{16} = \Omega$ and we have substituted definitions (5.1a) and (5.1b).* Combining this with the decomposition of states in terms of partial states,

$$\omega_{16} = \left\{ p\left[\bigcap_{i,j} \dot{\beta}_{ij}^u(t+\tau), \bigcap_h \varepsilon_{h0}^v(t+\tau) \middle| \bigcap_{i,j} \dot{\beta}_{ij}^r(t), \bigcap_h \varepsilon_{h0}^s(t) \right] \middle| u, r \in I_{ij}, \right.$$

$$\left. v, s \in J_{h0} \right\}$$

$$(5.6)$$

where I_{ij} and J_{h0} are index sets for the partial states of biota and environment respectively, and the intersections are taken over all subcompartments of the biota, including the biota itself (in the case of i and j), and over all regions of the environment, but not including the environment itself (in the case of h).

---

*Note again the assumption that $\alpha_{16}^w(t) = [\beta_{15}^r(t), \varepsilon_{25}^s(t)]$, where $\alpha_{16}^w(=\alpha^w)$ labels the complete ecosystem state. This is the minor assumption that there is no practical value in considering partial states of the ecosystem defined in terms of the relation between biota and environment.

Reexpressing the marginal scheme,

$$\omega_{15} = \left\{ p\left[\cap \dot{\beta}^u_{ij}(t+\tau) | \cap \dot{\beta}^r_{ij}(t), \cap \varepsilon^s_{h0}(t) \right] \right\} \tag{5.7a}$$

$$\omega^*_{25} = \left\{ p\left[\cap \varepsilon^v_{h0}(t+\tau) | \cap \dot{\beta}^r_{ij}(t), \cap \varepsilon^s_{h0}(t) \right] \right\} \tag{5.7b}$$

where $\omega_{15} = \omega$, $\omega^*_{25} = \omega^*$, the now-redundant star notation is retained for clarity, and, for simplicity, explicit reference to the index sets is dropped. However, the decomposition need not be stopped here, for it is now possible to expand ω and ω^* in terms of probabilities (and conditional probabilities) for the partial states.

To make this decomposition practical it is convenient to introduce some definitions. Just as it is possible to define marginal transition schemes for the biota and environment, it is possible to define (local) marginal schemes for any particular compartments. Thus, define

$$\omega_{hk} = \left\{ p\left[\beta^u_{hk}(t+\tau) | \beta^r_{15}(t), \varepsilon^s_{25}(t) \right] \right\} \tag{5.8}$$

as the complete marginal transition scheme of compartment c_{hk}. This may also be decomposed in terms of partial states, indeed using the right-hand side of equation (5.7a), but restricting the range of i and j to c_{hk} and its subcompartments. Likewise it is possible to define marginal transition schemes for every region of the environment.

We may finally define partial marginal transition schemes, i.e., transition schemes for a compartment in terms of its subcompartments at the next lower level. In this case

$$\dot{\omega}_{hk} = \left\{ p\left[\dot{\beta}^u_{hk}(t+\tau) | \beta^r_{15}(t), \varepsilon^s_{25}(t) \right] \right\} \tag{5.9}$$

where $\dot{\omega}_{hk}$ is the partial transition scheme of compartment c_{hk}.

The next step is to express the global scheme (for the biota as a whole) in terms of local schemes (for particular subcompartments) and to express the latter in terms of partial schemes (associated with individual levels of the compartments). To do this it is convenient to define products of schemes

$$\omega_{hk}\omega_{fg} \equiv \left\{ p\left[\beta^u_{hk}(t+\tau), \beta^v_{fg}(t+\tau) | \beta^s_{15}(t), \varepsilon^s_{25}(t) \right] \right\} \tag{5.10a}$$

$$\omega_{hk}\omega_{fg}\omega_{ij} \equiv \left\{ p\left[\beta^u_{hk}(t+\tau), \beta^v_{hk}(t+\tau), \beta^w_{ij}(t+\tau) | \beta^r_{15}(t), \varepsilon^s_{25}(t) \right] \right\}$$

$$\tag{5.10b}$$

and so on. This definition may also be extended to partial schemes and to "products" of partial and complete schemes. For example,

$$\dot{\omega}_{hk}\dot{\omega}_{fg} \equiv \left\{ p\left[\dot{B}_{hk}(t+\tau), \dot{B}_{fg}^v(t+\tau)|\beta_{15}^r(t), \varepsilon_{25}^s(t)\right]\right\} \qquad (5.10c)$$

$$\omega_{hk}\dot{\omega}_{fg} \equiv \left\{ p\left[B_{hk}(t+\tau), \dot{B}_{fg}^v(t+\tau)|\beta_{15}^r(t), \varepsilon_{25}^s(t)\right]\right\} \qquad (5.10d)$$

Thus, the rule is that the product of schemes for two or more compartments is the probability of the joint occurrence of the (complete or partial) state of the compartment at time $t + \tau$, given the complete state of the ecosystem at time t (the product of schemes is thus not the product of probabilities!).

With the above notation it is now possible to write the global transition scheme in terms of the partial scheme of the biota and the complete scheme of each population

$$\omega = \omega_{15} = \dot{\omega}_{15}\Pi\omega_{i4} \qquad (5.11)$$

where $i = 1,\ldots,n$, viz.

$$\Pi\omega_{i4} = \omega_{14}\omega_{24}\ldots\omega_{(n-1)4}\omega_{n4} \qquad (5.12)$$

This can be further decomposed into partial schemes:

$$\omega = \Pi\dot{\omega}_{ij} \qquad (5.13)$$

where i and j run over all subcompartments of the biota (including the biota) in the reference structure. Similarly, we can decompose the environment in terms of transition schemes for the region:

$$\omega^* = \omega_{25}^* = \Pi\omega_{h0}^* \qquad (5.14)$$

where h runs from $2n^4 + 1$ to $2n^4 + m$. There are, of course, a virtually indefinite number of other such forms.

Notice that in the decomposition of the biota into complete population compartments [equation (5.11)] there is one partial scheme which cannot be avoided. This is because the decomposition is contingent on our answering the question, what species are present?

5.5. THE CANONICAL REPRESENTATION

Recall the definition of the entropy of a transition scheme:

$$H(\omega_{hk}) = -\sum_{r,s,u} p\left[\beta_{15}^r(t), \varepsilon_{25}^s(t)\right] p\left[\beta_{hk}^u(t+\tau)|\beta_{15}^r(t), \varepsilon_{25}^s(t)\right]$$

$$\times \log p\left[\beta_{hk}^u(t+\tau)|\beta_{15}^r(t), \varepsilon_{25}^s(t)\right] \qquad (5.15)$$

where the sum is taken over all r, s, u (written explicitly to avoid ambiguity). This is exactly the same as equation (4.8), except that the entropy is here written for the arbitrary compartment c_{hk} rather than for ω (or equivalently ω_{15}). Rewriting in terms of partial schemes:

$$H(\Pi\dot{\omega}_{hk}) = -\sum_{r,s,u} p\left[\bigcap_{i,j}\dot{\beta}^r_{ij}(t), \bigcap_h \varepsilon^s_{h0}(t)\right]$$

$$\times p\left[\bigcap_{h,k}\dot{\beta}^u_{hk}(t+\tau)\middle|\bigcap_{i,j}\dot{\beta}^r_{ij}(t), \bigcap_h \varepsilon^s_{h0}(t)\right]$$

$$\times \log p\left[\bigcap_{h,k}\dot{\beta}^u_{hk}(t+\tau)\middle|\bigcap_{i,j}\dot{\beta}^r_{ij}(t), \bigcap_h \varepsilon^s_{h0}(t)\right] \quad (5.16)$$

where the convention is that h and k run over all the compartments of c_{hk}, as usual i and j run over all compartments of the biota, h runs over all regions of the environment, and the sum runs over all r, s, u. Suppose, for simplicity, that we are dealing with only two compartments, c_{10} and c_{20}. In this case

$$\Pi\dot{\omega}_{pq} = \dot{\omega}_{10}\dot{\omega}_{20} = \omega_{10}\omega_{20} \quad (5.17)$$

and the corresponding form of equation (5.16) can be expanded to give an identity in just the same way in which we developed our fundamental identity for the joint biota and environment schemes in Chapter 4. Thus

$$H(\omega_{10}\omega_{20}) = H(\omega_{10}) - H(\omega_{20}|\omega_{10}) \quad (5.18)$$

$$= H(\omega_{20}) - H(\omega_{10}|\omega_{20})$$

Adding the two expressions on the right and normalizing,

$$H(\omega_{1C}\omega_{20}) = H_e(\omega_{10}) + H_e(\omega_{20}) \quad (5.19)$$

where

$$H_e(\omega_{10}) = \tfrac{1}{2}\left[H(\omega_{10}) + H(\omega_{10}|\omega_{20})\right] \quad (5.20a)$$

$$H_e(\omega_{20}) = \tfrac{1}{2}\left[H(\omega_{20}) + H(\omega_{20}|\omega_{10})\right] \quad (5.20b)$$

will be called effective entropies (Conrad, 1976a). These are simply unique linear combinations of conditional terms, useful for expressing the entropy of a product of schemes in terms of the sum of normalized entropies and conditional entropies of the individual scheme.

The notion of effective entropies can be extended to higher-order products. Thus, for the general three-term expansion

$$H(\omega_{uv}\omega_{pq}\omega_{rs}) = H_e(\omega_{uv}) + H_e(\omega_{pq}) + H_e(\omega_{rs}) \qquad (5.21)$$

the effective entropies are

$$H_e(\omega_{uv}) = \tfrac{1}{3}\Big[H(\omega_{uv}) + \tfrac{1}{2}H(\omega_{uv}|\omega_{pq}) + \tfrac{1}{2}H(\omega_{uv}|\omega_{rs}) + H(\omega_{uv}|\omega_{pq}\omega_{rs}) \Big]$$

$$(5.22a)$$

$$H_e(\omega_{pq}) = \tfrac{1}{3}\Big[H(\omega_{pq}) + \tfrac{1}{2}H(\omega_{pq}|\omega_{uv}) + \tfrac{1}{2}H(\omega_{pq}|\omega_{rs}) + H(\omega_{pq}|\omega_{uv}\omega_{rs}) \Big]$$

$$(5.22b)$$

$$H_e(\omega_{rs}) = \tfrac{1}{3}\Big[H(\omega_{rs}) + \tfrac{1}{2}H(\omega_{rs}|\omega_{uv}) + \tfrac{1}{2}H(\omega_{rs}|\omega_{pq}) + H(\omega_{rs}|\omega_{uv}\omega_{pq}) \Big]$$

$$(5.22c)$$

This may be continued for higher-order expansions, except that the general formula is quite complicated. The main feature of these expansions, however, is that the leading term represents the behavioral uncertainty of the compartment (or level of compartment) in question, whereas the relative contribution of each of the remaining (smaller) terms describes the degree to which this behavior is independent of the behavior of other levels or compartments.

The behavioral uncertainty of a compartment (or level), considered in isolation, will appear as modifiability of that compartment or level. When other compartments and levels are taken into consideration it may happen that what we originally thought were independent modifications are quite contingent on the behavior of these other compartments or levels. For example, the relative position of body parts of an organism with a highly centralized nervous system may be quite modifiable, as in motor behavior. Considered in isolation, this modifiability may appear quite uncertain; but as soon as the state of the nervous system is taken into account this uncertainty may in large measure be dispelled. Thus the leading term (to be called the modifiability) gives an idea of the amount of variation exhibited by the level or compartment in question and the remainder terms (to be called the independence) give an idea of the degree to which this variation makes a bona fide contribution to variability.

Certain of the remainder terms shrink or disappear once the time order of events (i.e., the direction of information flow) is taken into account. The effect of considering time order is complicated if the flow of information is circular, as is often the case in biological systems. Thus we defer this point for the time being, returning to it in Chapter 9.

Effective entropies can be used for describing environment compartments as well as biota compartments. Also, conditional entropies, such as $H(\omega|\omega^*) = H(\Pi\omega_{ij}|\omega^*)$ can be expanded in terms of effective entropies, the only difference being that both the leading and remainder terms retain the conditioning on ω^*.

5.6. STATISTICAL LAWS

Now it is possible to express the fundamental inequality [equation (4.18)] in terms of partial schemes:

$$H\left(\Pi\hat{\omega}_{ij}\right) - H\left(\Pi\hat{\omega}_{ij}|\Pi\hat{\omega}_{h0}^*\right) + H\left(\Pi\hat{\omega}_{h0}^*|\Pi\hat{\omega}_{ij}\right) \geqslant H(\Pi\omega_{h0}^*) \quad (5.23)$$

where i, j run over all compartments of the biota and h runs over all regions of the environment. As in Chapter 4 the hats denote the scheme for which the environment is most uncertain, subject to the condition that the biota remains alive indefinitely. Thus $\hat{\omega}_{ij}$ denotes the statistically stressed complete scheme of compartment c_{ij} and $\hat{\omega}_{ij}$ denotes the statistically stressed partial scheme.

Rewriting equation (5.23) in terms of effective entropies,

$$\sum H_e\left(\hat{\omega}_{ij}\right) - \sum H_e\left(\hat{\omega}_{ij}|\Pi\hat{\omega}_{h0}^*\right) + \sum H_e\left(\hat{\omega}_{h0}^*|\Pi\hat{\omega}_{ij}\right) \geqslant \sum H_e(\omega_{h0}^*)$$

$$(5.24)$$

The left-hand side is again the adaptability of the biota, but expressed in terms of the modifiability and independence of each level of each compartment. The equation can also be expressed in terms of the transition schemes of the information transfer picture, simply by replacing $\hat{\omega}_{ij}$ by $\underline{\hat{\omega}}_{ij}$ and ω_{h0}^* by $\underline{\omega}_{h0}^*$.

If equation (5.24) were an equality rather than an inequality, it would be possible to make some very powerful statements about compensating changes in adaptability—for example, statements about the implications for changes in dependency or addition of restrictions on the modifiability of particular compartments or levels. In fact, I shall argue (in the next chapter) that communities tend to move in the direction of an equality in the course of evolution.

I take the opportunity here to sharpen our definition of adaptability, by adding a rider to condition A of Section 4.4. The modified definition is: $\Sigma H_e(\hat{\omega}_{h0})$ is a maximum over all possible transition schemes such that the half-life of the biota is not decreased *in any region of the environment*. The advantage of this modification is that it prevents the adaptability from growing beyond all bounds because of a localized wildness of environmental variability which destroys the biota in only one region.

5.7. INTERPRETATION OF THE TERMS

Recall that the adaptability has three components: behavioral uncertainty, tolerance for decorrelation, and tolerance for error (cf. also Section 4.4). This is also true for the new formulation, expressed in terms of partial states and local regions, except that we now distinguish a modifiability and independence contribution to each component. But even more important, each of the terms can now be interpreted in terms of variables concomitant to the level.

These concomitants are given in Table 5.2. Each of the terms is associated with the uncertainty of the partial state of the compartment, i.e., the uncertainty in the number and configuration of the subcompartments. The table is built by ascertaining the uncertainty associated with the answers to the same questions which were used to build up the table of biological observables associated with the various levels of organization (Table 5.1). Thus $H_e(\hat{\omega}_{15})$ is the uncertainty in the species composition and food web organization, $H_e(\hat{\omega}_{u4})$ is the uncertainty in the population size and spatial structure, $H_e(\hat{\omega}_{u3})$ is the uncertainty in the number of arrangements of organs in an organism, $H_e(\hat{\omega}_{u2})$ is the uncertainty in the number and topographic relations of the cells in the organ, $H_e(\hat{\omega}_{u1})$ the uncertainty in the relation between genome and subcellular phenome (including the uncertainty as to which genes are on or off), and $H_e(\hat{\omega}_{u0})$ is either the uncertainty in the state of the subcellular phenome or the uncertainty in the base sequence of the genome. It is important that these properties may not be uncertain once given the state transition of the environment; but this we are not given. Also, it is important to remember that the prior state of the biota and the prior state of the environment are given, so that not all *variations* in these properties are *variabilities*.

Now it is possible to identify the terms in Table 5.2 with terminologies which either are frequently used or could be used to describe the adaptabil-

TABLE 5.2
Modes of Adaptability

Compartment	Entropy	Modifiability component
Biota	$H_e(\hat{\omega}_{15})$	Migration, extinction, speciation; routability of matter and energy in food webs
Population	$H_e(\hat{\omega}_{u4})$	Culturability; social and topographical plasticity
Organism	$H_e(\hat{\omega}_{u3})$	Morphological plasticity (number of organs)[a]; morphological flexibility (including behavioral plasticity)
Organ	$H_e(\hat{\omega}_{u2})$	Morphological plasticity (size of organ)[a]; histological plasticity
Cell	$H_e(\hat{\omega}_{u1})$	Plasticity in cellular morphology[a]; differentiability[a]
Cellular phenome	$H_e(\hat{\omega}_{u0})$	Physiological plasticity of cells and body fluids
Genome	$H_e(\hat{\omega}_{(u-1)0})$	Gene pool diversity[b]

[a]All contribute to developmental plasticity.
[b]Gene pool uncertainty is expressed by $\Sigma H_e(\hat{\omega}_{(u-1)0})$, where u runs over the index set of the species. If somatic genetic variability is possible, u should run only over the germ cells.

ity of biological systems:

1a. *Potentiality for species change.* This is associated with uncertainty in the species composition. The biota may adapt to novel environmental situations through the disappearance of species (extinction) or the immigration of already existing species (migration). It may also adapt through speciation, but this is a slow process, by definition significant only on the long, evolutionary time scale.

1b. *Routability of matter and energy.* This corresponds to the uncertainty in the food web organization.

2a. *Culturability.* This is the uncertainty in the population size. For example, one of the ways in which bacteria can cope with environmental variability is to assume a spore form, but switch to rapid population growth in a suitable environment. In the absence of any information about the environment there will in general be uncertainty about changes in the population size. But, of course, such uncertainty may amount to fluctuation in a vertebrate population. The difference is the degree to which the uncertainty is reduced once the behavior of the environment is specified (i.e., by the difference between the behavioral uncertainty $H_e(\hat{\omega}_{u4})$ and the anticipation entropy $H_e(\hat{\omega}_{u4}|\Pi\hat{\omega}_{h0}^*)$. If the reduction is high the variation in population numbers contributes to culturability, whereas if it is low it contributes to undesirable fluctuation [therefore increasing the error term, $H_e(\hat{\omega}_{h0}^*|\Pi\hat{\omega}_{ij})$].*

2b. *Social and topographical plasticity.* The social plasticity is associated with unpredictable changes in territorial organization or other parameters which define the spatial arrangement of the population (unpredictable meaning unpredictable in the absence of information about the environment). The topographical plasticity is associated with the uncertainty of organism movements from one part of the environment to another.

3a. *Developmental plasticity.* This is uncertain variation in the morphology of the organism associated with (i) uncertainty in the number and arrangement of organs in the organism; (ii) uncertainty in the number and arrangement of cells in the organ; (iii) uncertainty in the relation between genome and subcellular phenome, including uncertainty about the induction and repression of genes; and (iv) uncertainty in the state of the subcellular phenome. For example, in case (i) the number of flowers or leaves on a higher plant may change, and in a way which is unpredictable in the absence of any information about the environment; or an organism with a genetically fixed number of organs (such as a vertebrate) may lose one but nevertheless continue to function (in which case we might properly speak of physiological reliability). In case (ii) the change in organism morphology may arise from growth in cell numbers or cell migration, and in case (iii) it may arise from switching into different modes of development in different environments. In case (iv) the change in organism morphology is due to change in cell size or shape. For example, plant growth may be due to change in cell number or change in cell size associated with the uptake of water. This is why it is virtually possible to watch plants increase in height after a period of cell multiplication.

3b. *Physiological control.* This is a special case of developmental plasticity in the sense that it is associated with the same uncertainties, but with no change in morphological structure (i.e., no change in either flexible or nonflexible constraints on the relative position of subcompartments of the organism). Needless to say, there are numerous physiological control mechanisms which regulate the body variables. To pick some random examples: changes in the amount of hemoglobin per blood cell, adjustments of heart beat and circulation, induction and repression of enzymes associated with

---

*An alternate measure of the degree to which the indeterminacy in population size contributes to culturability is given by $H_e(\hat{\omega}_{h0}^*|\underset{i,j\neq u,4}{\Pi}\hat{\omega}_{ij}) - H_e(\hat{\omega}_{h0}^*|\Pi\hat{\omega}_{ij})$, with the contribution being completely to undesirable fluctuation when the difference is zero and being increasingly to culturability as the difference increases. A large difference means that omitting the conditioning on the change in population numbers very much increases uncertainty about the environmental transition.

variations in food source, redundant occurrence of organelles (this subserves the function of adaptability as well as reliability), and occurrence of duplicate nuclei with different genetic potentialities (this increases the routability of matter and energy at the biochemical level).

3c. *Behavioral plasticity* (learning ability and motor flexibility). Learning ability is a special case of developmental plasticity and physiological control in which the plasticity and control are mediated by the nervous system and brain. This is normally coupled to motor behavior (such as posture change or movement). The motor flexibility is a special case of developmental plasticity and physiological control restricted to the rearrangeability of organs, but subject to the condition that the rearrangements do not interfere with the (fixed or flexible) constraints which define the morphological structure.

3d. *Competence for immunity.* This is a special case of developmental plasticity and physiological control mediated by the immune system.

4. *Gene pool diversity.* This is the uncertainty in the sequence of bases in heritable DNA.

Notice that extinction, migration, speciation (1a), and routability of matter and energy (1b) are all defined over the partial state of the biota, except that the latter is a configurational uncertainty. Culturability (2a) and social and topographical plasticity (2b) are defined over the partial states of the population, with the latter as the configurational component. The situation is different with the developmental, physiological, behavioral, and immunity modifiabilities. These are *phenotypic plasticities*, defined over the complete state of each organism, excluding the genome. This is because the organism is such an intertwined unity that biologists have to develop a terminology for the overt phenomena, such as development or learning, which includes contributions from components at different levels of organization. Moreover, the mechanism of contribution is not always known, and often the extent of modification at various levels is a matter of great controversy. For example, some authors propose that learning is an extension of development, involving changes in nerve cell connections (cf. Mark, 1974); others, that it arises from modifications in one or another physiological parameter of the neuron (cf. Hebb, 1949); still others, that it is based on conformation changes at the molecular level (cf. Conrad, 1976b) and molecular changes which mimic natural evolution (cf. Conrad, 1974).

For the above reasons, and to stick to the common usage, it is convenient to speak of the phenotypic plasticities as a whole. This does not mean that it is not possible to dissect them in terms of the partial state description, nor that such a dissection would not be useful from the mechanistic standpoint. What it does mean is that the normal usage, as it has developed through practice, is not necessarily definable in terms of uncertainties at any one level of phenotypic organization, and furthermore that different forms of the phenotypic plasticities (e.g., developmental or behavioral) are used to describe behavior associated with different phenotypic variables.

The price of retaining the normal usage is that we must use it with care, always remembering that each type of phenotypic plasticity receives a contribution from a number of compartments at different levels.

5.8. FURTHER BIOLOGICAL CORRELATES

So far we have considered only the modifiability terms. What is left are the independence terms, the modifiability and independence terms associated with the anticipation entropy, and the terms associated with tolerance for error.

The independence terms simply express the extent to which the modifiabilities of the different compartments and levels are independent, for example, the extent to which culturability is independent of physiological control, or the extent to which social and topographical plasticity is independent of behavioral plasticity. Thus as the independence of any two compartments or levels increases, the modifiabilities of these compartments or levels make a greater bona fide contribution to the adaptability of the system, since in this case the sum of all the (unnormalized) local modifiabilities will more nearly reflect the variability of the system considered as a whole.

Now consider the expression

$$H_e\left(\hat{\omega}_{ij}\right) - H_e\left(\hat{\omega}_{ij}|\Pi\hat{\omega}_{h0}^*\right) \tag{5.25}$$

This is the difference between the behavioral uncertainty and the anticipation entropy and is therefore the amount of environmental uncertainty which compartment c_{ij} can absorb at its top level. The modifiability component of the (effective) anticipation entropy indicates the extent to which the behavior of the compartment is correlated to the behavior of the environment; the independence component indicates the extent to which it is correlated given the behavior of other compartments at other levels. Thus the contribution of a particular kind of modifiability (e.g., developmental or behavioral plasticity) to system adaptability increases as the environment gives more information about the nature of the modification and as the *independence of the modifications at different levels given the behavior of the environment decreases.*

According to the discussion in Chapter 4, anticipation of the environment depends either on information transfer (and therefore on a finer set of
states and time delay) or on the fact that the community or population has
evolved in such a way that its behavior reflects regularities in the environment. If different compartments are independent apropos modifications but
not anticipation, this means that information is being transferred from one
compartment to another or from one level to another. In this case there is a
causative relation between the modifications at the different levels. This
introduction of causality (already discussed in Chapter 4) imposes more
detailed structure on the canonical (effective-entropy) representation of
adaptability of compartments in terms of subcompartments, but I defer
further discussion of this structure to Chapter 9.

Finally we turn to the indifference term, $H_e(\hat{\omega}^*_{h0}|\Pi\hat{\omega}_{ij})$. This may take
three basic forms: (1) *external despecialization*; (2) *restriction of spatial
range*, including avoidance reactions; or (3) *pathological variability*. For
example, an organism or population which is despecialized in the sense that
it has many sources of food, few metabolic requirements, or high tolerance
for or isolation from certain physical features in the environment does not
have to care about any of these factors. It also does not have to care about
environmental variability in regions of the environment in which it is not
located, so that restriction of spatial range, restriction to a particular
microhabitat, or avoidance reactions subserve the indifference function.
However, not all indifference is a bona fide contribution to adaptability. In
some cases the indeterminateness in the behavior of the compartment given
the environment behavior may mean that the system is exhibiting a pathological mode of behavior. In this case we will say that the indifference is
nonselective, therefore tantamount to error.

There are a number of points to note. The first is that there is no *a
priori* way to distinguish selective from nonselective indifference. However,
the distinction is a general feature of all statistical descriptions of information processing systems. If the system is simply a communication system,
not processing but only transmitting information, any uncertainty about the
environment (source) would be error. However, if the system actually
processes information (for example, computes), it ordinarily loses information about its past inputs (for example, consider a resetting process or an
"or" operation). Thus there is no way of distinguishing, without considering
function, between this kind of selective loss of information [a characteristic
feature of information processing (cf. Winograd and Cowan, 1963)] and
error.

The second point is that if the biota compartments anticipate the
environment well it is reasonable to assume that their behavior is suited to
the environment, for if the behavior were unsuited to the environment the

system would in fact suffer injury, which inevitably would be inconsistent with maintaining good anticipation. In other words, in biological systems the notion of unfit anticipation is self-contradictory.

The third point is: since $\Sigma H_e(\hat{\omega}_{ij}|\Pi \hat{\omega}_{h0}^*)$ and $\Sigma H_e(\hat{\omega}_{h0}^*|\Pi \hat{\omega}_{ij})$ both represent decorrelation between biota compartments and environment, why should we interpret the former as anticipation and the latter as indifference? The answer is that the biota and environment are not on an equal footing, since one is alive and the other not. It is this inherent asymmetry which breaks the symmetry between the potential uncertainty of the environment given the behavior of the biota and the potential uncertainty of the biota given the behavior of the environment.

One way to see this is to consider the case of a completely quiet environment. In this case the actual value of the indifference (or error) term would have to be zero, but the anticipation could be quite poor. This is possible, but the reverse situation is not.

REFERENCES

Conrad, M. (1974) "Evolutionary Learning Circuits," *J. Theor. Biol. 46*, 167–188.

Conrad, M. (1975) "Analyzing Ecosystem Adaptability," *Math. Biosci. 27*, 213–230

Conrad, M. (1976a) "Patterns of Biological Control in Ecosystems," pp. 431–456 in *Systems Analysis and Simulation in Ecology*, Vol. 4, ed. by B. C. Patten. Academic Press, New York.

Conrad, M. (1976b) "Molecular Information Structures in the Brain," *J. Neurosci. Res. 2*, 233–254.

Gause, G. F. (1934) *The Struggle for Existence*. Hafner, New York.

Hebb, D. O. (1949) *Organization of Behavior*. Wiley, New York.

Kolmogorov, A. N. (1956) *Foundations of the Theory of Probability*. Chelsea, New York.

Mark, R. (1974) *Memory and Nerve Cell Connections*. Oxford University Press, London.

Mayr, E. (Ed.) (1957) *The Species Problem*. AAAS Symposium 50, Washington, D.C.

Mayr, E. (1963) *Animal Species and Evolution*. Harvard University Press, Cambridge, Massachusetts.

Simpson, G. G. (1949) *The Meaning of Evolution*. Yale University Press, New Haven, Connecticut.

Sokal, R. R., and P. H. A. Sneath (1963) *Numerical Taxonomy*. Freeman and Company, San Francisco.

Volterra, V. (1931) *Leçons sur la théorie mathématique de la lutte pour la vie*. Gauthier-Villars, Paris.

Winograd, S., and J. D. Cowan (1963) *Reliable Computation in the Presence of Noise*. MIT Press, Cambridge, Massachusetts.

6

Evolutionary Tendency of Adaptability

Chapters 2 and 3 dealt with the general principles of dissipation, and in particular the peculiar features of dissipation in biological systems. For adaptability the main relevant point is that biological systems are capable of coping with disturbances if they are capable of forgetting these disturbances. In general forgetting disturbance means that disturbance is ultimately dissipated into a heat bath. The contribution to the total dissipation involved in this forgetting process may be small, but magnitude and importance are not the same thing. The component of dissipation associated with forgetting disturbance is all important for the control and stability of biological processes.

Actually the discussion was quite a bit more complicated than this. Sometimes biological systems cope with disturbances through isolation rather than dissipation. Also the system may change permanently under the impact of disturbance (e.g., may undergo evolution). This also involves dissipation in a way which is essential from the standpoint of adaptability, but in this case it involves the formation of new adaptations. The disturbance is not completely forgotten, but an observer who can distinguish only certain very general characteristics of the system (for example, whether or not it remains alive) will eventually lose sight of it.

From the microscopic point of view dissipation is a statistical process, so the general import of Chapters 2 and 3 is that the statistical characteristics of the ecosystem are fundamental from the standpoint of adaptability. In Chapter 4 we turned directly to these statistical characteristics, describing them in terms of entropy measures on transition schemes, or sets of transition probabilities governing the state-to-state behavior of the ecosystem. Then Chapter 5 described the ecosystem in terms of its subunits and reformulated the statistical description in terms of this subunit (or compartmental) structure. In this chapter I also characterized the hierarchy of classes of states (functionally distinct states, functionally equivalent states, macroscopically distinct but functionally equivalent states, macroscopically equivalent states) and the mechanism of information transfer from environment to biological system.

The result of the analysis was a set of very general statistical inequalities, both for the ecosystem described as a unit [equation (4.18)] and for the ecosystem described in terms of compartmental structure [equation (5.24)]. The plan in this chapter is to strengthen these inequalities by showing that efficiency considerations suggest that in general adaptability tends to fall to its lowest possible value. I begin with a very general argument; then I make the argument much more detailed and specific by classifying the various possible mechanisms and modes of adaptability and showing that in each case the general efficiency considerations hold, although not necessarily always with the same urgency. The strengthened inequalities make it possible to prove the self-consistency theorem of hierarchical adaptability theory and to define an operational procedure for measuring adaptability.

6.1. THE BASIC ARGUMENT

Recall the basic identity:

$$H(\omega) - H(\omega|\omega^*) + H(\omega^*|\omega) = H(\omega^*) \tag{6.1}$$

In the environment of maximum uncertainty we have

$$H(\hat{\omega}) - H(\hat{\omega}|\hat{\omega}^*) + H(\hat{\omega}^*|\hat{\omega}) = H(\hat{\omega}^*) \tag{6.2}$$

where $H(\hat{\omega}^*) \geqslant H(\omega^*)$. Subtracting (6.1) from (6.2),

$$[H(\hat{\omega}) - H(\omega)] - [H(\hat{\omega}|\hat{\omega}^*) - H(\omega|\omega^*)] + [H(\hat{\omega}^*|\hat{\omega}) - H(\omega^*|\omega)]$$

$$= H(\hat{\omega}^*) - H(\omega^*) \geqslant 0 \tag{6.3}$$

The inequality implies that one or more of the following conditions hold:

1. $H(\hat{\omega}) - H(\omega) > 0$
2. $H(\hat{\omega}|\hat{\omega}^*) - H(\omega|\omega^*) < 0$
3. $H(\hat{\omega}^*|\hat{\omega}) - H(\omega^*|\omega) > 0$

If condition (1) holds, the behavior of the biota is potentially more uncertain than it is ever required to be, i.e., it has a larger than necessary behavioral repertoire. If condition (2) holds, the biota is potentially better at anticipating the environment than it ever needs to be. And if condition (3) holds, the biota is potentially more indifferent than necessary, i.e., has a greater than necessary tolerance for behavior inappropriate to the environment or greater than necessary insensitivity to the environment.

The efficiency argument is that any one of these superfluities is an unnecessary cost for the biota and that therefore the ecosystem will tend to develop in directions which eliminate them. Under these circumstances the fundamental inequality [equation (4.18)] becomes

$$H(\hat{\omega}) - H(\hat{\omega}|\hat{\omega}^*) + H(\hat{\omega}^*|\hat{\omega}) \rightarrow H(\omega^*) \qquad (6.4)$$

where $H(\omega^*)$ is the uncertainty of the actual environment and the arrow represents an evolutionary tendency (in the direction of an equality).

The above tendency can also be expressed in hierarchical form:

$$\sum H_e(\hat{\omega}_{ij}) - \sum H_e(\hat{\omega}_{ij}|\Pi\hat{\omega}_{h0}^*) + \sum H_e(\hat{\omega}_{h0}|\Pi\hat{\omega}_{ij}) \rightarrow \sum H_e(\omega_{h0}^*) \quad (6.5)$$

and also in terms of the transition schemes in the information transfer picture

$$\sum H_e(\underline{\hat{\omega}}_{ij}) - \sum H_e(\underline{\hat{\omega}}_{ij}|\Pi\underline{\hat{\omega}}_{h0}^*) + \sum H_e(\underline{\hat{\omega}}_{h0}|\Pi\underline{\hat{\omega}}_{ij}) \rightarrow \sum H_e(\underline{\omega}_{h0}^*) \quad (6.6)$$

These equations, to be called the compensation equations, will play a most important role.

Note that each version of the equation is a single equation in four unknowns. Thus, by themselves, they cannot determine the actual values of the unknowns, i.e., the actual values of the entropies. These are determined in part by competition-generated uncertainty, discussed later in this chapter, and in part by the requirements for reliability and transformability of structure and function, discussed in Chapters 9 and 10. At that point it will become clear that the entropies formulated in terms of the functional transition schemes (i.e., in terms of $\hat{\omega}$) fall to their lowest possible values, but that in order for this to happen without undue increase in error it is in general necessary for the entropies formulated in terms of the transition schemes in the information transition picture (i.e., in terms of $\underline{\hat{\omega}}$) to increase.

The efficiency argument is clearly based on the intuitive idea that no form of adaptability is entirely without biological cost. Adaptability is an advantage, indeed a prerequisite for continued function; but it inevitably conjoins with this advantage a greater or smaller component of disadvantage, so that whenever excess adaptability is dispensible it tends to be dispensed with. What is required, to make this argument precise, is to show that this is true for each general mechanism of adaptability and for the manifestations of each of these general mechanisms at every level of organization. But more important, it is necessary to consider whether the mechanisms of evolution are in fact capable of dispensing with superfluities in each of the above forms of adaptability (both above and below the organism level), and also to form some idea of the rates at which the dispensing occurs. Finally it will be necessary to say more precisely what is meant by efficiency, or at least relate it more precisely to our earlier discussions of ecological thermodynamics and differential growth rates. However, I defer this to the next chapter, in the meantime regarding efficiency as a primitive biological concept.

It is worth pointing out at this stage that there are experiments which bear on the cost of adaptability and which suggest that superfluous adaptability in fact has a tendency to disappear in the course of evolution. One example is that bacterial mutants which forego even one biosynthetic function have a selective advantage on a medium in which the function is not necessary (Zamenhof and Eichhorn, 1967). This is true even if feedback inhibition or repression is available to reduce the cost of this function, indicating that the occurrence of the extra nucleic acid and indeed even any unnecessary biosynthetic steps leading to the unnecessary product contribute to the disadvantage. A similar but more direct example has been pointed out to me by L. Luckinbill. *E. coli* drawn from a nonminimal medium and cultured on a minimal medium of glucose and salts for 200 to 400 generations outcompete the parent strain on this minimal medium, but are rapidly displaced by the parent strain when returned to the nonminimal medium. Luckinbill has shown that the adaptation to the minimal medium involves the slow development of necessary biosynthetic capabilities on the one hand, and a marked loss of superfluous amino acid uptake on the other. The first development is adaptability-enhancing since it enables the strain to function in a wider variety of environments, whereas the second is adaptability-reducing. The minimal-medium-adapted strain thus possesses too much of one type of adaptability and too little of the other to compete in the richer medium. Such studies indicate that at least in microbial cultures unexercised adaptability is subject to negative selection even if its energetic cost is relatively small and that a strain with an inappropriate structure of adaptability is rapidly displaced in a competitive environment even if this structure is viable in the absence of competition. It is more difficult to perform such selection experiments on metazoans. But it is interesting that Levins (1968) has argued on an indirect basis that as the fitness curve of an organism, plotted against environmental parameters, becomes broader, its maximum height becomes lower. The argument is based in part on data which indicate that energy cost curves for insect development show this tradeoff when temperature is taken as the environmental parameter (Sacher, 1967). The implication of such a tradeoff is that organisms which can live in a wider variety of environments—hence which are more adaptable—could not function as efficiently as a less adaptable variant in any given environment.

6.2. GENERAL MECHANISMS

Each of the terms in equations (6.4)–(6.6) contributes to the ability of the community to contend with sources of dysfunction, whether internal or external. We already know that the processes which underlie these terms either involve physical isolation or selective dissipation into an ensemble of functionally acceptable states (or, more precisely, trajectories). These trajectories may be macroscopically or adaptively inequivalent, may initially be macroscopically or adaptively inequivalent and later macroscopically equivalent, or may never quite converge to equivalence. Thus there is a continuous spectrum of different possible situations. However, for the purposes of discussion it is convenient to classify them into five basic categories, namely physical reliability, organizational reliability, behavioral uncertainty, selective indifference, and failure. Physical and organizational reliability allow the system to combat noise or malfunction of internal origin. Selective indifference and behavioral uncertainty allow the system to cope with the unpredictability of the environment and therefore contribute to adaptability directly.

i. *Physical reliability.* The physical reliability mechanism allows the system to withstand noise because its components function with greater precision. The degree of precision is determined by quantum structure and the number of states which are equivalent from the macroscopic point of view (cf. Sections 2.7, 3.8, 4.11; also Table 4.1). In general it would be expected that precise components would be more unusual objects than less precise components, e.g., that a precise enzyme is either larger or allows for less variation in primary structure than an imprecise enzyme. Thus it is reasonable to suppose that physical precision always involves some cost. Note, however, that the degree of physical precision is not reflected in equation (6.4), for in this case the entropies are defined on the transition schemes of adaptively distinct states, and therefore would not reflect selectively equivalent variants, such as selectively equivalent amino acid sequences.

ii. *Organizational reliability.* The organizational reliability mechanism allows the system to withstand noise by virtue of its spatial or temporal organization (cf. Section 4.11). Recall that the general idea is that the organization allows for families of adaptively equivalent states (or trajectories) correlated to the environment in the same way (i.e., allows for the same chance of reproducing), along with a mechanism for maintaining the coherence of the family. This mechanism either may involve an external correcting (or coding) device, which must itself be reliable, or may be entirely internal, in which case all likely changes which endanger the coherence involve an increase in free energy. The second mechanism thus borders on physical reliability, except that we speak of organizational reliability when the ensemble of states is more readily associated with the entire system than with any of the individual components. The simplest example of organizational reliability is a cell with multiple copies of genetic material. If one of these copies is damaged this does not interfere with the function of the system. Furthermore, certain enzymes (the reliable components) are capable of returning the damaged system to its original state. As another example consider a structural change in a cell due to random perturbation. In this case it may be possible for the system to return to its original state through self-reassembly, i.e., through a nonenzymatic free energy minimization process (cf. Section 3.2). Here there is also an ultimate dependence on the specificity of system design in the sense that the components must fit to one another like the pieces of a jigsaw puzzle, but this is entirely a matter of the internal dynamics of the system.

The organizational reliability mechanism always involves either spatial or temporal redundancy of states, and therefore ultimately depends on either component redundancy, signal redundancy, or inherent specificities of system design. Thus, like physical reliability it involves a cost. Unlike physical reliability it does contribute to certain of the entropies, viz. to all those entropies defined over $\hat{\omega}$.

iii. *Behavioral uncertainty.* The behavioral uncertainty component of adaptability allows the system to cope with external disturbance (i.e., macroscopic variation of the environment) by assuming a mode of behavior which is functionally distinct and suited to the new environmental situation. There are three basic mechanisms:

1. *Constitutive adaptability.* In this case the repertoire of behavioral modes depends on the constitution of the system, i.e., on structures ("machinery") which are always present. The variations in the environment push the system into one or another of its functionally distinct trajectories, but the modifications disappear before too long and the system converges to what is essentially the same final situation. This transience is a general feature of constitutive adaptability, for if the modifications were permanent

it would be wasteful for the full machinery always to be available. Constitutive adaptability is thus particularly suited to disturbances which are significant on a relatively short time scale.

My choice of the term *constitutive adaptability* derives from what is perhaps the simplest example, viz. a microorganism constitutive for some enzyme. Such an organism may appear quite indifferent to the presence or absence of the substrate on which this enzyme acts. However, the presence or absence of the substrate does alter the detailed biochemical behavior of the organism. As another, more sophisticated example, imagine a system (e.g., a computer or a brain) which dissipates the effect of environmental disturbance according to a definite procedure, such as the procedure which allows a spider to repair its web. In other words, the system embodies an algorithm for coping with a certain class of possible environmental situations. The rule is always the same, but each different environmental situation produces a different detailed sequence of states in the system. The changes, however, are relatively transient and the machinery for handling the class of disturbances is always present.

2. *Inducible adaptability.* In this case the change in the environment pushes the system into an appropriate mode of behavior by inducing latent machinery in the system. Thus the changes in behavior are coordinated to structural change in the system and are capable of lasting for a relatively long time. Inducible adaptability allows biological systems to cope with long-lasting changes in the environment, or changes which have a good chance of lasting long enough to make it worthwhile for the system to undergo some structural change, but which are not so frequent that it would be economical to keep all machinery ready all the time.

The choice of the word *inducible* again derives from what is probably the simplest example, viz. a microorganism which is inducible for some enzyme. In this case the presence of a metabolite serves (indirectly) to activate the gene which produces the enzyme which acts on this metabolite, or, alternatively, the absence of a necessary metabolite serves to activate genes which produce enzymes which supply this metabolite. Another more dramatic example is provided by the different morphological patterns which are sometimes found among the same plant species in different environments. In this case the different environment induces different patterns of growth, either by pushing the genetic endowment of the organism into different patterns of activity or by altering cell growth or multiplication within the framework of a given pattern of genetic development.

3. *Selective adaptability.* This is adaptability through variation and natural selection. The most general thing which can be said is that the system which uses this form of adaptability contains a population of

subsystems which are rough replicas of one another. Changes in the environment change the distribution of replicas, thereby changing the behavior of the system as a whole. In general selective adaptability involves permanent changes in the system, although some degree of reversibility is in principle possible if the change in the environmental situation reverses itself.

The best example of the selective mechanism is genetic adaptability, i.e., the formation of new adaptations through variation and natural selection. In this case the species shuffles its genes, e.g., through mutation, recombination, or crossing over. The shuffling of course gives rise to an ensemble of genotypes. The corresponding phenotypes are functionally dissimilar, although often only slightly so. Thus the population consists of a variety of types, and if the variety is great enough it is highly likely that at least some of them will be able to cope with any given environmental situation. But even more important, the selective mechanism is the only mechanism which allows for the development of entirely novel adaptation. In this sense it is the master mechanism of adaptability, for it is through it that all other mechanisms of adaptability are ushered into the world.

There are probably other examples of the selective mechanism in biology. In particular, the evidence is now excellent that mechanisms of this type play a fundamental role in antibody formation and therefore in immunity (cf. Jerne, 1955; Burnet, 1959). The suggestion has also been made that the selective mechanism plays a role in behavioral learning. For example, protein molecules influencing the activity of neurons in the brain could be differentially cultured on the basis of their functional value for the behavior of the organism, either by culturing the nucleic acids coding for them or by culturing inducers which initiate their production (cf. Conrad, 1974).

The difference between the induction and selection mechanisms is that in the former the essential feature is environmental control of the release or suppression of the construction of latent machinery. In the latter the essential feature is environmental influence over the amplification of preexisting machinery through reproduction or replication processes. The amplifiable machinery can all preexist, in some cases in small numbers, or it can arise *de novo* through noise in the replication process. The selective mechanism is thus costly, not because the system carries potentially necessary machinery in latent form, but because it carries it in overt form. In this sense it is more expensive than the induction mechanism, especially as some variations are likely to be nonfunctional in any situation. Moreover, even if all the varieties are functional it is unlikely that they have equal functional value in any given environment, so that a population with a greater than necessary number of varieties is never as efficient as it could be. Indeed, this is the basic content of the so-called competitive exclusion principle (or

Gause hypothesis), according to which no two species can coexist in the same niche. Presumably the two species are in some way phenotypically distinctive and therefore inevitably not precisely matched in their "struggle" for any given type of existence (cf. Section 5.1).

The induction and selection mechanisms, while distinct as general mechanisms, frequently operate together in biological control processes. A simple example is sporulation and culturability (cf. Section 5.7). The spore form of a microorganism germinates as a result of an environmental change. This goes into the induction mechanism category, assuming that something latent (e.g., machinery present only in the blueprint form) is brought out of latency. The culturing process, however, makes it possible for the environment to influence (select) which types are reproduced and therefore which machinery is amplified. To the extent that this is the case it properly belongs to the selection mechanism category. If the reproduction process involves variation there is also the possibility for the evolution of *de novo* adaptations, in which case the culturing process necessarily falls into the selection mechanism category.

Notice that both selective adaptability and organizational reliability are based on redundancy. However, in organizational reliability the redundancy is used to eliminate the detectable effects of noise, whereas noise helps generate the ensemble of rough replicas in selective adaptability and therefore plays the critical role in the development of new adaptations.

The constitutive, inducible, or selective mechanisms are all mechanisms of behavioral uncertainty. Thus they always contribute to $H(\hat{\omega})$. They may also contribute to $H(\hat{\omega}|\hat{\omega}^*)$, but the extent to which they contribute to adaptability depends on the difference between $H(\hat{\omega})$ and $H(\hat{\omega}|\hat{\omega}^*)$. Thus the maximum contribution occurs when there is no contribution to $H(\hat{\omega}|\hat{\omega}^*)$, which is tantamount to saying that the maximum increase in adaptability occurs when anticipation is complete.

iv. *Selective indifference.* The selective indifference mechanism allows the system to cope with macroscopic disturbances either by dissipating these disturbances into the ensemble of macroscopically equivalent states or by isolating the system from these disturbances. In general the physical isolation mechanism is based on spatial isolation and therefore involves niche breadth. If the geographical range of the organism is restricted it is clear that it will be unaffected by events outside this range, or, more precisely, that the effect of these events will be dissipated by the environmental medium before they reach the organism. Spatial isolation, however, is not restricted to geographical isolation. The organism may also restrict itself to certain temporal or even spatial microenvironments, in which case certain physical features of the environment are used to dissipate the effects of geographically local disturbances. Alternatively it may restrict itself to certain temporal microenvironments, in which case it combines spatial isolation (or niche narrowing) with prepatterned anticipation of the environment, or it may retreat to particular locations in the environment under appropriate circumstances (the avoidance reaction mechanism), in which case it combines spatial isolation with behavioral uncertainty.

The organism may also develop its own "external medium" for dissipating disturbance. In this case, instead of using or constructing a natural feature of the environment, it grows its own feature (shell, hide, insulating device, and so forth) for dissipating the disturbance into an ensemble of macroscopically equivalent states. Actually, this form of selective indifference is quite similar to the constitutive adaptability mechanism, except that it is natural to think of selective indifference as involving static structures, such as shells or bone encasements, and of constitutive adaptability as involving dynamic processes, such as the occurrence of enzymes which make it possible to metabolize a greater number of food sources or to synthesize necessary components from a smaller number of sources. The association of constitutive adaptability with $H(\hat{\omega})$ and that of selective indifference (arising from protective structures) with $H(\hat{\omega}^*|\hat{\omega})$ are thus in some measure a matter of convention, for if we chose to look at our biological system in sufficient detail, we would almost certainly detect some transient but physically measurable change in response to external perturbation of the protective structure.

Constructing protective mechanisms, either in the course of development or by manipulating the environment, is clearly a cost to the organism. Spatial isolation (or niche narrowing) is also a cost, for it is clear that in this case the organism forgoes using certain resources in the environment. Arbitrary indifference is thus clearly no solution to the problem of adaptability; indeed, it is quite inconsistent with the essential processes of life.

v. *Failure (or nonselective indifference).* There is also the possibility that the biological system (community, population, organism...) either fails to cope with the environment or fails because of internal breakdown. In the first case the environmental variation drives the system into states which are in some way nonfunctional or subfunctional, e.g., states of injury or disease in the organism, death of an unusual number of organisms in the case of a population, eutrophism or other disruption at the level of the community, or pathological organization or behavior at the level of the cell. In the second case (internal breakdown) the failure is an expression of system unreliability. Note, however, that failure in this sense should be distinguished from senescence and death resulting from senescence. This is a natural process, part of the organism's normal repertoire of behavior, and in fact we could even regard failure to die in organisms in which death is the normal mode of behavior as a form of failure, except that this never seems to occur. Furthermore, there is no ambiguity as to the class of organisms for which aging in the direction of death is the normal model of behavior. This includes all the multicellular organisms and also individual cells which do not undergo binary fission. If the cell undergoes binary fission, the result is in some sense two refurbished daughter cells, and no parent whatsoever. In the case of multicellular organisms, however, it is only the germplasm which has no definite half-life (to the extent that there is no definite extinction time for the species). Since the somaplasm develops from the germplasm it is really inevitable that the parent organism must eventually experience death, for there is no way it can refurbish its individual self through self-reproduction, the universal mechanism of self-repair. Even if it could refurbish itself through more specialized mechanisms of self-repair, however, the wages of eternal life would be the absence of adaptability through evolution, for the entire living space would soon be filled up with old organisms, the birth rate would have to drop to the accident rate (which would be relatively lower in the absence of aging), and therefore the genetic variability (or, more precisely, the uncertainty in behavior at the genetic level) would drop to zero. In other words, aging and expected lifetime (or expected doubling time in the case of unicellulars) are properly regarded as part of the spectrum of biological adaptabilities, not as failure.

How does failure enter into the formalism? If the behavior of the community (or population, or organism, ...) is improperly correlated to that of the environment, it is pathological—indeed, this could almost serve as a definition of pathological. The problem is that the equations tell us nothing about whether correlations are proper or not, but only how much correlation there is. However, it is reasonable to suppose that pathological behavior is in general less correlated to the behavior of the environment, since it could hardly be expected that an improperly functioning organism could continue to process information about the environment as effectively as a normally functioning organism.

There are, in the equations, two terms which express decorrelation, viz. $H(\omega|\omega^*)$ and $H(\omega^*|\omega)$. Thus the question arises as to which of these entropies increases in the case of failure. In the case of the first (or the anticipation) term both increase or decrease are logically possible. To see this, suppose that the organism (or other biological system) is less effective at anticipating the environment (or less effective at processing information about the environment). It is initially possible for its behavioral uncertainty to be just as great or even greater than before. In this case $H(\omega|\omega^*)$ will increase. If the organism dies, however, the behavior will be definite (in the sense that the organism goes into a state which we call dead and stays there). In this case $H(\omega|\omega^*)$ must decrease. Thus, for the organism, and for all other biological systems, the final outcome is always a decrease if the pathology is not corrected.

In the case of the second of the correlation terms (the indifference term) the only really plausible direction of change is increase. This is again implicit in the meaning of subfunctional or pathological behavior, for it would be hard to call the behavior of an organism subfunctional if the organism was more effective at processing information about the environment; and furthermore, it is unlikely that interference with other aspects of the organism's behavior (e.g., metabolism, self-maintenance or repair capabilities, ability to move or otherwise affect the environment) could result in any improvement in these information processing capabilities. Thus the conclusion is that failure always expresses itself in the indifference term, except that it does so in a way which is of no advantage to the system, hence in a way which is nonselective.

Note that the inherent entropies, $H(\hat{\omega}|\hat{\omega}^*)$ and $H(\hat{\omega}^*|\hat{\omega})$ do not themselves change with injury; if the organism is incapable of coping with disturbance this just means that its inherent entropies are too small. Thus there is a certain inherent amount of unreliability (or ineffectiveness at anticipating) which is tolerable; there is also a certain amount of nonselective indifference which is tolerable. A natural question is: is it possible to distinguish this tolerable component of nonselective indifference from selective indifference? It would seem that in the absence of any reference system we can never unambiguously distinguish error from normality in biological systems, and that therefore the distinction between selective and nonselective indifference must be regarded as a conceptual distinction, demarcating indifference which is concomitant to overriding disadvantage from indifference which is concomitant to overall advantage. From the practical point of view, however, it is probably possible to make the distinction in many instances, e.g., by determining whether increasing the particular form of indifference has favorable or unfavorable consequences for the survival curve of the system.

6.3. CORRELATION AND DECORRELATION MECHANISMS

The decorrelation term, $H(\hat{\omega}|\hat{\omega}^*)$, is important, not because it reflects mechanisms concomitant to the dissipation of or isolation from external disturbance, but because it reflects the effectiveness of these mechanisms. For example, in a purely deterministic environment complete anticipation is in principle possible, in which case there will be no need for dissipation of unexpected disturbance (although the deterministic behavior of the organism,

including behavior associated with the performance of essential life functions and behavioral change anticipating the deterministic changes of the
environment, is itself necessarily dissipative). On the other hand, if the
anticipation is incomplete the demand on mechanisms of behavioral uncertainty and indifference will be much greater, so that the component of
dissipation associated with adaptability will necessarily be positive.

By itself this component may not be large; certainly it is not in general
large relative to the total dissipation (cf. Section 2.2). However, it is
associated with the mechanisms discussed in the previous section and
therefore clearly critical from the functional standpoint, i.e., from the
standpoint of the system's "fitness" to the environment. Thus, anticipation
is important from the standpoint of biological economy. As with the other
mechanisms (behavioral uncertainty, indifference) it also involves a cost, or
perhaps more aptly, an investment.

The basic issues have already been covered in the discussion of the
information transfer picture (cf. Section 4.7). The essence of anticipation is
either preestablished correlation with the environment or ability to utilize
certain features of the environment as cues for potentially significant
environmental events. Preestablished correlations are of course based on the
past experience of the organism or species, i.e., on learning or evolution.
Examples are life-cycle phenomena and circadian, seasonal, or other
rhythms. All are clock phenomena, or, to draw out the clock metaphor, all
are ordered by the setting of alarms of the time pieces which all organisms
somehow carry. Maintaining the settings, once evolved, either requires a
certain amount of selection (therefore the elimination of variant organisms)
or genetic mechanisms which prevent variations in the first place (e.g.,
inversions, agamy, or high gene flow within the population). In the absence
of selection pressure and conservative genetic mechanisms, the precision of
evolved correlation will certainly erode. Moreover, in the absence of any
such pressure the conservative mechanisms will also eventually loosen,
partly because of the absence of any selection for maintaining the particular
ratchets which act on the correlations in question, but also because all of
these ratchets are concomitant to structures or modes of behavior which
involve some cost to the organism, although perhaps a relatively small cost.
Thus we can expect unusable ability to anticipate (due to evolved correlation) to fall off, for there can be no selective pressure for maintaining an
unusable ability.

The same considerations hold for anticipation based on the ability to
cue in on features of the environment insofar as this cueing ability is based
on evolution. The cueing ability and also preestablished correlation may
also arise through learning processes, e.g., through classical conditioning, via
trial and error learning, or on the basis of more elaborate mechanisms for

forming associations between cues and appropriate reactions to likely future states of the environmert, or, more generally, mechanisms for putting together behavioral sequences correlated to the sequences of environmental states. For obvious reasons I call this mechanism for the sharpening of anticipation the *associative* mechanism. There are perhaps many such mechanisms in biology. The associative capacity of the central nervous system is of course the most well known. The immune system is also in a certain sense an associative learning system, at least to the extent that antibody–antigen specificity is a form of association. The common feature of all such mechanisms is the presence of a memory store, a device for putting memories into storage, and a device for retrieving them. The storage medium and devices may vary in different types of systems, but excess capacity of this type (e.g., superfluous neural tissue) always has to go into the cost category. Thus it can be expected that superfluous ontogenetic capabilities for sharpening anticipation to the environment will also fall away in the course of evolution, for all that can be expected from such superfluously sharp capability is superfluously sharp correlation. Moreover, there is a definite formula for distinguishing the superfluous from the nonsuperfluous, for once the behavioral uncertainty and indifference are given, any decreases in the anticipation term which make the adaptability greater than the environmental uncertainty must be regarded as superfluous. (But recall that if the behavioral uncertainty and indifference are not fixed it is only possible to say that some portion of one or more of the three terms is superfluous.)

6.4. APPARENT PARADOX OF COMPETITION

Recall that the decorrelation terms reflect not only anticipation (or rather lack of anticipation) but also decorrelation arising from internal sources of noise. When the ensemble of states generated by the noise process are functionally equivalent the situation is one of organizational reliability [cf. Section 6.2(ii)]. Otherwise the system is unreliable. It is of fundamental importance that this unreliability reflects the *statistical aspect of competitive processes*. We have already considered one example at some length, viz. genetic error and its role in evolution (which is certainly in the broad sense a competition process). However, there are many other aspects of competition, especially at the level of intra- and interspecies interactions.

Precise definition of competition is not easy. Roughly speaking, however, it is reasonable to say that the two species in a community cooperate when they specialize in the utilization of different resources in a complementary way, and that they compete when resources are usable by both and in such a way that use by one interferes with use by the other. This definition actually extends to predator–prey interactions, in which case the usable resource is prey. The predator uses the prey for food and the prey uses itself to produce more prey. In this case the interaction is competitive if the prey never actively sacrifices itself to the predator, whereas it is cooperative if the prey actively contributes to regulation of its numbers through the predation process, or if the predator actively refrains from overfishing the prey (this is the prudent predator and efficient prey notion of Slobodkin, 1961). The definition also applies to intraspecies competition, in which case it is the different members of a single population or family group whose resource utilization is either complementary or overlapping.

In the competitive interaction the actions of one of the competitors (e.g., organism, population) disturb the other. If these actions are indeterminate, the disturbances will be indeterminate, therefore concomitant to internal sources of noise within the population or community. In short, biological systems must contend not only with the unpredictable behavior of the external environment but also with the unpredictable behavior of other biological systems.

Notice that as the uncertainty of the biological environment of a particular biological system (e.g., organism or population) increases, the adaptability of this system must increase also, assuming that the adaptability tends to fall to its minimum possible value (i.e., assuming the consequence of all the foregoing arguments). At first sight this appears to imply that the adaptability of the larger unit (e.g., population in the case of the organism, community in the case of the population) must increase, and therefore that our whole argument leads to a fatal contradiction; for if the adaptability of each organism in a population increases, due to intrapopulation competition, it would seem that the adaptability of the population would increase, in which case the latter adaptability would exceed the uncertainty of the external environment of the population (which is unchanged by the internal competition process).

More careful examination, however, shows that there is in fact no contradiction. To see this consider the first two terms in our adaptability equation, viz. $H(\hat{\omega}) - H(\hat{\omega}|\hat{\omega}^*)$. As the level of internally generated competition increases, both these terms increase equally, so that there is no net increase in adaptability. In other words, if the uncertainty is internally generated it clearly increases the behavioral uncertainty of the higher-level unit (e.g., population if the uncertainty is associated with interactions of organisms in this population). Also, if it is internally generated this means that it is uncorrelated to the behavior of the environment, so the decorrelation entropy will increase by the same amount as the behavioral uncertainty. The only possible exception is the case in which the competitive interactions are parametrically controlled by the environment; but in this case the unit serving as the source of uncertainty can only surprise another unit if this other unit is unaware of the parametric environmental influence. However, this unawareness must appear in the anticipation and indifference terms, so that again there can be no increase in overall adaptability.

The noncontribution of competition-generated uncertainty to overall community adaptability does not at all mean that it is biologically unimportant or invisible. It is certainly visible at inner levels of organization and indeed may be regarded as serving as a spur to the adaptability of populations and organisms and therefore as a mechanism for maintaining a reserve of inner adaptabilities which can always be converted to forms capable of coping with bona fide external disturbance in the event that the quantity or nature of this disturbance increases or changes. For example, if the level of competition-generated uncertainty in a population increases, this may result in an increase in any of the modes of adaptability of the constituent organisms, ranging from the genetic to the behavioral. If the level of environmental uncertainty increases, the population will definitely suffer disturbance in a way which it cannot completely forget (assuming that its adaptability has fallen to the minimum possible value). However, if the behavioral, developmental, and genetic adaptabilities used for protecting organisms from other organisms in this population are particularly high, there is an extremely good likelihood that they can fairly readily be redirected toward the novel disturbances of the external environment. This is particularly true for genetic adaptability and also learning ability, neither of which are in general precisely directed toward particular classes of disturbances in any case.

The reserves of internal adaptability maintained by competition-generated adaptability are mechanistically no different than forms of adaptability maintained by externally generated environmental uncertainty. Thus the inescapable conclusion is that competition-generated uncertainty is a cost. However, the minimal tendency arguments do not apply in this case, for these are evolutionary tendencies and evolution is inherently a matter of variability and competitive selection. Certainly it is untenable to argue that competitive evolution always reduces the level of uncertainty associated with competition, so it must also be untenable to argue that reserves of adaptability concomitant to such competition always tend to disappear. Furthermore, it is clear that such reserves are often advantageous and clearly play an important role in the dynamics of evolution itself. However, the existence of these inner adaptabilities is completely consistent with all the statements about the minimal tendency of adaptability, for they in effect cancel one another out.

It is sometimes useful to think of these intra- and interpopulation interactions in game theory terms. For example, imagine that each organism is a player with certain possible moves. The other player may be the external environment or both the external environment and other organisms. The moves of the environment are in general uncertain, hence the requirement for adaptability. The moves of the other organisms may also be uncertain, hence the requirement for adaptability directed to uncertainty of biological origin. In some instances the two organisms are better regarded as being in coalition, in others, as opponents. These correspond to the situations of cooperation and competition. The analogy can also be extended to higher levels, for example by regarding species as players in a game of evolution.

It is useful to analyze the situation in terms of the hierarchical formalism. The simplest case is a system with two subcompartments, e.g., a community with two populations. The equation is

$$H\left(\hat{\omega}_{15}\hat{\omega}_{14}\hat{\omega}_{24}\right) - H\left(\hat{\omega}_{15}\hat{\omega}_{14}\hat{\omega}_{24}|\hat{\omega}^*\right) + H\left(\hat{\omega}^*|\hat{\omega}_{15}\hat{\omega}_{14}\hat{\omega}_{24}\right) \rightarrow H(\omega^*) \quad (6.7)$$

where $\hat{\omega} = \hat{\omega}_{15} = \hat{\omega}_{15}\omega_{14}\omega_{24}$. The expansion of the decorrelation term is

$$H\left(\hat{\omega}_{15}\hat{\omega}_{14}\omega_{24}|\hat{\omega}^*\right) = H_e\left(\hat{\omega}_{15}|\hat{\omega}^*\right) + H_e\left(\hat{\omega}_{14}|\hat{\omega}^*\right) + H_e\left(\hat{\omega}_{24}|\hat{\omega}^*\right) \quad (6.8)$$

where, according to equation (5.22),

$$H_e\left(\hat{\omega}_{14}|\hat{\omega}^*\right) = \tfrac{1}{3}\Big[H\left(\hat{\omega}_{14}|\hat{\omega}^*\right) + \tfrac{1}{2}H\left(\omega_{14}|\hat{\omega}_{15}\hat{\omega}^*\right)$$

$$+ \tfrac{1}{2}\left(\omega_{14}|\hat{\omega}_{24}\hat{\omega}^*\right) + H\left(\omega_{14}|\hat{\omega}_{15}\hat{\omega}_{24}\hat{\omega}^*\right)\Big] \quad (6.9)$$

$$H_e\left(\hat{\omega}_{24}|\hat{\omega}^*\right) = \tfrac{1}{3}\Big[H\left(\hat{\omega}_{24}|\hat{\omega}^*\right) + \tfrac{1}{2}H\left(\hat{\omega}_{24}|\hat{\omega}_{15}\hat{\omega}^*\right)$$

$$+ \tfrac{1}{2}\left(\hat{\omega}_{24}|\hat{\omega}_{14}\hat{\omega}^*\right) + H\left(\omega_{24}|\hat{\omega}_{15}\hat{\omega}_{14}\hat{\omega}^*\right)\Big] \quad (6.10)$$

Suppose that the modifiability terms increase but that the independence terms do not, i.e., $H(\hat{\omega}_{14}|\hat{\omega}^*)$ and $H(\hat{\omega}_{24}|\hat{\omega}^*)$ increase but all other terms stay the same. In this case the ability of the individual populations to anticipate the external environment decreases, so that the decrease in reliability should be associated with a decrement in community anticipation. It is also possible for the independence terms to increase, but not the modifiability terms. In this case terms such as $H(\hat{\omega}_{14}|\hat{\omega}_{24}\hat{\omega}^*)$ and $H(\hat{\omega}_{24}|\hat{\omega}_{14}\hat{\omega}^*)$ increase, but subject to the obvious condition that they never exceed $H(\omega_{14}|\hat{\omega}^*)$ or $H(\hat{\omega}_{24}|\hat{\omega}^*)$. In this case the increase in unreliability is associated with increasing decorrelation between the two populations rather than between the individual populations and the environment. In general, however, the independence component of the behavioral uncertainty [e.g., $H(\hat{\omega}_{14}|\hat{\omega}_{24})$ and $H(\hat{\omega}_{24}|\hat{\omega}_{14})$] increases by the same amount, so that there is no decrease in adaptability due to increasing unreliability or increase due to increasing behavioral uncertainty.

The above hierarchical considerations can be condensed into a simple, general statement, namely: *as the uncertainty associated with competition increases the independence terms increase*. However, the converse is not always the case, i.e., it is not always reasonable to say that a relative increase in the independence part of the decorrelation term is associated with our intuitive sense of the word *competition*. There are clearly many cases in which internally generated uncertainty associated with decorrelation among subunits does not play a functional role which can be identified with competition, or even any functional role at all. However, in no case does this lead to any violation of our general statements about evolutionary tendencies.

It is appropriate to close out this section with some further critical comments on the words *cooperation* and *competition*, in particular insofar as these words draw meaning from game-type situations. The game theory analogy has been pursued, both informally and formally, starting I believe with a suggestion of Waddington (1957). Lewontin (1961) has formulated in game theoretical terms the problem of determining (relative to an arbitrary measure of fitness) the optimal choice of evolutionary strategies (for example, what is the degree of sexuality which might be optimal in a particular instance?). Slobodkin and Rapoport (1974) have dealt with evolution in terms of what they call existential games, in which the effective goal of the species is only to stay in the game. The idea clearly goes outside the bounds of formal game theory. However, it certainly has biological merit, for it is quite reasonable to suppose (reversing the argument) that it is only the properties of those species which actually stay in the game which in fact have any significant reality.

The game theory analogy is suggestive and perhaps a useful metaphor for capturing the adaptability theory notion of competition-generated uncertainty. However, the formal structure of game theories is too restrictive for the much more variegated and generalized situations of biology. To the extent that the notions of competition and cooperation are tied to this framework, they must also be regarded as merely suggestive, and necessarily limited in applicability. To the extent that these words carry with them the connotations which their service in ordinary language earns for them, they must be regarded with even more caution, for in many cases it is certainly unclear whether we are dealing with a situation of competition, one of cooperation, a mixture of these, or a situation which is well outside the framework of these words. What is objectively observable and measurable is only the degree of internally generated uncertainty in a system, i.e., the degree of uncertainty associated with the lack of information of one part of a biological system about the behavior of another part. This is perhaps more accurately expressed in terms of the degree of integration (or rather pluralism) among the parts of the system. In some instances this approaches a form which can be expressed in terms of an (informal) game theory framework and therefore a form to which the words *competition* and *cooperation* can be given fairly definite meanings. However, it should always be remembered that the use of these words implies a definite functional interpretation of behavior. In particular, when we speak of competition-generated uncertainty we are making a definite functional interpretation for internally generated uncertainty, and one which may not always be applicable. However, we can always safely talk about unexpected disturbances of internal origin and associate these with either (advantageous) lack of integration or (advantageous) decentralization, depending on whether or not the survival curve of the system would be improved by increasing the degree of integration.

The game theory models developed so far do not seem to lead to testable predictions except in very locally defined situations. This is partly because the formal structure of games is too restrictive for biological systems, but much more importantly because they require the arbitrary introduction of fitness variables in order to define payoff functions. This is not the case for the adaptability theory developed in this book. Indeed, the adaptability theory may be regarded as a theory of a certain aspect of fitness, namely its statistical aspect.

6.5. MECHANISMS AND MODES OF ADAPTABILITY

The classification of mechanisms developed in Sections 6.2 and 6.3 applies to all levels of biological organization. Thus it encompasses all the various modes of adaptability discussed in the previous chapter and listed in Table 5.2. Actually this table shows only the modifiability component of various forms of behavioral uncertainty (e.g., gene pool diversity, morphological and behavioral plasticity, and culturability). There are numerous other components (associated with the independence, anticipation, and indifference terms), many of which, indeed most of which, have not been christened in the biological literature and about which it would therefore be difficult to speak outside the framework of the formalism of the adaptability theory. For example, consider the term

$$H\left(\hat{\omega}_{u4}|\prod_{j<4}\hat{\omega}_{ij},\prod_{v\neq u}\hat{\omega}_{v4}\right) \tag{6.11}$$

where i runs over all subcompartments of c_{u4}. This is the behavioral uncertainty of population u (e.g., its culturability, social modifiability, topographical plasticity) given all information about its subcompartments and also the behavior of all other populations in the community in terms of their complete states. Now consider

$$H\left(\hat{\omega}_{v4}|\prod_{j<4}\omega_{ij},\prod_{v\neq u}\omega_{v4},\prod\hat{\omega}_{h0}^{*}\right) \tag{6.12}$$

and

$$H\left(\hat{\omega}_{u4}|\prod_{j<4}\hat{\omega}_{ij},\prod_{v\neq u,r}\hat{\omega}_{v4},\prod\hat{\omega}_{h0}^{*}\right) \tag{6.13}$$

The first of these is the behavioral uncertainty of population u given the same information as above and in addition the contemporaneous behavior of the environment. The second is the same as the first, except that the behavior of population r is not specified. If the second term is much greater than the first, this means that it is the coincidence of events in population r and in the environment which are significant apropos the ability of population u to anticipate the environment and therefore for the extent to which (6.12) contributes to its adaptability. If certain of the $\hat{\omega}_{h0}^{*}$ are removed but this makes no difference, the implication is that events in these regions of the environment provide no information about the population and are therefore not anticipated; similarly, if the $\hat{\omega}_{h0}^{*}$ are further expressed in terms of transition schemes associated with particular features of the environment, it is possible to see which features can and cannot be anticipated, or rather, to measure the degree to which they can be anticipated. Conversely, one can write

$$H\left(\hat{\omega}_{h0}^{*}|\prod_{i,j}\hat{\omega}_{ij}\right) \tag{6.14}$$

and

$$H\left(\hat{\omega}_{h0}^{*}|\hat{\omega}_{u4}\right) \tag{6.15}$$

which specify, in the first instance, how much the behavior of the community is capable of telling about the particular region of the environment (therefore how sensitive it is to this environment) and, in the second instance, how much population u by itself is capable of telling. If the two terms are practically the same, this means that population u (at the population level) is as sensitive to the particular environment (or to the particular feature of the environment) as the whole community, i.e., as sensitive as possible. If the second term is very large this means that population u is insensitive to the particular region of the environment.

Expressions of the above type provide a useful formal tool for thinking about the flow of influences in an ecosystem, or within lower-level biological systems such as organisms or cells. The resulting description is actually completely phenomenological, for determining the existence or even the strength of an influence is not the same as determining its mechanistic basis, although it is not at all unlikely that there are certain patterns of influence which are generally associated with certain mechanisms, or combinations of mechanisms. However, what is important at the moment is only that all the various mechanisms are more or less costly, and that therefore all the different combinations of terms of the above type are more or less costly. This is why the argument about evolutionary tendency extends to both the nonhierarchical and hierarchical forms of the fundamental inequality [cf. equations (6.4) and (6.5)].

6.6. DISPENSING WITH ADAPTABILITY

The argument so far is that all mechanisms of adaptability are in principle costly (for the time being taking cost as an intuitive, primitive concept of biology) and that therefore all the various modes of adaptability, including modes of modifiability, independence, anticipation, and indifference, are more or less costly. Furthermore, it is supposed that features of biological systems which are costly and entirely superfluous tend to atrophy in the course of evolution and therefore that superfluous forms of adaptability tend to disappear. This supposition is clearly based on the idea that Darwinian evolution is the fundamental biological optimization process and that it therefore automatically diminishes superfluous adaptability.

It is important to consider the validity of this supposition in more detail. In recent years a great deal of attention has been paid to the level at which selection acts and to whether evolution arguments carry hidden assumptions about selection acting on groups of individuals, and indeed to whether it is legitimate to consider such group selection at all. I have learned from a number of discussions that it is important to clarify this point as early as possible. Careful analysis shows that the minimal tendency of adaptability in evolution can occur solely on the basis of selection on the individual, although it would also be mediated by selection at higher levels to the extent that this occurs.

There are two major situations: at and below the organism level, and above the organism level. The assumption that superfluous adaptability tends to disappear in the first situation depends only on the very plausible assumption that the reproduction probabilities of organisms are diminished by superfluous but costly properties. The argument applies to the organism in its entirety (at the organism level and at the level of all its subcompartments) since the whole system is acted upon in a unitary way by natural selection.

What about adaptability above the organism level, for example, social plasticity, culturability, topographic plasticity? The critical point is that the adaptability at one level is never completely independent of processes taking place at other levels. For example, culturability is dependent on physiological events such as sporulation and germination. This dependence is not only a cross-level dependence at any given time (i.e., expressed in the independence terms), but also a temporal dependence. This is an inherent feature of the formalism, for the original transition schemes [equations (4.2) and (4.33)] express the probabilities of the state of the population at time $t + \tau$ given the state of the whole ecosystem at time t. This means that all of the entropies used to characterize the population are defined over variables of the organism as well as variables of the population. Thus it is not accurate to suppose that adaptabilities at the population level (or at any other level above the organism level) will not be affected by selection at the organism level.

One way to see the mechanism here is to look at it in the framework of evolutionarily stable strategy (ESS), developed by Maynard Smith and Price (1973). A trait is evolutionarily stable if the carriers of this trait would have a larger chance of reproducing than would a variant which dispensed with this trait, but was otherwise identical. But this is just the situation with superfluous adaptability. Since adaptability is costly and since this cost is mixed into the organism level, the reproduction probabilities of the organisms in this population are less than they might be. The argument can also be phrased in the language of populations. If the reproduction probabilities are less than they otherwise might be, the population is smaller than it otherwise might be. A second population, without the superfluous adaptability, will consist of a larger number of organisms. The net effect is that the second population will predominate over the first in the course of evolution; however, this outcome is in no way dependent on group selection (i.e., selection on the population as a unit). Far to the contrary, the organisms in the second population have more favorable reproduction probabilities and therefore will inevitably replace the organisms in the former. All selection is on the individual organism.

The population phraseology is more suggestive in one respect. It exposes the "reaction–diffusion" dimension of the process. This dimension becomes important if organisms from the area occupied by one population immigrate into the area occupied by the other, or alternatively if populations in nature are broken up into smaller, intergrading populations (sometimes called demes) whose properties are not quite the same. This assumption is quite realistic, for unless gene flow over a larger area were so complete that it allowed for thorough mixing, it would be expected that the properties of populations (including adaptability) would exhibit some statistical variation in space. The tendency to migrate is also a genetically controlled property. If there is a significant enough advantage associated with migration, the tendency to migrate could easily become more pronounced, with the consequence that the diffusion gradient mechanism would become more important as its potential contribution to the minimization of adaptability became greater (i.e., as the superfluity of adaptability among certain local populations became greater).

The above process is very likely equivalent to the concept of interdeme (or, alternatively, intergroup) selection, as described, for example, by Wright (1932). Some caution is necessary, however, since the terms *interdeme selection* and *group selection* have been used with a variety of meanings in the literature (cf. Wilson, 1975; Maynard Smith, 1976) and sometimes interdeme selection is treated as a form of or equivalent to group selection. Interdeme selection (as described above) is definitely a matter of natural selection acting on organisms, in conjunction with the fact that it is more likely for organisms to migrate from larger to smaller populations than the reverse, whereas group selection involves selection on the society, population, or community as a group. The latter mechanism has mainly been invoked to account for the presence of coadaptive, altruistic traits in populations and communities at the expense of the individual. Forms of adaptability, for example at the organism level or below, are certainly not coadaptive in this sense. Others may or may not be. However, our problem is not to account for the origin of adaptability, whether or not concomitant to altruism, but to account for its elimination. Thus, the intra- and interdeme processes (involving only selection on individuals) are sufficient for the minimization tendency. In this sense the controversial and difficult question of group selection is moot from the standpoint of dispensing with adaptability.

The origin of adaptability is, however, a different question. A population or community without adequate adaptability will certainly eventually suffer catastrophic change (this is the way adaptability is defined). The result will be reduction of population numbers, thereby automatically increasing the importance of the interdeme mechanism. Thus population and community adaptabilities can expand as well as contract through individual selection coupled with migration. Actually, the expansion process is potentially much faster than the contraction process, as the decrement in population numbers in an insufficiently adaptable community would be much more rapid than in an overly adaptable one. Adaptabilities at the organism level and below can develop without migration processes (that is, through selection on the individual alone), but this does not mean that migration processes cannot play a role. In order for higher-level adaptabilities to develop, migration or other mechanisms (to be discussed) are necessary.

The argument is sometimes made that if the actual migration rates observed in nature are small (as they often are), the mechanism of interdeme selection must be rejected. This argument is circular. Migration rates (determined, for example, by behavioral binding of populations to particular areas) are themselves the consequences of evolution and may themselves be regarded as adaptations or coadaptations. The existence of such biological impediments to spatial dispersion is, however, contingent on adequate adaptability, for in the absence of adequate adaptability, the community would become extinct, then be replaced by communities whose populations have high migration rates. Low migration rates are thus latter-day characteristics of communities which already have well-developed adaptability mechanisms, so that it cannot be concluded that the interdeme mechanism is inoperable on the basis of observations on communities on which it has already played a decisive role. By preventing mixing, the reduction in migration rate serves to maintain a greater variability of adaptability characteristics among neighboring communities, so that it actually increases the effectiveness of the migration which does occur. But if the adaptability of any such community in fact became so inadequate that catastrophic change ensued, the immediate result would be its replacement not by such a similar but slightly more adaptable population, but rather by colonizing populations, which necessarily have high adaptability, or at least high migration rates (which are themselves a form of adaptability). Thus measurements on the migration between contiguous populations would again be irrelevant, for it is the migration rates of the colonizing populations which are significant under these circumstances.

The above argument goes through for coadaptations involving disadvantages to the individual but advantages to the population or community as well as for forms of adaptability at the population and community level. Again the reason is that population will be larger in the properly coadapted community than in a similar one which is not so well coadapted, so that the interdeme mechanism automatically comes into play. For example, this would be the case in a situation in which overly high intrinsic growth rates resulted in severe oscillations with resulting depression in the time averages of population size (due to lags in regrowth). Communities which develop governor mechanisms capable of reducing such oscillation would have the advantage in interdeme competition, although ungoverned individuals would have the advantage in each community.

The interdeme argument does not prove that group selection never plays a role, but only that it is not in principle necessary to invoke it. A number of critical expositions have been written on the question, with some authors (e.g., Wynne-Edwards, 1962) taking the pro position and others (e.g., Williams, 1966) taking a completely contrary position. One of the sources of difficulty is undoubtedly that it is not always clear that any particular adaptation is altruistic. All adaptations must be thought of in terms of both organism and environment, where the environment includes other organisms. Thus all are coadaptations in the sense that they are concomitant to complementary adaptations on the part of other organisms in the community and therefore more or less advantageous to the community. But are any advantageous to the community at the expense of disadvantage to the individual, i.e., are there any bona fide cases of altruism? In fact one of the arguments has been that all so-called altruistic coadaptations, when properly described, turn out to be advantageous to the individual and therefore are explainable in terms of selection on the individual. However, this interpretation seems overly restrictive, as the interdeme mechanism (which involves only the addition of spatial considerations to individual selection) is capable of accounting for coadaptations disadvantageous to the individual. Another argument (on the other side) is that since we are dealing with complementary adaptations there is a certain sense in which the whole community is involved in the selection process. In this case it is simpler to regard the community as determining the forces of selection acting on many individuals, as opposed to supposing that the interrelatedness of the selection forces implies any unitary reproduction process on the part of the population or community (for which there is no mechanistic justification). However, there are undoubtedly transitional situations (corresponding to organizations intermediate between organism and society) in which the unity of the reproduction process and therefore the actual unit of selection are ambiguous. These are just the situations in which the definition of organism and society become ambiguous.

A number of other mechanisms have also been suggested to account for coadaptive traits involving some sacrifice on the part of the individual. One is kin selection (Hamilton, 1964). The basic idea is that traits, even those disadvantageous to the individual, will be propagated if they increase the chance that genetically related individuals will succeed in reproducing. As a result, altruistic adaptability would be evolutionarily stable in organism A if it increased the chance of A's genes being propagated through B. If these related individuals carry copies of the same genes, the chances that these "gene species" will flourish increases. The real fitness of a gene—its inclusive fitness—is determined by how many offspring genes it contributes, not simply by how many offspring organisms particular individuals which carry that gene produce. This type of analysis has been much used in sociobiological argumentation. A corollary is that the gene is at least to some extent the unit of selection, an idea expressed by some exponents of the theory in terms of the paradigm (or slogan) of "the selfish gene." Genes are viewed, in effect, as using (or abusing) individuals to reproduce themselves. The weakness of the theory, in this exaggerated form, is that it loses sight of the obvious fact of a coadapted community of genes and of the integrity of the organism. One might think in terms of the selfish genome rather than the selfish gene, but then the number of possible genomic varieties becomes so large that the idea of A propagating its genome through B loses clarity. Even for the individual gene locus, with its numerous allelic variations, the notion of a definite class of genes as entities more fundamental than the organism becomes awkward. The above considerations suggest that caution should be exercised in making arguments on the basis of what would appear to be advantageous for the gene. But to the extent that a notion of relatedness can be defined, the mechanism of kin selection should certainly underlie important biological phenomena and could give rise to forms of adaptability whose occurrence would otherwise be perplexing.

A different mechanism which is most interesting involves the interaction of genetic events and population dynamics. Suppose that increase in population size leads to environmental overexploitation, causing efficient organisms to lose some of their advantage over inefficient organisms. The probability for reproduction of both will decrease and become more equal, with the result that the inefficient organisms make a relatively greater contribution to the next generation. As the population grows back, the more efficient organisms increase more rapidly, leading to a repeat of the entire process and therefore to a stable limit cycle (rather than an asymptotic increase) of population efficiencies. This situation has in fact been observed in simulations of ecosystem evolution (Conrad, 1969, 1981; Conrad and Pattee, 1970). In these simulations the genetic load, that is, the complement of fitness-reducing genes, could be increased or decreased by increasing or decreasing the extent of the population oscillation. It was not an advantage to the individual to be less efficient, but a variety of less efficient forms did increase the adaptability of the population. In the specific situations studied the decrement in individual efficiency actually increased the biomass of the population by preventing overfishing of the environment. The genetic load which developed was coadaptive in the strict sense and at the same time evolutionarily stable.*

I now terminate the discussion of the potential role of group selection. The theory outlined in this book is unlikely to be affected by the outcome of the controversy, except insofar as the issue bears on the rates with which different modes of adaptability are dispensed. These rates clearly depend on the costs of the particular modes of adaptability, for as the cost of superfluous adaptability increases, its significance in terms of differential growth rates must increase. However, it is to be expected that adaptabilities requiring the interdeme process for their reduction would decrease at a much slower rate. If bona fide group selection processes have any reality, this rate could be increased.

6.7. PHYSIOLOGICAL TENDENCIES

The mechanisms of adaptability minimization described so far are all mediated by evolution. Actually biological systems also exhibit physiologically mediated adaptability expansion and contraction. Organisms are generally capable of developing physiological adaptations to environments more extreme than usual, provided that the extreme is turned on adiabatically rather than suddenly and that it does not overreach the limits of adaptation of the particular organism. The newly developed adaptations are presumably concomitant to a physiological expansion of adaptability, provided that the organism does not lose the capacity for living in either the original or other possible environments in the process of gaining the capacity to function in the new one. Conversely, as the organism is removed from the influence of such extreme environments it in general gradually loses the adaptations which developed in response to the influence of the environment, thereby undergoing a contraction in adaptability.

The above argumentation is indirect, since to the author's knowledge there has been no systematic study of adaptability expansion and contraction apart from work which has been done on specific physiological adaptations and the universal experience with the impermanent effect of exercise. However, this experience implies the existence of such contraction

*M. Gilpin has pointed out to me that this effect is tantamount to an environment-mediated global effect of the behavior of the population on the individual and is therefore a form of group selection as the term is defined in Gilpin (1975), except that the mechanism is statistical rather than purely dynamic in nature.

and expansion, under the assumption (which is mild, but probably difficult to test) that development or loss of adaptation to a particular environment does not imply loss or gain of adaptation to some other possible environment. Expandable adaptability is based on either the inducible or selective mechanisms since these allow for the development of adaptations which are not always present. However, it only becomes significant when the time delay associated with the development of the adaptation is significant, so that the full adaptability of the organism is not immediately accessible. The complete elimination of this time delay would be expected to involve some biological costs. Thus expandable adaptability may be regarded as a way of lessening the costs conccmitant to adaptability in environments whose statistical characteristics allow for this device (i.e., environments which make it possible to develop adaptation to extremes in gradual stages).

Expandable adaptability means, in effect, that some adaptability can be dismissed under appropriate circumstances. Thus it is clear that the average adaptability (obtained by weighting each possible quantity of adaptability by the frequency of its occurrence) is smaller than the maximum possible adaptability. However, this is entirely consistent with the proposition that adaptability tends to fall to the actual uncertainty of the environment, for this uncertainty is itself decreased by the condition that certain sudden changes in environmental variables are not possible.

6.8. UPPER BOUND OF ADAPTABILITY

The argument so far has dealt with the lower bound of adaptability. There are also some considerations which suggest that there are upper limits, albeit somewhat indefinite ones. The key point is that regulation of population numbers (birth control) is a necessary condition for ecological stability and therefore the persistence of the community in a given form. This control may be internal, for example involving density-dependent mechanisms. However, it also may be mediated by environmental disturbances, in which case the populations in the community are so adapted that they are "prunable" by either periodic or aperiodic variations in the environment. This is, in particular, a possibility in environments of roughly constant uncertainty. Reduction in the actual uncertainty of the environment in this case would be accompanied by decontrol of population numbers and consequent disturbance to the community, accompanied by successional changes to a new (but possibly identical) climax. Conversely, any analogous expansion in the adaptability in the community would have the same result, indicating the existence of an upper bound (except that in this converse case there is no plausible evolutionary mechanism for the expansion).

There is some evidence that certain populations in nature are in fact controlled, or partially controlled, by environmental disturbance. Flask ecosystems cultured under a high degree of statistical stress and then placed in a quieter environment often become extremely green, resulting in an increase in oxygen tension, and sometimes a consequent transition to a gray (or eutrophic) state (personal observation). Many other mechanisms may contribute to this, including imbalances in energy or materials input (as in pollution resulting from growth induced by the addition of phosphates). However, subjecting such a system to occasional disturbance may push it to an earlier successional stage, so that in this case a certain level of environmental uncertainty serves to stabilize the system, which in turn means that the adaptability of the system cannot be arbitrarily high. This technique is in fact used and is called "pulse stability" (cf. Odum, 1971).

6.9. SELF-CONSISTENCY OF HIERARCHICAL ADAPTABILITY THEORY

The evolutionary tendency of adaptability has an important implication, expressed in the

SELF-CONSISTENCY PRINCIPLE.   The assumption of the hierarchical and compartmental structurability of the state description of biological systems is self-justifying within the framework of adaptability theory in the sense that such structurability is a necessary condition for efficient adaptability.

ARGUMENT.   To make matters concrete, consider the simple, three-compartmental decomposition of a single organism (organism one, here taken at level one, with genome and phenome compartments one and two at level zero). For the behavioral uncertainty component of adaptability

$$H(\hat{\omega}_{10}\hat{\omega}_{20}\hat{\omega}_{11}) = H_e(\hat{\omega}_{10}) + H_e(\hat{\omega}_{20}) + H_e(\hat{\omega}_{11}) \qquad (6.16)$$

where

$$H_e(\hat{\omega}_{10}) = \tfrac{1}{3}\{ H(\hat{\omega}_{10}) + \tfrac{1}{2}H(\hat{\omega}_{10}|\hat{\omega}_{20}) + \tfrac{1}{2}H(\hat{\omega}_{10}|\hat{\omega}_{11}) + H(\hat{\omega}_{10}|\hat{\omega}_{20}\hat{\omega}_{11}) \}$$

$$(6.17a)$$

$$H_e(\hat{\omega}_{20}) = \tfrac{1}{3}\{ H(\hat{\omega}_{20}) + \tfrac{1}{2}H(\hat{\omega}_{20}|\hat{\omega}_{10}) + \tfrac{1}{2}H(\hat{\omega}_{20}|\hat{\omega}_{11}) + H(\hat{\omega}_{20}|\hat{\omega}_{10}\hat{\omega}_{11}) \}$$

$$(6.17b)$$

$$H_e(\hat{\omega}_{11}) = \tfrac{1}{3}\{ H(\hat{\omega}_{11}) + \tfrac{1}{2}H(\hat{\omega}_{11}|\hat{\omega}_{10}) + \tfrac{1}{2}H(\hat{\omega}_{11}|\hat{\omega}_{20}) + H(\hat{\omega}_{11}|\hat{\omega}_{10}\hat{\omega}_{20}) \}$$

$$(6.17c)$$

As always, effective entropy consists of a modifiability term and a number of conditioned modifiability terms, called independence terms. For given modifiabilities the effective entropies are largest when each of the conditioned terms are equal and equal to the modifiability term (since they cannot be larger). Thus the conditions for a maximum uncertainty component of adaptability for given observable modifiabilities are

$$H(\hat{\omega}_{10}) = H(\hat{\omega}_{10}|\hat{\omega}_{20}) = H(\hat{\omega}_{10}|\hat{\omega}_{11}) = H(\hat{\omega}_{10}|\hat{\omega}_{20}\hat{\omega}_{11}) \quad (6.18a)$$

$$H(\hat{\omega}_{20}) = H(\hat{\omega}_{20}|\hat{\omega}_{10}) = H(\hat{\omega}_{20}|\hat{\omega}_{11}) = H(\hat{\omega}_{20}|\hat{\omega}_{10}\hat{\omega}_{11}) \quad (6.18b)$$

$$H(\hat{\omega}_{11}) = H(\hat{\omega}_{11}|\hat{\omega}_{10}) = H(\hat{\omega}_{11}|\hat{\omega}_{20}) = H(\hat{\omega}_{11}|\hat{\omega}_{10}\hat{\omega}_{20}) \quad (6.18c)$$

where $H(\hat{\omega}_{10})$, $H(\hat{\omega}_{20})$, and $H(\hat{\omega}_{11})$ are the observable modifiabilities. To the extent that these conditions are not realized, the observable modifiabilities of different compartments will be correlated, implying that the extent of modification required to achieve a given degree of adaptability increases.

There are two considerations. The first is that modification increases the cost of adaptability. A reasonable assumption is that the cost is proportional to the required amount of modification, hence inversely proportional to the independence. The second consideration is that in order for two compartments to be independent, either there must be a constraint which prevents their interaction or the system must be built so that it is capable of supporting modes of dissipation which enable the two compartments to disregard each other's behavior. This is the cost of independence. A reasonable assumption is that the cost is very low and that maintaining an isolating constraint is facilitated by the independence. Initially the cost must be very low, since isolating constraints and extra mechanisms of dissipation are small fixed costs. But complete independence must be very costly since it is incompatible with the coordination necessary for life. The conclusion is that there should be a tendency for independence to develop to the point where the decreased cost in terms of modifiability of adding to the independence is exceeded by the increased cost in terms of constructing and maintaining a functional system with this added independence. A necessary concomitant of efficient adaptability is that systems be broken up into levels of organization and compartments at different levels which are as independent as possible, subject to this condition. Furthermore, since independence has a basis in the construction of the system, it is itself an aspect of organization which must be protected from disturbance, implying that the whole hierarchy of adaptabilities contributes to the maintenance of the independence which increases their efficiency. The initial assumption that adaptability of biological systems is describable in terms of compartmental and level structure is thus consistent with the fact that such structure is necessary for efficient adaptability.

The argument generalizes straightforwardly to an arbitrary number of compartments or levels, therefore to the community as a whole. It should be emphasized that certain strong correlations are inherent in fundamental biological constraints, such as the genotype–phenotype relationship. However, to the extent that there are compartments and levels the ramification of disturbance throughout the system can be attenuated, lessening the cost of absorbing and dissipating disturbances. To the extent permitted by the necessary constraints these different levels and compartments should appear as independent (autonomous) as possible.

6.10. SEGREGATION OF GENOTYPE AND PHENOTYPE

It is worth considering further the special relationship between genotype and phenotype. Complete independence is not possible since the genotype determines the possible phenotypes. Through chemical signals the phenotype can control the expression of the genotype. But this is compatible with a high degree of independence provided that this control is mediated by chemical signals rather than by grossly distinct changes in the phenotype.

There is another kind of independence between genotype and phenotype which is connected to time scale. By definition the heritable modes of adaptability are those which involve the transmission of traits from parent to offspring, while the nonheritable modes are those which involve activities restricted to the lifetime of the individual organism. In practice the genotype is identified with nucleic acid sequences which are transmitted from parent to offspring. But it is evident that in certain cases acquired adaptations are in fact transmitted to the offspring through the cytoplasm. This is a recognized phenomenon in single-celled organisms, where it is difficult to prevent cytoplasmic adaptations from being divided among both daughter cells. Thus even though all the evidence suggests that genes, taken as DNA sequences, are not modified in an adaptive way by the environment, heritable features of the cytoplasm may be. From the functional point of view the occurrence of this phenomenon should in principle blur the distinction between genotype and phenotype. This is because the genome is the part of the organism which specializes for storage of information transmitted to the offspring, whereas the phenome is the part which acts on the environment and on the genome to produce a new phenome. But in those instances in which the phenome mediates the transmission of adaptations to the offspring it is acting as a genotype. Thus the functional distinction between genotype and phenotype does not necessarily imply a physical demarcation of genome and phenome, and it is conceivable that the descriptive and active functions could be so thoroughly intermixed that no experimental procedure could distinguish genome and phenome as definite physical objects.

In practice the problem of distinguishing genome and phenome, both functionally and physically, appears not to be a major one. In part this is undoubtedly connected to the advantages of a sharp genotype–phenotype distinction for faithful and efficient reproduction at the cellular level. But this cannot be the sole reason since "nongenetic" inheritance does occur in some species. There are two other factors. Both are connected with the cost of "phenotypic" modes of inheritance and with the increase in this cost as the complexity of the organism increases. The first factor is that as the reproduction time becomes longer the time scale of genetic change becomes long in comparison to the time scale of physiological change. As a consequence it becomes less likely that any physiological adaptation of the parent will be useful to the offspring and less desirable for it to have the stability needed to make a significant contribution to inheritance. The second factor is that inherited physiological adaptations are less likely to be reversible in complex multicellular organisms than in simpler organisms. Due to the influence of the initial cells on the development of other cells, the parent's adaptation is likely to propagate itself independently of the environment, and possibly into a form very different from the original. In multicellular organisms which reproduce by fission or budding some inheritance of acquired characteristics will inevitably occur, as in microorganisms. However, as organism complexity increases, mechanisms which enforce a sharper distinction between genotype and phenotype become increasingly evident. It is in principle possible that germ cells could carry adaptive cytoplasmic features which could propagate into the offspring. But it is not in principle inevitable and it appears that the effects—at least the known effects—are negligible. It is not a mechanistic necessity for the independent life of these complex organisms to begin with a single cell or for this single cell to carry in its cytoplasm only what is needed for motility or initial nutrition. But by beginning in this way the extent to which it is possible to isolate genetic and phenotypic modes of transmission from parent to offspring is increased to a maximum.

The restrictions on the time scale of phenotypic adaptabilities in complex organisms reduce the uncertainty of the phenotypic state at birth, and therefore reduce the contribution of these adaptabilities to $H(\hat{\omega})$. Such restrictions may be compensated by other modes of adaptability. In complex organisms they are often compensated by behavioral modes which allow an alternative functionally acceptable form of transmission to offspring. As another example, in mammals there is some short-term transmission of immunity through the mother's milk. But in general enforcing different time scales on modes of adaptability specifically connected to the conception of new offspring and modes connected with other activities of the organism reduces error by enforcing more appropriate response times. Enforcing these different time scales confers the maximum possible independence on genetic and phenotypic modes of adaptability, thereby reducing the total amount of variation which must be exhibited in order to achieve the required level of adaptability.

The above considerations have an important bearing on the major assumption of modern evolutionary theory, namely the assumption that the inheritance of acquired traits is not an important factor in evolution. This is sometimes called the strong principle of inheritance. Given the modern identification of genes with nucleic acid sequences, the strong principle can be taken as asserting that the inheritance of acquired characteristics through the genes does not occur and that their inheritance through phenotypic mechanisms is not a significant factor in evolution. Since there are no known effects through the genes, the major open possibility for violation of the strong principle is through nongenetic mechanisms. But the cost considerations described above suggest that as phenotypic heritability increases, the capacity to evolve diminishes, especially in complex species. As a consequence deviations from the strong principle of inheritance are small in the more complex forms and comparatively ineffective in terms of their contribution to evolution in the less complex forms. Moreover, the known mechanisms of phenotypic heritability do not provide a source of novel variability, in contrast to mechanisms involving the genes proper.

The connection between the above discussion and Lamarckian models of evolution should also be mentioned. My concern has been with cytoplasmic transmission of acquired adaptations and possibly of cultivated adaptabilities. This is a fact in some instances and I have argued that it is not an important factor in evolution. Lamarckian models are usually concerned with the possibility of transmitting acquired traits to the offspring through the genes. There is no evidence for this and many strong arguments against it, totally apart from adaptability theory. There is no known mechanism which would allow the unfolding of a protein and its backward reading into DNA. Even if an alternative mechanism of back-reaction were possible it would be necessary for the organism to filter out the small subset of environmentally induced modifications which are beneficial, otherwise all the information accumulated in the course of evolution would be degraded. On these grounds alone the isolation of genotype from phenotype is a desirable feature. If the organism embodied mechanisms for modifying the DNA depending on its needs, this would require the earliest organisms to embody a deductive system which would beg the issue of evolution. The strong principle of inheritance asserts the nonoccurrence of these mechanisms in all cases. Once this form of the strong principle of inheritance is admitted, it must be admitted that discrete genes exist which can be recombined to form novel descriptive patterns. The alternative, mixing model of inheritance does not provide the variation necessary for selection to act on. Historically, the existence of such discrete units of inheritance was the major prediction of the Darwin–Wallace theory of evolution. At the present time the fact of discrete genes can be taken as implying an important role for variation and selection in evolution, while the tendency to suppress inheritance based on the division or mixing of cytoplasmic components suggests that the role of these mechanisms is secondary and increasingly reduced in importance in the more complex forms.

6.11. OPERATIONAL DEFINITION OF ADAPTABILITY

A major advantage of the statistical definition of adaptability (as ability to cope with the uncertainty of the environment) is that it can be made operational. I here use *operational* in the usual sense of an actual procedure for measurement. This procedure is: culture identically prepared laboratory ecosystems (such as the flask ecosystem) under controlled environmental conditions. According to the definition, the adaptability of such a cultured community is equal to the uncertainty of the most uncertain environment which does not cause unacceptable change in, e.g., its species composition. Suppose that we prepare a large ensemble of such microcosms. Each undergoes a sequence of changes which depends on environmental conditions (i.e., initial distribution of substrate and living matter). The environmental conditions can be completely controlled, the initial abiotic conditions fairly well controlled, and the initial biotic conditions made more identical by exposing the system to high stress (e.g., a long period with no light and periodic low temperatures, cf. Conrad, 1976). By imposing higher and higher degrees of environmental uncertainty, systems are cultured whose adaptability is higher and higher and whose actual adaptability is higher if the evolutionary tendency is valid (since each environment brings forth certain potentialities in the system and an uncertain environment necessarily brings forth those potentialities which enable the system to function in this environment). Thus a series of cultures can be produced, each ordered according to adaptability. The absolute value of the adaptability of none is known, but the adaptability relative to one another (or to any one taken as a standard) is known. In principle, therefore, it is possible to study the relationship between changes in adaptability and the particularities of biological organization, the relation between given degrees of adaptability and the particularities of organization under different degrees of environmental stress, and the extent to which adaptability is optimized in the course of evolution. To determine the different forms of adaptability which actually develop, more detailed examination of the community or placement of constraints on the forms of community adaptability is necessary. It is also possible to determine the relative importance of the different components of adaptability (e.g., anticipation) by introducing greater or lesser degrees of correlation into the sequence of environmental changes.

What if there is no unique most uncertain environment? The idea that there is a potential ensemble of trajectories, therefore a quantity $H(\hat{\omega})$, is still meaningful. It may be that this apparently most uncertain environment does not evoke all the trajectories in this ensemble. Then another environment could be constructed which would. What if there is a trajectory which can be evoked by an environment which has not been considered—in practice not all could be considered. Of course one expects such trajectories to exist. But according to the evolutionary tendency argument trajectories never evoked by the culturing procedure should tend to disappear.

REFERENCES

Burnet, F. M. (1959) *The Clonal Selection Theory of Acquired Immunity*. Vanderbilt University Press, Nashville, Tennessee.

Conrad, M. (1969) *Computer Experiments on the Evolution of Co-adaptation in a Primitive Ecosystem*. Ph.D. Thesis, Biophysics Program, Stanford University, Stanford, California.

Conrad, M. (1974) "Evolutionary Learning Circuits," *J. Theor. Biol. 46*, 167–188.

Conrad, M. (1976) "Biological Adaptability and Human Ecology," in *Proceedings of the International Meeting on Human Ecology*, pp. 467–473. Georgi, Switzerland.

Conrad, M. (1981) "Algorithmic Specification as a Technique for Computing with Informal Biological Models," *BioSystems 13*, 303–320.

Conrad, M., and H. H. Pattee (1970) "Evolution Experiments with an Artificial Ecosystem," *J. Theor. Biol. 28*, 393–409.

Gilpin, M. E. (1975) *Group Selection in Predator–Prey Communities*. Princeton University Press, Princeton, New Jersey.

Hamilton, W. D. (1964) "The Genetical Evolution of Social Behavior," *J. Theor. Biol. 7*, 17–52.

Jerne, N. K. (1955) "The Natural Selection Theory of Antibody Formation," *Proc. Natl. Acad. Sci. USA 41*, 849–857.

Levins, R. (1968) *Evolution in Changing Environments*. Princeton University Press, Princeton, New Jersey.

Lewontin, R. C. (1961) "Evolution and the Theory of Games," *J. Theor. Biol. 1*, 382–403.

Maynard Smith, J. (1976) "Group Selection," *Q. Rev. Biol. 51*, 277–283.

Maynard Smith, J., and G. R. Price (1973) "The Logic of Animal Conflict," *Nature 246*, 15–18.

Odum, E. P. (1971) *Fundamentals of Ecology*. W. B. Saunders, Philadelphia.

Sacher, G. (1967) "Complementarity of Entropy Terms for the Temperature-Dependence of Development and Aging," *Ann. N. Y. Acad. Sci.* *138*(2), 680–712.

Slobodkin, L. B. (1961) *Growth and Regulation of Animal Populations.* Holt, Rinehart and Winston, New York.

Slobodkin, L. B., and A. Rapoport (1974) "An Optimal Strategy of Evolution," *Q. Rev. Biol.* *49*, 181–200.

Waddington, C. H. (1957) *The Strategy of Genes.* Macmillan, New York.

Williams, G. C. (Ed.) (1966) *Adaptation and Natural Selection: A Critique of Some Current Evolutionary Thought.* Princeton University Press, Princeton, New Jersey.

Wilson, E. O. (1975) *Sociobiology.* Harvard University Press, Cambridge, Massachusetts.

Wright, S. (1932) "The Roles of Mutation, Inbreeding, Crossbreeding and Selection in Evolution," *Proc. Sixth Int. Cong. Genet.* *1*, 356–366.

Wynne-Edwards, V. C. (1962) *Animal Dispersion in Relation to Social Behavior.* Hafner, New York.

Zamenhof, S., and H. H. Eichhorn (1967) "Study of Microbial Evolution through Loss of Biosynthetic Functions: Establishment of 'Defective' Mutants," *Nature 216*, 456–458.

7

The Meaning of Efficiency

The concept of efficiency (or cost in terms of efficiency) has been used in a completely intuitive way, with the strategic motive of deferring the frustrating problem of seeking a fundamental definition. In the opinion of the author there is no hope of finding such a definition outside the circle of biological concepts themselves, that is, outside the framework of the theory of evolution. However, within this framework the idea of efficiency translates into the concept of fitness, and therefore into a form which is as elusive and primitive as fitness.

Despite this elusiveness it is possible to transform all statements in the previous chapter concerning the relation between superfluous adaptability and unnecessary cost (or unnecessarily low efficiency) into statements which are formulated in terms of still more primitive, physical concepts, and without making any assumptions about the organization of the organism, the relation between organism and environment, or the dynamics of populations. The reason is that it is possible to specify all the thermodynamic processes which determine efficiency and to make statements about how efficiency changes whenever any of these change, all else constant. This semantic transformation puts the theory on a logically firmer basis and at the same time gives considerable insight into the fundamental nature of fitness and efficiency. The crucial point is that factors which determine efficiency are processes, not variables which characterize the system in question. These processes are themselves determined by the organization of the organism and its relation to the environment.

7.1. THE CONNECTION BETWEEN EFFICIENCY AND FITNESS

We recall (from Chapter 1) that the theory of evolution is not a theory of fitness *per se* but rather a theory of the mechanism of the historical development of systems which we call fit. The idea is that the environment does the selecting (or influences the differential growth rates of populations) and therefore the results of the selection process must fit (in a metaphorical sense) the environment. Thus fitness is a function of the properties of the biological system and those of the environment, where the properties of the former are determined by past reproduction probabilities and the function itself determines the reproduction probabilities in the immediate future, that is, determines the expected contribution to the next generation. In this sense it is an expression of the peculiar circumstance that from the evolution theory a bona fide concept of value emerges. Within the framework of the theory it is convenient to associate value with expected contribution to the future and by extension to the biological properties which determine this contribution. However, the theory offers no prescription for establishing this extension and can be formulated without explicit reference to it. We cannot even say that evolution always tends to optimize (increase) fitness, as is sometimes done, for populations almost inevitably eventually become extinct, in which case their fitness runs to zero.

This zeroing tendency reflects a fundamental difficulty in the population genetics definition of an organism's fitness as its expected contribution of offspring to the next generation—a difficulty which the analysis in this chapter is designed to circumvent so far as the structure of adaptability theory is concerned. As pointed out by Thoday (1953), a definition of fitness might validly consider the long-term as well as the immediate future. As an example, if the reproduction probabilities are too high, the population may overfish the environment, thereby producing a major decline in these probabilities. As another example, the environment might change, in which case adaptability would enhance long-term fitness even though it subtracts from short-term fitness. As a consequence it may not be an advantage to the species to be overly fit, an apparent contradiction in terms. One might attempt to repair this difficulty by averaging fitness over a number of generations. But there is no well-founded way of determining how far into the future one should look. If one looks far enough fitness will always average to zero since after a large enough number of generations one almost certainly expects zero reproduction probability. If one looks the right amount ahead one may have to admit that lower fitness has higher value.

The usefulness of associating value with reproduction is well-founded, however. It is clearly based on the fact that it is only reproductively successful systems that can long exist and which are therefore of any interest. The association is therefore nonarbitrary as long as existence is supposed to have more value than nonexistence. But it is important to recognize that this does not mean that the value concept can be reduced to any simple principle, such as fidelity of reproduction, or number of genes transmitted. Imperfect reproduction and restraint on number of genes transmitted are values for continued existence as well. Nor does it mean that the value concept is derivable from physical theory. It is still based on the existence of constraints which could not be deduced from physical theory without reference to a historical development which presupposes these constraints and which is dependent on all of them.*

Everything so far said about fitness could be said about efficiency. However, rather than thinking in terms of organisms fitting to the environment, we can think in terms of the efficiency of their operation in this environment. The metaphor has changed from lock–key in the case of fitness to effectiveness of utilization in the case of efficiency. The change is potentially fruitful, however, once it is recalled that biological systems are thermodynamic systems and therefore that their reproduction probabilities are necessarily bounded by the effectiveness of energy utilization. Restricted in this way, we can regard efficiency as the energy side of fitness.

7.2. THERMODYNAMIC PARAMETERS OF EFFICIENCY

We already have much of the apparatus for pursuing the above consideration. This is the thermodynamics of open systems developed in Chapter 2. Formulation of the efficiency concept in this framework gives considerable insight into the special peculiarities of biological efficiency and also makes it possible to give sharper formulation to statements about the relation between changes in the amount of superfluous adaptability and changes in efficiency (or biological cost).

The first task is to seek a definition of efficiency in terms of the quantities which appear in the thermodynamic equations. What is needed is a ratio between some actual utilization of the environment and the optimal utilization. One possibility is actual versus maximum possible biomass. However, it is hard to know what the maximum possible biomass is in this case. A more practical means of definition is to compare the food harvested from the system (under some circumstance) to the food (or light energy) available to it. If the system is a population this in fact corresponds to the commonest usage of the word *efficiency* (cf. Slobodkin, 1960). In this case the food extracted is the energy per unit time taken from the prey as yield (under natural circumstances) and the food available is the energy per unit time ingested by the prey population. In the case of the community as a

*I have recently found that a related view concerning the impossibility of complete reduction of the value concept inherent in the constraints of evolution theory to physical concepts outside this framework is argued in a book by Ebeling and Feistel (1982).

whole the food available is the light energy entering the community (or chemical energy entering from the outside) and the food extracted would be the maximum amount of food that could be removed without significant change in the nature of the community (which means that the basic chemical components of this extracted food must be recycled in degraded form). Defining efficiency for an organism is a bit more difficult, since in this case we are dealing with a system which grows and dies. Also, the organism is often the quantum of food transfer, or, in plainer language, it is not always possible to eat an organism in fractional amounts and still have a living organism. However, since the organism is the unit of food, it is safe to assume that if the population is less efficient the average organism will be also. In this way it is possible to define an unambiguous average efficiency for the organism without adding extra complications or generalizing the definition. The efficiency ratio is thus

$$\varepsilon = \frac{Y}{I} \tag{7.1}$$

where Y is the yield, I is the food or light energy absorbed, and ε is the efficiency (sometimes called the gross ecological efficiency in the population case, cf. Phillipson, 1966). At the organism level ε is an average quantity.

Note that it is not workable to define the efficiency in terms of energy input and output, for clearly it would be improper to include heat energy and in any case we would always end up with an efficiency of one in the steady state under this definition. Nor is it useful to define it in terms of work performed, since from the standpoint of organic evolution the end significance of all activities of an organism is solely in terms of reproduction. (But later the relation to the work terms will emerge.) I do not assume that biomass is optimized (in the first possible definition), or that extractable food, ingested food, or even efficiency is optimized.

Recall the general thermodynamic equation of an open system, equation (2.5):

$$\frac{dE}{dt} - T\frac{dS}{dt} = \frac{d_e W}{dt} + \frac{d_e R}{dt} + \sum_{i=1}^{n} e_i \frac{d_e n_i}{dt} - T\sum_{i=1}^{n} e_i \frac{d_e n_i}{dt} - T\sum_{i=1}^{n} s_i \frac{d_e n_i}{dt} - \phi$$

$$\tag{7.2}$$

This may be simplified (making the realistic assumption of constant temperature and pressure) by substituting

$$e_i = \mu_i - pv_i + Ts_i \tag{7.3}$$

where $\mu_i = (\partial G/\partial n_i)_{T p n_i'}$ is the chemical potential of component i, $v_i = (\partial V/\partial n_i)_{T p n_i'}$ is the molar volume of that component, and e_i is just the Gibbs free energy ($G = E + pV - TS$). This gives

$$\frac{dE}{dt} - T\frac{dS}{dt} = \frac{d_eW}{dt} + \frac{d_eR}{dt} + \sum_{i=1}^{n} \mu_i \frac{d_e\mu_i}{dt} - p \sum_{i=1}^{n} v_i \frac{d_e n_i}{dt} - \phi \quad (7.4)$$

which has the advantage that the flow of energy into and out of the population is expressed in terms of chemical potentials.

The next task is to express the efficiency ratio in terms of equation (7.4). To do this, however, it is necessary to accommodate the following biological restrictions:

1. Assimilable energy input (I) must be in the form of chemical matter or light.
2. Yield (Y) is only in the form of chemical matter.
3. Some chemicals exported to the environment represent waste rather than a contribution to yield.

Restrictions (1)–(3) mean that it is necessary to split the observed flows into an imported part and into both high- and low-grade exported parts (i.e., potential food and waste products). Thus define

$$\sum_{i=1}^{m} \mu_i \frac{d_e n_i}{dt} = \sum_{i=1}^{r-1} \mu_i \left| \frac{d_e n_i}{dt} \right| - \sum_{i=r}^{s-1} \mu_i \left| \frac{d_e n_i}{dt} \right| - \sum_{i=s}^{m} \mu_i \left| \frac{d_e n_i}{dt} \right| \quad (7.5)$$

where $(n_1, \ldots, n_{r-1})$ are the chemicals which enter the system, $(n_r, \ldots, n_{s-1})$ are the high-energy chemicals leaving the system, $(n_s, \ldots, n_m)$ are the low-energy chemicals leaving it, and the absolute values and choice of signs indicate the direction of observed flow. With this definition, which makes the above restrictions explicit, we can reexpress equation (7.1) as

$$\varepsilon = \frac{Y}{I} = \frac{\displaystyle\sum_{i=1}^{s-1} \mu_i \left| \frac{d_e n_i}{dt} \right|}{\displaystyle\sum_{i=1}^{r-1} \mu_i \left| \frac{d_e n_i}{dt} \right| + \frac{d_e R}{dt}} \quad (7.6)$$

where the yield is identified with the high-grade (usable) exports and the inputs with imported matter (taken as low-grade in the case of photosynthetic plants) and radiation (taken as zero in the case of animals).

Combining equations (7.4) and (7.5):

$$\frac{dE}{dt} - T\frac{dS}{dt} = \frac{d_eW}{dt} + \frac{d_eR}{dt} + \sum_{i=1}^{r-1} \mu_i \left| \frac{d_e n_i}{dt} \right| - \sum_{i=r}^{s-1} \mu_i \left| \frac{d_e n_i}{dt} \right|$$

$$- \sum_{i=1}^{m} \mu_i \left| \frac{d_e n_i}{dt} \right| - p \sum_{i=1}^{m} \nu_i \frac{d_e n_i}{dt} - \phi \qquad (7.7)$$

which is the combined law, but with explicit consideration of the splitting of matter flows. Combining this with equation (7.6),

$$\varepsilon = 1 - \frac{\dfrac{dE}{dt} - T\dfrac{dS}{dt} + \sum_{i=s}^{m} \mu_i \left| \dfrac{d_e n_i}{dt} \right| - \dfrac{d_e W}{dt} + p \sum_{i=1}^{m} \nu_i \dfrac{d_e n_i}{dt} + \phi}{\sum_{i=1}^{r-1} \mu_i \left| \dfrac{d_e n_i}{dt} \right| + \dfrac{d_e R}{dt}} \qquad (7.8)$$

which is the expression for efficiency in terms of the combined law (Conrad, 1977). Taking the partial derivatives (e.g., $\partial \varepsilon / \partial \phi$), we see that the efficiency increases as the chemical potential of waste exports decreases, as the quantity of waste exports decreases, as the work performed by the system decreases (since work performed is negative), as the dissipation decreases, as the chemical potential or quantity of food inputs increases, and as the radiation input increases (in the case of plants).* The efficiency increases if the free energy of the populations decreases (i.e., if its total energy decreases or its entropy increases), but in this case the increase is at the expense of a reduction in biomass (or standing crop). This type of increase is thus impossible in the steady state, i.e., when $dE/dt = 0$ and $dS/dt = 0$. Under these conditions the efficiency equation reduces to

$$\varepsilon = 1 - \frac{\sum_{i=s}^{m} \mu_i \left| \dfrac{d_e n_i}{dt} \right| - \dfrac{d_e W}{dt} + p \sum_{i=1}^{m} \nu_i \dfrac{d_e n_i}{dt} + \phi}{\sum_{i=1}^{r-1} \mu_i \left| \dfrac{d_e n_i}{dt} \right| + \dfrac{d_e R}{dt}} \qquad (7.9)$$

*The increase in efficiency with increase in energy input (e.g., $d_e R/dt$) may at first sight appear to contradict the original definition [equation (7.1)]. However, recall that we are dealing with $\partial \varepsilon / \partial(d_e R/dt)$, so that all other quantities are assumed constant. This means that the numerator is constant and therefore that the yield must show a relative increase (e.g., because the radiation input increases but the dissipation does not).

The efficiency also increases as the specific molar volume of food inputs or food outputs decreases, but in general this can be ignored. Also, we ignore the fact that certain components, for example water, flow through the system, that others, such as trace elements, contribute a catalytic rather than an energy function, and that some of the compounds taken as output may be the same as those taken as input. However, these complications could be incorporated into the formalism by splitting input components into two parts, thereby distinguishing low-energy and support elements from energy carriers *per se*. Finally, we ignore any contribution associated with either heat or work of transfer (arising from excess kinetic energy of matter exchanged with the environment or temperature differences between this matter and the environment). This contribution drops out automatically because of the simplifying assumptions of constant temperature and pressure.

The question mary arise: is the splitting of waste products and yield an *ad hoc* device or does it have some more fundamental justification? The answer is that it is required for consistency, for in the absence of such a splitting the efficiency would increase as the chemical potential and quantity of waste exports increased. Furthermore, such waste export is almost inevitable in any system which runs itself on the basis of importing high-grade energy in the chemical form and degrading this to lower forms of chemical potential energy. In addition, the export of the degraded chemicals provides a convenient way of carrying entropy to the environment, and therefore for maintaining and developing organization. The only alternative is simple heat export, which is of course important as well.

7.3. FITNESS AND EFFICIENCY IN THE LIGHT OF THERMODYNAMICS

The critical conclusion from the previous section is that efficiency has certain definite relations to the thermodynamic parameters which characterize populations or even communities [cf. equations (7.8) and (7.9)]. Thus, the basic principles of thermodynamics justify the statement: efficiency increases if dissipation can be decreased without any concomitant increases in energy input, work performed, or chemical potential and quantity of waste products. However, this does not mean that this is possible in any particular case. Whether or not it is possible depends on the functional relations among the various properties (food input, materials exported, work performed, dissipation). The problem is that these functional relations depend on the detailed constraints which govern the behavior of the system—that is, on the details of path—and are therefore beyond the reach of purely thermodynamic analysis.

Consider, for example, an arbitrary (but nonbiological) physical system. The maximum efficiency of such a system might be limited by certain fundamental principles, but certainly its minimum efficiency could be small, even zero. The system could be so constructed that any slight decrease in work performed would render it completely nonfunctional, with efficiency zero, whereas a slight increase might have no effect, might increase the efficiency, or might decrease it. Everything depends on the design parameters and on the detailed conditions under which the machinery is operating. The same could be said for the relation between efficiency and any of the other properties (e.g., dissipation or high energy input). In other words, thermodynamics, like logic, says all that is possible and not possible, but it does not say (without supplement) what actually is the case.* This is why it serves to specify precisely the parameters relevant to efficiency and the ways in which they can change relative to one another, but without specifying the actual functional interdependencies among these parameters (and therefore the particular way they in fact change).

The situation is exactly the same for biological systems, except that in this case the interdependencies among the various factors are not entirely arbitrary. This is because such systems are the product of long evolution, so that the details of path and the way these details are reflected in the parameters entering into the efficiency (regarded as the energetic aspect of fitness) is both a primitive concept of biology and at the same time subject to a certain degree of analysis in terms of thermodynamics. It is a primitive concept because the interplay among the factors affecting efficiency is constrained by evolution, so that it is inevitably a historical property which could not be deduced in principle from more primitive concepts of physics without subtracting from the generality of these concepts by adding to them the peculiar constraints which make evolution possible.† On the other hand it is subject to a certain degree of analysis in terms of more primitive concepts because the efficiency equation applies to all open systems satisfying very broad assumptions, whether or not these systems are biological.

---

*The logic metaphor has been used by Katchalsky (1969). Many authors have of course pointed out that the laws of thermodynamics can be regarded as impotence principles, therefore principles which allow us only to deduce the boundaries of the possible.
†Medawar (1975) has suggested that biological systems are properly regarded as more specialized than physical systems and that it is this specialization which gives them their interesting properties. The comparison is to mathematics, in which it often happens that adding more restrictions to a structure (e.g., that it have a unique inverse, identity element) produces greater mathematical richness up to a certain point, beyond which point the structure becomes increasingly uninteresting. This point has also been discussed by Rosen (1973) in the context of the mathematical analogy, by Pattee (1973) in the context of physical constraint, and by Stravinsky (1970) in the context of music.

This justifies the earlier claim that efficiency is a primitive concept of biology, but that it is nevertheless possible to make statements about it in terms of more primitive physical concepts. I emphasize that the concept of efficiency discussed here is not the same as the thermodynamic concept of efficiency as applied to heat engines, which is the ratio between the work performed by an actual system and the maximum amount of work which could be performed by that system (given by the amount of work performed by a reversible system operating between two temperatures). What is of interest in the biological case is not mechanical work *per se* (which is only one condition for reproduction), nor is it possible in principle for biological systems to be reversible. For one thing, reproduction is inherently irreversible (since it requires energy dissipation to drive a system far from equilibrium and in any case certain initial details of the environment must inevitably be destroyed or else the offspring would reflect these as well as the properties of the parent, which contradicts the definition of self-reproduction). Rather, what is relevant are the details of path, which is inevitable once one admits the fundamental role of the dissipation, i.e., the fundamental role of a path or history-dependent differential, one that is not a function of the state of the system and which therefore is controlled by details of both system and environment. This justifies the earlier claim that efficiency, like fitness, is not a function of the state of a biological system but rather an indicator of the relation between system and environment.

7.4. REFORMULATING STATEMENTS ABOUT EFFICIENCY

Fortunately the limited degree of discourse about efficiency allowed by its general thermodynamic analysis makes it possible to reformulate all our earlier statements about cost (or about increases or decreases in efficiency) in terms of statements which make no reference to cost or efficiency *per se*. For example, consider the statement: superfluous biological adaptability always involves an unnecessary biological cost (or an unnecessary decrease in efficiency). This statement can be transformed to: any increase in the difference between the tolerable and actual uncertainty of the environment of a biological system is always accompanied by increases in its dissipation, increases in the chemical potential of waste exports, increases in the quantity of waste exports, increases in the work it performs, decreases in the chemical potential of food inputs, decreases in the quantity of food inputs, or decreases in the utilization of radiation input. Some of these changes are of course much more likely than others, e.g., increases in dissipation, increases in work performed, or decreases in the quantity of food inputs. However, any single one or any combination of the possible changes will do the job, so long as the change in this one (or combination) is not counterbalanced by a greater than compensating change in the others.

The above reformulation makes it possible to rephrase the entire argument of the previous chapter concerning evolutionary tendencies in completely operational terms, i.e., in terms of the operational definition of adaptability and in terms of well-defined, measurable physical quantities. I want to emphasize again, however, that this does not mean that a fundamental definition of fitness (or of the equivalent concept of efficiency) is in hand or even possible. What is in hand is a description of the energy and entropy aspects of efficiency which necessarily follow from the fundamental laws of thermodynamics. These aspects (embodied in the efficiency equation) reflect the detailed operation of the system in its given environment (or how the system fits to the environment), but by itself the equation does not allow deduction of efficiency from these details.

Does the new reformulation allow the minimum adaptability argument to escape the circle of evolutionary concepts? The answer has to be no. We must still make an evolutionary assumption, namely: all other things being equal, those populations will predominate in the course of evolution which are less dissipative, which perform less work, which produce less waste export, which produce lower potential energy waste export, which import more food, which import higher-potential-energy food, or which utilize radiation more effectively. The assumption is now very precise and I think mild when formulated in these terms. Naturally, it can also be formulated directly and equivalently in terms of the biological efficiency concept [as defined in equation (7.8)], except that it must be remembered that this concept now has a precise, thermodynamic meaning.

7.5. BIOMASS AND TURNOVER IN THE CONTEXT OF EFFICIENCY

According to equation (7.8) changes in efficiency are accompanied by changes in energy and entropy, assuming all other terms constant. However, the equation gives no information about the actual values of either energy or entropy. Thus it gives no information about the actual value of the biomass (as the energy and entropy are extensive properties, therefore proportional to biomass). This is consistent and indeed a necessary feature of any definition of efficiency, for it is clear that many values of the total energy, entropy, and biomass are consistent with given rates of energy inflow and outflow. Thus these properties are historical properties of the system, whereas efficiency is a non-history-dependent property since it is a property, or fitting relation, between biological system and environment at a given slice of time.

The above consideration is relevant, as it implies that we cannot in general take biomass to be an index of efficiency. If biomass increases, all other things constant, the efficiency increases, but this does not mean that increases in efficiency are necessarily associated with increases in biomass. For example, the efficiency of a population may increase, but nevertheless conditions may change so that it is contracted to a smaller region of its terrain. Clearly we do not want efficiency to depend on population size, since this is ordinarily a historical property.

The purely thermodynamic independence of efficiency and biomass does not preclude extra thermodynamic relations between biomass concentration and components of efficiency. On the contrary, this would in general be expected since biomass concentration is the analog of population density. For example, suppose that the dissipation is an increasing function of biomass (i.e., if biomass is increased, all other things constant, more energy is turned over) and also that it is an increasing function of inherent turnover rate (i.e., if birth and death rates increase, this means more rapid cycling of energy through the system). Under these circumstances biomass and turnover rate are complementary quantities, in the sense that an increase in either one of them is always compensated by a decrease in the other. Thus, if all birth and death rates are increased, there is an increase in dissipation which necessarily destroys the current steady state and creates a new one in which the biomass is lower; similarly, if the biomass is pushed to a different level, the dissipation inevitably changes in such a way that it is pushed back to its former level.

It is not difficult to put the above argument in formal terms (cf. Conrad, 1973, 1974). The essential idea is simply to write (if we are considering a whole community)

$$\phi_j = \phi_j(k_j, b_j, v_j) \qquad (7.10)$$

where ϕ_j is the dissipation in region j, b_j is the biomass concentration in this region, k_j is the average turnover rate (or average over birth rates and death rates), and v_j are other relevant variables which characterize the region. The total dissipation is given by $\sum_{j=1}^{m} \phi_j$, assuming m regions. The assumptions about the dissipation function are

1. $\partial \phi_j / \partial k_j > 0$

2. $\partial \phi_j / \partial b_j > 0$

which are reasonable over a wide variety of kinetic schemes. In the steady state equation (7.2) becomes

$$\sum_{j=1}^{m} \phi_j(k_j, b_j, v_j) = \frac{d_e W}{dt} + \frac{d_e R}{dt} \tag{7.11}$$

where we retain the simplifying assumption that we are dealing with the whole community (for a single population it would be necessary to add the matter flow terms). This equation, along with properties (1) and (2), directly implies the complementarity of biomass concentration and turnover rate, but to complete the argument it is necessary to show that the steady state condition is not an extra assumption, i.e., that any violation of this condition produces change in the energy and entropy which eliminates this violation. To do this, decompose the free energy of the entire system

$$F = F_A + F_B + F_{AB} \tag{7.12}$$

where F_A is the free energy of the nonliving phase, F_B is the free energy of the biotic phase, and F_{AB} is the free energy associated with the interaction of the two phases. This may be rewritten

$$F = af_a + bf_b + F_{AB} \tag{7.13}$$

where $f_a = F_A/a$ is the free energy of the nonliving phase per unit concentration of substrate (denoted by a); f_b is the free energy of the living phase per unit concentration of biomass (denoted by b); and for simplicity we suppose that the concentrations are the same in all regions (this makes it possible to drop the subscript j). Differentiating with respect to time,

$$\frac{dF}{dt} = (f_b - f_a)\frac{db}{dt} + (1-b)\frac{df_a}{dt} + \frac{b\,df_b}{dt} + \frac{dF_{AB}}{dt} \tag{7.14}$$

where $a = 1 - b$ because of conservation of mass. Combining this with the combined first and second law [equation (7.2)] gives

$$\frac{dW}{dt} + \frac{d_e R}{dt} - \sum_{j=1}^{m} \phi_j(k_j, b, v_j) = |f_b - f_a|\frac{db}{dt} + (1-b)\frac{df_a}{dt} + \frac{b\,df_b}{dt}$$

$$+ \frac{dF_{AB}}{dt} + \frac{S\,dT}{dt} \tag{7.15}$$

where we utilize the fact that the system is closed, that by definition $dF/dt = dE/dt - T\,dS/dt - S\,dT/dt$, and that the functional form [equation (7.10)] can be used for the dissipation (with $b_j = b$). The absolute sign is used to express the fact that $f_b > f_a$ (since the free energy of the living phase per unit concentration of living matter is greater than the free energy of the nonliving matter). Equation (7.15) implies that db/dt has the same sign as the left-hand side of the equation, provided there are no major changes in temperature, interaction energies of the two phases, or free energy per unit concentration of matter in each phase. The identity of signs [along with property (2)] means that the dissipation always changes in such a way as to make the left-hand side tend to (or oscillate around) zero, thereby implying that b has a stable unique value for given k_j and v_j.

The argument outlined above is based on the simplistic but not entirely outrageous assumption that the living matter of an ecosystem can be treated as a collection of autocatalytic entities, with the inorganic environment as substrate. This is almost true by definition since chemical self-reproducing systems are necessarily autocatalytic (with inner cross-catalyses). The *almost* is a necessary qualifier, however, as we are dealing with a system competent to evolve and therefore one which is more accurately described as quasi-autocatalytic in the sense that its properties are amenable to change. The argument is thus only good so long as these changes are restricted to changes in birth rates and death rates (where changes in death rates may be thought of in terms of changes in birth rates of substrate), and in such a way that there are no fundamental changes in orgainization, either in terms of community structure or functional capabilities of the organism. Even with these rather serious reservations and despite the rather unusual setting for such a thermodynamic argument, I think the main conclusion of the argument is safe (though amenable to qualification in detail). This is: the dissipation along with turnover rate does determine biomass and therefore the thermodynamic state of the biotic community (since total energy and entropy are proportional to biomass). This is certainly not true for ordinary thermodynamic systems, where the total dissipation has no self-regulating connection to rate of dissipation. To increase the biomass for a given value of the turnover rate and dissipation it would be necessary to change the presumptive functional relation between dissipation and biomass and turnover rate. Without any such fundamental changes, the result implies that there can be no orthogenetic evolutionary tendencies as regards either biomass or turnover rate, since any orthogenesis in one of these quantities would have to be compensated by an opposite orthogenesis in the other (for which there is not evidence).*

---

*Prigogine and co-workers (e.g., Prigogine *et al.*, 1972) have developed a model of prebiological evolution in which instabilities in nonequilibrium systems (associated with an initial increase in dissipation) can result in their movement to new asymptotically stable states which are even further from equilibrium (with the movement being accompanied by a decrease in the dissipation). This is a situation which, from the present point of view, would involve change in efficiency.

The argument has another interesting and relevant consequence. Suppose that the biomass concentrations and turnover rates are locally heterogeneous, i.e., that $b_u \neq b_v$ and $k_u \neq k_v$. It is inevitable that this will be concomitant to flows (or migrations) of populations from one local region to another, so that no evolutionary advantage of high turnover rate would result in an orthogenetic reduction in biomass and conversely. For example, any local region in which biomass is maintained higher than usual at the expense of a more than compensating decrease in turnover will inevitably suffer an influx of organisms from neighboring regions, since there will be more offspring in these regions. The conclusion is that both biomass and turnover must be considered together and that their complementarity buffers local regions from local evolutionary extremal tendencies as regards either one (apart from fundamental increases in efficiency).

The above argument is essentially a thermodynamic formulation of the interdeme selection argument discussed in the previous chapter. It suggests an interesting notion, viz. that there is some kind of balance or equality principle as regards biomass and turnover rate (taken together) in comparable communities. However, it is possible to generalize this and state it with greater simplicity and more long-run applicability in terms of efficiency, for it is reasonable that populations and communities with higher yields for given energy inputs are more competent to invade the space of other populations or communities. This means that more efficient populations should tend to dominate in the course of evolution (which should be the case if fitness is efficiency) and that the only stable situation is one in which equality of efficiencies obtains. Furthermore, this is independent of the particular cause of the higher yield, i.e., whether it is based on change in the dissipation function (allowing for increase in biomass and turnover) or on some other component of efficiency. The next section discusses and adds qualification to this extension.

7.6. EVOLUTION OF EFFICIENCY

First I want to reemphasize that no assumption has been made about the optimization of efficiency, only that changes which increase efficiency without having any negative effect will be favored in the course of evolution. This does not mean that efficiency always tends to increase. In fact, there is evidence that this is practically never the case. The evidence is the phenomenon of extinction, which eventually infects practically all populations. Thus, if we really wanted to make a generalization about the long run it would have to be that efficiency almost always goes to zero (recall the same comment as regards fitness in Section 7.1).

The immediate conclusion is that there are no orthogenetic tendencies as regards the efficiencies of populations. What about the (biotic part of) the ecosystem as a whole? This question must be divided into two parts. First, there is the very long run, in which the basic details of organization of the system change (for example, origin of photosynthesis, evolution of other fundamental biochemical pathways, development of the capacity to exploit altogether new environments). It is reasonable to suppose that this generally involves increases in radiation input, decreases in dissipation, or decreases in work required for successful reproduction, and therefore for increase in efficiency. Second, there is the short run, in which no fundamentally new reactions or processes develop, but in which rate constants may be tuned. According to the theory of adaptability, changes in environmental uncertainty should always be reflected in changes of community efficiency of this type. Thus, if the uncertainty of the environment increases, the efficiency should decrease (which is just the other side of the argument about the minimization of adaptability). However, efficiency depends on details as well as statistical properties of the environment and therefore would also be expected to decrease if the environment changes in such a way that the system must perform more work, must be more dissipative, must export more waste products, and so forth [cf. equation (7.8)].

How does the relation between adaptability and efficiency coexist with the notion that there is some kind of equality principle governing the efficiencies of different populations in nature? The equality idea (which has been discussed in various alternate forms) actually derives from analogy to economics, at least from those idealized cases where the equilibrium (and optimal) situation is one in which all companies are equally profitable.* In the case of populations in nature this would mean that populations would divide resources in such a way that they were "equi-efficient," which in terms of the definition given here must mean that they are capable of producing comparable yield for comparable input. Thus if a population were extremely efficient it would tend to expand its niche until its efficiency decreased to a point where it no longer provided a competitive advantage. Conversely, an inefficient population would be contracted into a smaller niche, one which it can hold on to if the contraction is in fact accompanied by increased competitiveness in terms of efficiency and if no other expanding or contracting population can use this niche more efficiently.

The equality idea is simple and therefore attractive. However, it must be significantly amended to have any claim to validity. The first point is that the maximum efficiencies may be quite different in different environ-

*For a discussion see Samuelson (1961). A discussion of feeding behavior from the standpoint of an economic analogy can be found in MacArthur (1972).

ments. The maximum possible ratio of yield to input is in general less in a very harsh environment (either because such environments are inherently less favorable to life or because living matter did not initially evolve in these environments or because living matter has not yet evolved the basic mechanisms to exploit them efficiently).* It is in general also different at different trophic levels since at each higher trophic level the total free energy available becomes significantly smaller and therefore in general more work (e.g., motion of the organisms) is required to assimilate this free energy. But, more important in the present context, the efficiency is even smaller in a statistically more uncertain environment (since increases in adaptability are complemented by reverse changes in efficiency). Thus the amendment must be that efficiency is equally allocated among populations which can expand into one another's niches and in which the niches are comparable from the standpoint of stress, trophic level, and statistical perturbation. In other words, fluidity of population movement, stress, trophic level, and uncertainty are extra constraints on the allocation process. There are undoubtedly other constraints as well.†

The adaptability constraint is different from these other constraints, however, since it is itself often contractable and expandable under biological control (for example, by contractions and expansions of the niche space). If the uncertainty of the environment remains the same but all adaptability (aside from indifference) decreases, then niche breadth must contract. The result is a gain in efficiency, provided that efficiency is measured against the reduced availability of input energy in the reduced niche. According to the equality principle, the contraction in niche space should stop as soon as the population's efficiency becomes equal to that of any populations contending for this space (provided that the original niche is such that the contraction serves as a mechanism for decreasing adaptability). Thus the equality principle governs the allocation of adaptabilities to different populations in the community. The rule is that insofar as possible the adaptabilities will be divided up in such a way that the efficiencies are roughly equal

---

*Fortunately it is not necessary for us to address the difficult question of whether any special physical–chemical nature of the environment is more hospitable to life or whether the hospitability of certain earthly environments is the result of the fact that these environments have in fact been the rearing chamber of life. The problem has been discussed by Henderson (1913), who argues that in fact certain physical–chemical conditions are especially propitious, although of course one may still argue about how narrow the range of propitiousness is.

†The equality principle predicts a high degree of regularity among ecological efficiencies in nature, the constrained equality principle much less so. Slobodkin (1972) has argued that generalizations about ecological efficiency have been overplayed and that the regularities are in fact much less than previously thought.

(factors which determine how efficiency is related to adaptability in different types of populations are discussed in Chapter 11).

The principle of equal allocation of efficiencies is a useful notion, but at most it can be only very roughly applicable in practice. In an economic situation the principle may fail because of constraints (knowhow, availability of materials, habit, lack of information) which prevent movement of capital from one sector of the economy to another, or which prevent expansion in a given sector. Likewise in a biological situation there are many constraints which may prevent a system from utilizing a higher than ordinary efficiency to expand into another environment. Not least of these is the time required for evolution (if genetic changes are required for the expansion). Also, an entrenched population may have a number of advantages even if it is inefficient, e.g., from the standpoint of probability of reproduction, from the standpoint of coadaptation to other populations, or even from the standpoint of a population structure which may provide strong defenses, but which may involve paying for this defense in terms of efficiency. Also, the first population moving into an unexploited niche is likely to utilize it very inefficiently for some time. However, once the population extends itself to this niche (e.g., an extreme environment, a new trophic level) the step is likely to be irreversible. It only requires a speciation process to split it into two populations, i.e., a more efficient one for the original niche and a less efficient one for the new niche (alternatively, the unsplit population would be driven out of the original niche). Furthermore, in some cases the low efficiency of the low-efficiency population may be permanent, for under certain environmental conditions the maximum possible efficiency obtainable by biological systems (at any particular stage of evolution) may be limited and less than the efficiency under optimal conditions. Nevertheless, any population which develops adaptations enabling it to function under such conditions would continue to function, in clear violation of any global law of equality of efficiencies (as opposed to the amended, local equality principle).

There is another critical point, emanating from the analogy to economics. In the economy of humans money plays a fundamental role. Equality of profitability, to the extent that it is a valid economic principle, is considerably facilitated by the fact that flow of labor, capital, and raw materials is to a considerable extent controllable by an entity (money) which is symbolic in the sense that its physical nature is almost completely independent of energy (labor, capital, raw materials). Thus, credit is possible, and also creation of money to expand credit, thus speeding the flow of capital and labor between unequally profitable sectors of the economy or speeding the exploitation of new possibilities. There is no corresponding phenomenon in ecology or anywhere in biology. Sometimes it is stated that ATP is biological currency.

This is entirely misleading, for ATP is hardly symbolic of energy, but rather is a unit of biological energy *per se*. A better analogy is perhaps between ATP and gold, not between ATP and paper money. Unsupplemented by paper money, however, gold allows for no credit expansion, which is why a barter economy may be more stable than a money economy, but which is also why it is unsupportive of investment and therefore the development of new technologies. Furthermore, even this analogy breaks down, because gold (as opposed to ATP) has little direct use and therefore really does serve a symbolic function. Its relation to energy is properly regarded as arising from constraints on its production and decay which stabilize it for expressing the value of other commodities from the supply side (while its limited use stabilizes it from the demand side). Because of this inert and noninteractive nature gold subserves a catalytic function which, peculiarly, is more similar to that of an enzyme than that of a biological energy store. These considerations are important, for they mean that the amendments to the equality principle would be much more serious in the "economy" of an ecosystem than in a money- or even a gold-based economy.

REFERENCES

Conrad, M. (1973) "Thermodynamic Extremal Principles in Evolution," *Biophysik 9*, 191–196.

Conrad, M. (1974) "Thermodynamic Correlates of Evolution," *BioSystems 6* 1–15 [printer's errata in following issue].

Conrad, M. (1977) "The Thermodynamic Meaning of Ecological Efficiency," *Am. Nat. 111*, 99–106.

Ebeling, W., and R. Feistel (1982) *Physik der Selbstorganisation und Evolution*. Akademie Verlag, Berlin (GDR).

Henderson, L. J. (1913) *The Fitness of the Environment*. Macmillan, New York.

Katchalsky, A. (1969) "Chemical Dynamics of Macromolecules and Its Cybernetic Significance," pp. 267–298 in *Biology and the Physical Sciences*, ed. by S. Devons. Columbia University Press, New York.

MacArthur, R. H. (1972) *Geographical Ecology*. Harper and Row, New York.

Medawar, P. (1975) "Geometric Models of Reduction and Emergence," pp. 57–63 in *Studies in the Philosophy of Biology. Reduction and Related Problems*, ed. by F. J. Ayala and T. Dobzhansky. University of California Press, Berkeley.

Pattee, H. H. (1973) "Physical Problems of Decision-Making Constraints," and ensuing Discussion, pp. 217–225 in *The Physical Principles of Neuronal and Organismic Behavior*, ed. by M. Conrad and M. Magar. Gordon and Breach, New York and London.

Phillipson, J. (1966) *Ecological Energetics*. Arnold, London.

Prigogine, I., G. Nicolis, and A. Babloyantz (1972) "Thermodynamics of Evolution," Parts I and II, *Physics Today*, November and December, pp. 23–28, 38–44.

Rosen, R. (1973) Discussion, p. 225 in *The Physical Principles of Neuronal and Organismic Behavior*, ed. by M. Conrad and M. Magar. Gordon and Breach, New York and London.

Samuelson, P. A. (1961) *Economics, An Introductory Analysis*, 5th Ed. McGraw-Hill, New York.

Slobodkin, L. B. (1960) "Energy Relationships at the Population Level," *Am. Nat. 94*, 213–236.

Slobodkin, L. B. (1972) "On the Inconstancy of Ecological Efficiency and the Form of Ecological Theories," *Trans. Conn. Acad. Arts. Sci. 44*, 293–305.

Stravinsky, I. (1970) *Poetics of Music in the Form of Six Lessons.* Harvard University Press, Cambridge, Massachusetts.

Thoday, J. M. (1953) "Components of Fitness," *Symp. Soc. Exp. Biol. 7*, 96–113.

The Connection between
Adaptability and Dynamics

I originally wrote down symbols such as $\hat{\omega}$ to express the dynamics of biological systems notationally. But the analysis has been in terms of the statistical quantity, $H(\hat{\omega})$. Now it is necessary to investigate the connection between these two objects. One important question is, what is the connection between the predictability of biological systems and the global structure of adaptability? I employ the answer to establish a correspondence between components of adaptability and the various notions of stability and instability used in dynamical descriptions of biological systems. This correspondence provides a natural interpretation of the functional significance of stability and instability in terms of the adaptability theory framework. An important implication is that the evolutionary tendency to economize adaptability implies a corresponding economics of stability and instability. A deep question is, does the structure of $\hat{\omega}$ along with the actual history of the environment impose evolutionary tendencies on $H(\hat{\omega})$ or do evolutionary tendencies of $H(\hat{\omega})$ impose structure on $\hat{\omega}$?

8.1. AUTONOMY, PREDICTABILITY, AND THE BATH OF UNREPRESENTED ADAPTABILITIES*

First I want to examine the connection between the structure of adaptability and the dynamics of an individual compartment or level. The compartment or level is autonomous to the extent that its behavior is independent of other compartments or levels. From the standpoint of the transition scheme, $\hat{\omega}$, a compartment or level is autonomous to the extent

*Some of the material in Sections 8.1 and 8.2 was originally presented in Conrad (1979a), and has been adapted by permission.

that the transition probabilities need not be conditioned on other compartments or levels. The degree of autonomy can be expressed by the ratio

$$\xi_{ij} = H_e(\hat{\omega}_{ij})/H(\hat{\omega}_{ij}) \tag{8.1}$$

where compartment (i, j) is completely autonomous if $\xi_{ij} = 1$ and not autonomous if $\xi_{ij} < 1$ (since the effective entropy is less than the unnormalized modifiability).

If compartment (i, j) is completely autonomous ($\xi_{ij} = 1$), $\hat{\omega}_{ij}$ is as good a predictor of the behavior of this compartment as the global community scheme $\hat{\omega}$. In this case only properties (variables) of the compartment need enter into the predictor. If $\xi_{ij} < 1$, the "law will be broken," with the degree of breaking depending on the degree of autonomy (or degree of dependence). In this case variables from other compartments must be added to the predictor to restore its predictive power. (If the modifiability is high and the anticipation term low, reflecting good anticipation, a still-better predictor could of course be constructed by including environmental variables.)

In and of itself the predictivity condition is basically trivial—all it says is that if other components give information about the compartment of interest, it is possible to construct a more predictive but less local transition scheme. However, in conjunction with the compensation equation this condition is highly nontrivial and leads directly to the following key statement: *As the predictive value of a transition scheme (transition function, rule, law) used to describe the behavior of any given compartment (level) of a global biological system depends less on the incorporation of variables associated with other compartments (levels), the adaptabilities of these other compartments (levels) become in general more important for maintaining this predictive value.* This statement follows directly from the fact that predictivity means high independence, but high independence means that environmental disturbances affecting other compartments (levels) are absorbed and dissipated in those compartments or levels, with minimum ramification to the compartment (or level) of interest. There are four possible patterns of modifiability and independence (assuming an uncertain environment):

A. *Independence high* (comparable to the modifiability)

   1. *Modifiability low.* In this case adaptabilities elsewhere in the system enable the compartment to support arbitrary but determinate and predictable dynamics despite environmental uncertainty, i.e., the adaptabilities are unrepresented in these dynamics but are critical for supporting them.

2. *Modifiability high.* In this case the compartment of interest makes an efficient contribution to total adaptability, providing the anticipation entropy is relatively low. An optimal predictor can be constructed without reference to other compartments, but the best predictions are necessarily probabilistic. If the transition scheme of the environment is added to the predictor it will be better, provided that the coordinated anticipation entropy is smaller than the modifiability. The predictor can be deterministic to the extent that the anticipation entropy is small. If the adaptability of the compartment of interest is less than the total required adaptability (in general the case), adaptabilities unrepresented in its dynamics are critical for maintaining these dynamics even if it has high modifiability.

B. *Independence low* (smaller than the modifiability)

3. *Modifiability low.* In this case adaptabilities elsewhere in the system absorb most of the environmental uncertainty, but nevertheless the modifications of these other compartments ramify to the compartment of interest, making its behavior less than optimally predictable without enlargement of the set of variables. Thus fewer unrepresented adaptabilities would play a role in supporting a predictable dynamics.

4. *Modifiability high.* In this case the compartment of interest makes a less than maximally efficient contribution to total adaptability even if the anticipation entropy is relatively low. An optimal predictor cannot be constructed without enlargement of the set of variables, thereby implying that fewer unrepresented adaptabilities would play a role in supporting the (in general stochastic) dynamics.

As a somewhat contrived example, imagine a population undergoing phyletic evolution. In the case of low modifiability and high independence for the population level, all the uncertainties would be absorbed at the genetic and organismic levels. More realistically, these modifications would also appear in the population dynamics. This would be the case of low modifiability and low independence. In the case of high modifiability and high independence, the lower-level adaptabilities would not appear, but the population dynamics would itself contribute to the adaptability. A more realistic case is the one of high modifiability and low independence. In this situation the genetic and organismic modifiabilities would manifest themselves at the population level, but the population dynamics would itself contribute to adaptability.

The conclusion is that any apparently autonomous compartment or level of biological organization is really highly controlled by a very much larger system of unviewed compartments or levels that are protecting it from environmental disturbance. This is not the same as saying that metabolism, repair, growth, irritability, and movement underlie the dynamics of, say, a population. It is only insofar as these processes undergo either adaptive or maladaptive change that they can produce modifications in the dynamics of the population, and it is only insofar as any adaptive changes are unrealistically efficient that they can fail to produce modifications in these dynamics. In general the learning at one level can only rarely be completely hidden from the standpoint of the behavior of another.

8.2. BIOLOGY OF STABILITY, INSTABILITY, AND BIFURCATION

Now it is possible to address the question, what is the relation between the various forms of stability or instability exhibited by biological systems and their adaptability? According to the previous section, as the behavior of any particular level of biological organization appears more autonomous, the contribution of unrepresented adaptabilities at other levels becomes in general more critical for maintaining this behavior. The key to using hierarchical adaptability theory to study the functional significance of stability and instability is to look at this statement from the other side, viz. from the side of the represented adaptabilities. From this standpoint, as the behavior of any particular level of biological organization appears more autonomous, its potential contribution to maintaining the behavior patterns at other levels increases (provided that its modifiability term is large).

The problem is that basic stability concepts (e.g., weak stability, asymptotic orbital stability, structural stability) are ordinarily used in biology within the context of differential dynamical models [as in Rosen (1970)], whereas hierarchical adaptability theory has been formulated in a discrete-time, finite-state representation. The most convenient way to make a cross-correlation is to define analog concepts for the representation used here, basing these on the notion of a tolerance [originally discussed in the context of biological models by Zeeman and Buneman (1968) and most prominently developed for automaton models by Dal Cin (1974)]. The idea of a tolerance corresponds roughly to an allowable difference between states. For the present purposes the criterion for *allowable* can be established arbitrarily. But if *allowable* is interpreted as a least noticeable difference, systems whose moves are always between states in tolerance will appear to be moving continuously. Formally, a tolerance is the same as a topology, but is not transitive. That is, it is a binary, symmetric, reflexive, but nontransitive relation on a set of states, to be denoted by " $\sim$ ". Analog definitions are:

A. *Weak stability.* A trajectory of a deterministic transition scheme is weakly stable if and only if it is possible to define a nontrivial tolerance such that for all n if $\alpha_{ij}^s(t) \sim \alpha_{ij}^r(t)$, then $\alpha_{ij}^u(t+n) \sim \alpha_{ij}^v(t+n)$, where α_{ij}^u and α_{ij}^v are the states at time $t+n$ and α_{ij}^s and α_{ij}^r are the states at time t. A scheme all of whose trajectories are weakly stable will be called weakly stable. Such schemes are analogous to differentiable dynamical systems with a constant of the motion (since a constant of the motion means that nearby states always remain nearby but never become identical).

B. *Strong stability.* A state α_{ij}^r of a (probabilistic or deterministic) transition scheme is strictly stable if and only if it is possible to define a tolerance set such that the system returns to α_{ij}^r by time $t+n$ (n sufficiently large) if $\alpha_{ij}^s(t)$ is in this set, and furthermore if for all $k < n$ states at $t+n-k$ are either identical to α_{ij}^r or connected to it by tolerance. Strong stability could also be defined for a trajectory if it is only required that by time $t+n$ the states of the perturbed and unperturbed schemes be identical, but not necessarily identical to the initial state. For deterministic systems, strong stability is analogous to asymptotic orbital stability, where nearby states change to more and more nearby states (or where trajectories converge, as for stable limit cycles).

C. *Structural stability.* Consider the joint transition scheme $\omega_{pq}\omega_{rs} = \{ p[\alpha_{pq}^a(t+1), \alpha_{rs}^b(t+1)|\alpha(t), \beta(t)] \}$. The term α_{rs}^b will be called a parameter (and ω_{rs} a parametric scheme) if it remains the same for all t, while α_{pq}^a in general changes (i.e., is the variable). The joint scheme will be called structurally stable to a change from α_{rs}^b to α_{rs}^c if it is possible to define a tolerance on the states of compartment (p, q) such that $\alpha_{pq}^u(t+n) \sim \alpha_{pq}^v$ $(t+n)$, where n is arbitrary and the α_{pq}^u are states of the compartment with parameter α_{rs}^b and the $\alpha_{pq}^v(t+n)$ are states of the compartment with parameter α_{rs}^c. If the scheme is not structurally stable to this particular change in parameter, the change will be called a bifurcation-producing change and the scheme will be said to bifurcate. The definition of structural stability is analogous to the dynamical notion of qualitative invariance to change in parameter and the notion of bifurcation to the notion of qualitative change in response to slight change in parameter. [If $\alpha_{pq}^u(t+n) = \alpha_{pq}^v$ $(t+n)$, then rather than calling the joint scheme structurally stable it will be said that the two compartments do not interact.]

These notions of stability play a fundamental role in relation to all the basic biological capabilities—metabolism, repair, growth, reproduction, irritability, and movement. Adaptability is also a fundamental biological capability, but in a higher-order sense, in that it involves variation and modulation of these first-order capabilities so as to absorb and dissipate disturbance. Here the concern is only with this connection. In fact it is decisive as regards the forms of dynamics which develop in the course of evolution, hence the particular forms stability and bifurcation take in relation to processes involving metabolism, repair, growth, and so forth. *A major conclusion is that instability and bifurcation make a fundamental contribution to the stability of biological systems, or at least to the stability of the most essential processes in such systems.* From adaptability theory each of the notions of stability acquires a natural functional interpretation:

A'. *Weak stability.* Weakly stable models are frequently criticized (as candidates for descriptions of biological systems) on the grounds that they are not structurally stable to a change in parameter that would represent the addition of even the smallest amount of dissipation. Nevertheless a compartment accurately describable by a weakly stable model would be capable of making a highly efficient contribution to adaptability and could under suitable conditions persist in a form so describable. The suitable condition is that the compartment be completely protected by a bath of adaptabilities provided by other compartments from all disturbances that could introduce a source of dissipation into its dynamics. The efficiency of the contribution to adaptability would result directly from this necessarily high autonomy together with the fact that slight perturbations, excluding those which introduce sources of dissipation, produce only slight modifications in the system and an enormous number of slight perturbations could therefore be absorbed without significant cost. The likelihood of the necessary conditions being met is in general not compelling; nevertheless the above considerations should be taken into account if it appears that a system is accurately describable by a weakly stable model.

B'. *Strong stability.* The argument generally made in favor of strongly stable models is that they are necessarily dissipative (since they forget perturbation), and therefore not necessarily structurally unstable to a perturbation that introduces more dissipation. The fundamental equations of physics are conservative, therefore weakly stable, so that ultimately the dissipation of disturbance is equivalent to the absorption of disturbance in a weakly stable heat bath. This heat bath is thus the unrepresented physical adaptability that supports the strongly stable dynamics of the compartment in question. However, the particular form of these dynamics is also maintained by other, biological adaptabilities, which are necessary to prevent different types of dissipative perturbation from mixing in, and which play an increasingly important role insofar as the degree of autonomy increases.

The above considerations suggest how strongly stable dynamics (with asymptotically stable states or limit cycle behavior) might be significant from the standpoint of a system performing particular biological functions, how the strong stability contributes to adaptability by representing the absorption of disturbances in a heat bath, and how the appearance of some degree of dynamical autonomy (or predictability in terms of a restricted set of variables) can more or less be supported by unrepresented adaptabilities. Taking the other point of view, however, instabilities of a strongly stable system involving transitions among multiple steady states or multiple limit cycles potentially provide adaptabilities that may be unrepresented in the dynamical descriptions of other compartments, but which support these dynamics. The costs of such adaptabilities depend on the biological structures and processes required, on the modifications required to move from one stable state (organization) to another, on the degree of independence of such modifications from modifications of other compartments, and on the compatibility of the different states with the efficiency of the compartment and other compartments.

C'. Structural stability. The argument for structurally stable models is the same as above, except that the dynamical structures can be so chosen that wider classes of perturbation can be mixed in and also so chosen that the bifurcations (catastrophes) occur in a structurally stable way. The cross-correlation to the components of adaptability is different at and away from the bifurcation points. Away from the bifurcation points it is the stable behavior which must be structurally stable. At the bifurcation points it is the unstable behavior which must be structurally stable. The basic connections are:

1. *Away from the bifurcation points.* If the chain of imperceptible parameter changes (that is, changes in tolerance with each other) required to produce perceptible change in the variable compartment is longer, the structural stability of the system can be said to increase. The structural stability of the dynamics thus increases as the independence of the variable compartment increases, provided that the variable and parametric compartments interact. The basic mechanism (aside from the case of noninteraction) is the ensemble of fine, imperceptibly different states of the variable compartment. As this ensemble increases, the capacity of the variable compartment to dissipate the perceptible changes of the parametric compartment increases. The magnitude of both the modifiability and independence terms must be large when defined over these states in order for the independence to be large when defined over the functionally distinguishable states. The magnitude of the latter may be large or small as long as it is comparable to the modifiability.

2. *Near the bifurcation points.* Near the bifurcation points an imperceptible change in the parametric compartment can produce a perceptible change in the variable compartment. The variable compartment appears to be independent of the parametric compartment, so the independence terms are high relative to the modifiability. But this is true whether or not the unstable behavior is structurally stable. To be structurally stable, the actual magnitudes of both the independence and modifiability terms must not be too high, or else the possible functionally different states into which the variable compartment can switch will be high. Each distinctive imperceptible change (distinctive in the information transfer picture) should be coordinated to one possible switch given the previous state of the parametric compartment. The chance of limiting the number of switches increases as the number of microscopic states compatible with the functional state into which the system switches increases, implying that the magnitude of both the modifiability and independence terms must again be large when defined over the functionally equivalent states.

For a system whose stable behavior has low structural stability, adaptive or maladaptive changes in parametric compartments (generally represented in terms of a single lumped parameter) cause any law written solely in terms of the variables describing the compartment of original interest to be broken. The correlated quantitative changes in the compartment of interest may also be adaptive or maladaptive, but the contribution to adaptability is not independent of the contribution coming from the parametric compartments. Such a limitation on independence is inherent and obviously desirable in, for example, the relation between genotype and phenotype. But even here keeping the independence high is advantageous. By increasing the structural stability of the stable (also the unstable) behavior, the effect of genetic changes on the phenotype can be smoothed out, thereby greatly facilitating the evolutionary process. This is the important phenomenon of transformability, to be discussed at length in Chapter 10. From the standpoint of the present cross-correlation between dynamics and adaptability theory, the important requirements are that the magnitudes of the independence and modifiability terms defined over the functionally distinguishable states be large away from the bifurcation points, that these magnitudes be comparable but not too high near the bifurcation points, and that the magnitudes be high in terms of the finer states in both cases.

Another situation in which lack of independence is fundamental is information processing. Low independence increases adaptability by increasing anticipation, that is, by decreasing terms such as $H(\hat{\omega}_{pq}|\hat{\omega}^*\omega_{rs})$, where $\hat{\omega}_{rs}$ is the transition scheme of the parametric compartment. But in order for this to be advantageous a parametric compartment must be organized in such a way that its modifications are really inexpensive. In that case the parametric compartment can be viewed as a sensor or control element. In the most advantageous case the functional states of the variable compartment are correlated to informationally distinct, but functionally equivalent states of these sensors or control elements. The situation merges into the information transfer picture and one can write $H(\hat{\omega}_{pq}|\hat{\omega}^*\hat{\omega}_{rs}) \approx H(\hat{\omega}_{pq}|\hat{\omega}^*) > H(\hat{\omega}_{pq}|\hat{\omega}^*\underline{\hat{\omega}}_{rs})$. If the parametric compartment were excluded from consideration, the quantitative variation of behavior of the compartment of interest would appear as a form of unstable behavior.

For a system whose unstable behavior has low structural stability any law written solely in terms of the compartment of interest will be stochastic. Changes in the parametric compartment will cause this stochastic law to be broken unless its variables are included in the description. Insofar as structurally stable bifurcation allows for different modes of behavior (for example, different modes of development) it provides a mechanism for absorbing disturbance similar to that provided by weak stability or multiple steady states. But features of such unstable behavior would appear quite incomprehensible if an attempt were made to interpret it solely in terms of the variables connected to the compartment of interest. As with structurally stable behavior, the correlation between compartments of interest and parametric compartments can be used for anticipation. Again the parametric compartment should be specialized as an inexpensively modifiable sensor or control, which is much easier near to the bifurcation points than away from them.

In sum the modifiability terms cross-correlate to instabilities, associated with jumps to alternate weakly or strongly stable states (absorption of disturbance) or to stable behavior involving the direct return to an initially strongly stable state subsequent to disturbance (dissipation of disturbance). The independence terms cross-correlate with either structural stability or weakening of interaction. The translations have been developed for discrete-time, finite-state dynamical systems. But it is not unreasonable to expect essential carryover to more general situations.

8.3. INTERPRETATION OF A CLASSICAL MODEL AND SIGNIFICANCE OF CHAOS*

The simplest example which illustrates these points is the classical Lotka–Volterra equations. For a two-species system without crowding these

*Some of the material in this section was earlier discussed in Conrad (1979b), and has been adapted by permission.

are

$$dN_1/dt = a_1 N_1 - b_1 N_1 N_2 \qquad\qquad (8.2\text{a})$$

$$dN_2/dt = -a_2 N_2 + b_2 N_1 N_2 \qquad\qquad (8.2\text{b})$$

where N_1 is prey, N_2 is predator, a_1 is the growth rate of prey in the absence of predation, a_2 is the death rate of predator in the absence of prey, and b_1 and b_2 express the encounter and utilization rates which lead to decrease of the prey population (in the case of b_1) and increase in the predator population (in the case of b_2). As is well known, separation of variables gives a functional relationship between number of predator and number of prey (see, e.g., Rescigno and Richardson, 1973). Thus the system is conservative, that is, has a constant of the motion and forever pursues a periodic oscillation not describable in terms of elementary functions. It is important that this oscillation is not of the limit cycle type. Rather it is weakly stable in the sense that perturbation of either the number of predator or the number of prey results in a shift of the trajectory by a distance which will forever be periodically reestablished.

The biological and mathematical unrealisms of the Lotka–Volterra equations have frequently been discussed. Here, however, our interest is restricted to illustrating the cross-correlation between adaptability theory and dynamical models and in its use for interpreting such models. The main points are:

1. Only population variables enter, no environmental variables enter, and all other properties (e.g., organismic properties) are reflected in constants. Thus the modifiability and anticipation terms are zero and the condition for the applicability of equations (8.2a, b) is that all environmental uncertainty be absorbed by adaptability at other levels or by indifference (or that there be no environmental uncertainty at all). Thus the population dynamics are completely controlled and appear autonomous and independent of the environment.

2. Suppose that an extra condition is added to equations (8.2a, b) which expresses the effect of environmental variability, e.g., by imposing a probability distribution on the constant of motion. The behavioral uncertainty at the population level (i.e., modifiability plus independence terms) is then the measure of the uncertainty of the ensemble of constants. If the modifiability is high this means that environmental disturbances are being absorbed (but not dissipated), that this absorption is preventing disturbances from ramifying to other levels of organization, and that no

correlated lower-level changes occur which are not reflected in these constants. Thus adaptability is achieved with a minimum of total observable modifiability and the dynamics of the population level appears autonomous (but not independent of the environment). As before, modifiability at other levels might absorb or dissipate other types of disturbance.

The existence of a constant of the motion in the Lotka–Volterra equations makes it possible to construct a statistical mechanics of many-species systems (Kerner, 1957). From the standpoint of adaptability theory the implication of such models is that the ensemble of possible weakly stable trajectories is providing an isolated bath for the absorption of environmental disturbances. Another example of such a model is Goodwin's (1963) statistical mechanics of gene–protein interactions in a cell. Here again the ensemble of weakly stable trajectories can be regarded as producing a bath for the absorption of disturbances.

3. Suppose that the constants (a_1, b_1, a_2, b_2) are treated as variables, e.g., that extra lower-level equations are added that contain these variables. In this case the modifiability of the population is dependent on the modifiability at the organismic level, but not conversely. This breaks the appearance of autonomy at the population level. However, the dynamical behavior of the population is structurally stable in the limited sense that its qualitative character does not change as long as the "constants" do not change sign or become equal to zero. As before, disturbances may be absorbed in modifiability of the population sizes, in modifiability at lower levels, or in indifference. If the absorption is in terms of organismic modifiability, however, it will entail modifiability at the population level as well, implying that the adaptability is achieved with a greater than minimum total modifiability. In order for the adaptabilities at other levels to play as effective a support role as possible, this dependence must be small. But on the other hand such dependence allows the population dynamics to adapt to varying circumstances. Also, dependence on information processes at the lower level allows for the use of environmental cues to adjust the dynamics.

Conservative models such as the Lotka–Volterra model are, of course, not structurally stable in the sense that the addition of even slight damping will radically alter their qualitative behavior. The addition of such damping terms (e.g., crowding terms of the form $c_1 N_1^2$ and $c_2 N_2^2$) may allow the system to be asymptotically orbitally stable or to have limit cycle behavior. The essential added element is that such systems are capable not only of absorbing but also of dissipating environmental disturbances. As previously discussed, this ultimately means that the disturbance is absorbed into a physical heat bath (ensemble of microstates) from which the behavior of the system is independent. The heat bath is thus the hidden, controlling system in gradient-type dynamics and is the basis, along with support by adaptability at other levels, for the appearance of autonomy of the system (including eventual autonomy relative to environmental disturbances which directly affect the population level).

With the addition of a dissipative term the possibility of using weak stability is lost, since the conservative character of the system is broken. However, absorption is still possible if the system has multiple steady states. Unless the population dynamics are really highly protected from disturbance by adaptabilities at other levels this is the only plausible basis for absorption. This is due to the fact that only dissipative systems are potentially structurally stable to a realistic class of perturbations.

If the deterministic dynamics is replaced by a stochastic dynamics—again the more realistic situation—the trajectorial and the anticipation entropies will increase. If only population or population and environmental variables are considered, the situation is the same as in the deterministic case, except that the magnitudes of the terms are higher. But if other variables enter it is possible for the stochastic dynamics to increase the adaptability, for example by disentraining the age structure and hence increasing the uncertainty at the organism level. The increase in adaptability will occur if this spreading out of the age structure either enhances the adaptability at the organism level or is coupled to improved anticipation at the population level.

The addition of time lags to determinate dynamical models is known in some cases to produce chaotic behavior which is indistinguishable from stochastic behavior (see May, 1976). Such aperiodic pseudorandom solutions can be found in discrete models and in completely continuous models with at least three variables (Rössler, 1977). These results make it very difficult to determine whether phenomenologically stochastic behavior is generated by a random or a deterministic mechanism. The issue is virtually undecidable. It is therefore of some interest that from the standpoint of adaptability theory a decision is not necessary. The dynamics has the same adaptability theoretic significance regardless of whether it is generated by a deterministic or a random mechanism.

8.4. $H(\hat{\omega})$ VERSUS $\hat{\omega}$

The connection between adaptability theory and dynamics suggests that corresponding to the evolutionary development of adaptability there is an evolutionary development of stability and instability. Insofar as selection acts on biological systems to produce an effective structure of adaptability, it acts to produce a dynamics with suitable stability characteristics. But this raises a deep question as to which is more fundamental, $\hat{\omega}$ or $H(\hat{\omega})$.

Initially one is inclined to suppose that the dynamics, therefore $\hat{\omega}$, is fundamental and that $H(\hat{\omega})$ is a derived quantity. However, changing the structure of adaptability means either changing $\hat{\omega}$ or using a time-dependent $\hat{\omega}$ which maps reality sufficiently well that the changes in the structure of adaptability are all derivable from the initial conditions and history of environmental events. The problem is that if $\hat{\omega}$ is taken as a law arrived at through the standard procedures of physics, it will certainly be unmanageable, even to the point where the applicability of these procedures becomes an undecidable issue. But if it is not based on these laws, phenomena will be ignored which should be relevant to the evolution of the system. High-level dynamical models of the type discussed in the previous section are of this nature. The epistemological status of the transition scheme is thus inherently vague. From the practical and operational point of view it is not meaningful. This is not the case for $H(\hat{\omega})$. The existence of an ensemble of possible behaviors compatible with the life of the community at any given point in time is plausible and operationally meaningful. From this point of view $H(\hat{\omega})$ is the fundamental object and $\hat{\omega}$ the inferred object.

REFERENCES

Conrad, M. (1979a) "Ecosystem Stability and Bifurcation in the Light of Adaptability Theory," pp. 465–481 in *Bifurcation Theory and Applications in Scientific Disciplines*, ed. by O. Gurel and O. E. Rössler. Annals of the New York Academy of Sciences, New York.

Conrad, M. (1979b) "Hierarchical Adaptability Theory and Its Cross-Correlation with Dynamical Ecological Models," pp. 131–150 in *Theoretical Systems Ecology*, ed. by E. Halfon. Academic Press, New York.

Dal Cin, M. (1974) "Modifiable Automata with Tolerance: A Model of Learning," pp. 442–458 in *Physics and Mathematics of the Nervous System*, ed. by M. Conrad, W. Güttinger, and M. Dal Cin. Springer-Verlag, New York and Heidelberg.

Goodwin, B. C. (1963) *Temporal Organization in Cells*. Academic Press, New York.

Kerner, E. (1957) "A Statistical Mechanics of Interacting Biological Species," *Bull. Math. Biophys. 19*, 121–146.

May, R. M. (1976) "Simple Mathematical Models with Very Complicated Dynamics," *Nature 261*, 459–467.

Rescigno, A., and I. W. Richardson (1973) "The Deterministic Theory of Population Dynamics," pp. 283–359 in *Foundations of Mathematical Biology*, ed. by R. Rosen. Academic Press, New York.

Rosen, R. (1970) *Dynamical System Theory in Biology*, Vol. 1. Wiley Interscience, New York.

Rössler, O. E. (1977) "Continuous Chaos," pp. 184–199 in *Synergetics: A Workshop*, ed. by H. Haken. Springer-Verlag, New York and Heidelberg.

Zeeman, E. C., and O. P. Buneman (1968) "Tolerance Spaces and the Brain," pp. 140–151 in *Towards a Theoretical Biology*, Vol. 1, ed. by C. H. Waddington. Edinburgh University Press, Edinburgh.

9

The Connection between Adaptability and Reliability

A fundamental point, raised but not pursued, is that each of the evolutionary tendency equations is a single equation in four unknowns, which of course means that the actual magnitudes of the unknowns are undetermined. To make the issue more precise, recall the evolutionary tendency equation in nonhierarchical form:

$$H(\hat{\omega}) - H(\hat{\omega}|\hat{\omega}^*) + H(\hat{\omega}^*|\hat{\omega}) \to H(\omega^*) \qquad (9.1)$$

In the optimal situation the arrow can be replaced by an equality. There is no doubt that it is possible to satisfy this equality overall (since it can always be replaced by the original identity simply by increasing the uncertainty of the environment). Not so obvious, however, is whether it is possible to approach an equality but at the same time maintain the indifference term small. If this is impossible, lowering the adaptability is clearly not feasible and we would have to admit that it might be more efficient for biological systems to pay for maintaining low error by maintaining excessive adaptability.

To satisfy tendency (9.1) without undue indifference, it is clearly necessary to insure that the reduction in adaptability not be incompatible with the reliable transfer or processing of information. Classical information theory deals with a closely related question for technical information processing systems. Reliability can be achieved at the expense of redundancy as long as the information transmitted does not exceed the channel capacity (Shannon, 1948; Winograd and Cowan, 1963). If this transmitted (or processed) information is used to control the future behavior of the system, the ensemble of possible functionally significant modes of behavior could be made as small as allowed by the uncertainty of the environment, provided that enough spatial or temporal redundancy is introduced to overcome the effects of noise. Clearly the expansion in the ensemble of finer states associated with the increase in redundancy means an increase in the actual magnitudes of $H(\hat{\omega})$ and $H(\hat{\omega}|\hat{\omega}^*)$. The technical analogy (to be discussed further below) is in general applicable to biological information processes only with some stretching. The situation and problem around which it is built are really too special to address all or even the most fundamental processes. The most fundamental problem faced by biological systems is not merely to perform various functions reliably, but rather to undergo graceful function change. After all, transformation of function is the essence of the evolution process. The basis for such transformation of function, however, is the transfer of information from genotype to phenotype, a process possessed of attributes (such as protein folding) which so far have not been embodied in any technical system. Transformability, like reliability, also depends on the actual magnitudes. For example, a protein with more redundancy of weak bonding may change its shape, hence function, more gradually with single changes in primary structure. It may be a larger, more expensive protein to build. However, the advantage is that more variations on it are functional and hence progeny with new, useful adaptations are most likely to derive from organisms with such proteins. The increase in the number of variations means an increase in $H(\hat{\omega}^*)$ and $H(\hat{\omega}|\hat{\omega}^*)$. Thus transformability, like reliability, requires an increase in the actual magnitudes, except that in the case of transformability nonselective indifference can never be eliminated entirely, even in principle. Rather the increase in the magnitudes allows for a more favorable balance between alterations of the system which decrease fitness and those which increase it.*

The present chapter considers reliability, a topic of active research for some time, with the objective of showing in principle that the evolutionary tendency can go to completion in the special case in which the technical analogy is applicable and that it can move in the direction of completion even when the analogy breaks down. Transformability requires different techniques and will be treated thoroughly in the next chapter.

9.1. EMBEDDED COMMUNICATION NETWORK AND THE IN-PRINCIPLE SOLUTION

Recall that there are two pictures, one in terms of adaptively distinct states [transition scheme (4.2)] and a second information transfer picture in

---

*This means that $H(\hat{\omega}^*)$ and $H(\hat{\omega}|\hat{\omega}^*)$ will also increase to some extent, since not all the variations are functionally equivalent, and in fact the distinction between functionally equivalent and functionally distinct becomes blurred. Note, from equation (4.37), that terms such as $H(\hat{\omega}|\hat{\omega}^*)$ can be replaced by $H(\hat{\omega}|\omega^*)$.

terms of finer states [transition scheme (4.33)]. For a system to use adaptability effectively it must transmit and process information about the
environment reliably. This is why the magnitudes which increase are not the
entropies defined over the functionally significant states [e.g., the $H(\hat{\omega})$], but
rather the entropies defined over the finer states of the information transfer
picture [e.g., the $H(\underline{\hat{\omega}})$]. Stated formally, what we want to establish is the
following:

PROPOSITION (tendency of the magnitudes). It is possible to satisfy
the tendency

$$H(\hat{\omega}) - H(\hat{\omega}|\hat{\omega}^*) + H(\hat{\omega}^*|\hat{\omega}) = H(\hat{\omega}^*) \to H(\omega^*) \qquad (9.2)$$

subject to the condition that the value of the nonselective component of
$H(\hat{\omega}^*|\hat{\omega})$ is as small as it could be for any greater value of the adaptability.
However, this is possible only if in the equation

$$H(\underline{\hat{\omega}}) - H(\underline{\hat{\omega}}|\hat{\omega}^*) + H(\hat{\omega}^*|\underline{\hat{\omega}}) = H(\hat{\omega}^*) \to H(\omega^*) \qquad (9.3)$$

the actual magnitudes of the first two terms become large.

In essence the argument involves an identification between compartments of a biological system as described in the information transfer picture
and the elements of an embedded communication network. This makes it
possible to specify the relation between selective components of indifference
and errors of communication and computation. Once this relation is established, Shannon's noisy coding theorem can be invoked to show that it is
possible to eliminate errors caused by noise, though at the expense of an
increase in the actual magnitudes associated with either spatial or temporal
redundancy.

In the information transfer picture the state of the environment at time
$t + \tau_E$ is correlated to the state of the environment at time $t + \tau$ and is
therefore capable of serving as a message to the biological system about the
environment state at the later time. This message is subject to noise
perturbation, in both its transmission and its processing. This noise perturbation has no necessary logical relation to $H(\omega^*)$, the uncertainty of the
environment in terms of its macroscopic states. However, any failure of
transmission or processing caused by it will reduce the likelihood of a
functional response appropriate to the environment, therefore will increase
the nonselective component of indifference. Thus the problem collapses to
the problem of combating the noise.

Suppose, as an initial simplification, that the noise only affects transmission processes. The situation is pictured in Figure 9.1. The environment is the source of the ensemble of messages, the sensors transduce these messages, the encoders put them in a form appropriate for transmission over some channel, the decoder puts the transmitted message in a form suitable for determining the next states of all other compartments in the system (which may in fact include the sensors, encoders, channel and decoders). To the extent that the environment at time $t + \tau_5$ is correlated to the environment at time t, it is possible for these next states to be appropriate to the environment. To the extent that there is no such correlation, some nonselective indifference is inevitable. However, a greater value of the adaptability could never reduce this indifference. The indifference will also be increased by error in the transmission of the message due to noise in the channel. Under these circumstances an increase in the ensemble of functionally distinct states [with an increase in $H(\hat{\omega}) - H(\hat{\omega}|\omega^*)$] might reduce the indifference by increasing the likelihood of a response resulting from an error having functional value. Nevertheless the minimum possible indifference can still be obtained without such an increase in adaptability. This is because, according to Shannon's noisy coding theorem, *it is possible, even for a noisy channel, to assign an error-independent maximum to the amount of information which can be transmitted per unit time, where by an error-independent maximum is meant that there is a lower limit to the amount of spatial or temporal redundancy of signals required to transmit this or any lesser amount of information with arbitrarily small frequency of error (true in the limit of long sequences of signals).* This justifies the first part of the proposition, that it is possible to complete the tendency as expressed in equation (9.1), since this error-free maximum can always be made sufficiently high by using a suitable channel.

This also justifies the second part of the proposition, that the actual magnitudes of the entropies in terms of informationally equivalent, functionally distinct states must in general increase. This follows from the fact that the number of signals required to send the message must increase, by increasing either spatial or temporal redundancy. The situation can again be pictured with the help of Figure 9.1, but with $\tau_1 = \tau_2 = \tau_3 = \tau_4 = \tau_5$. In the case of spatial (or component) redundancy, the number of signals representing the message is increased by sending more at any given instant of time. Since this increases the number of possible states, $H(\cap \hat{\underline{\omega}}_{mn})$ increases, which means that $H(\hat{\underline{\omega}})$ must increase as well. $H(\cap \hat{\underline{\omega}}_{mn}|\hat{\omega}^*)$ and therefore $H(\hat{\underline{\omega}}|\hat{\omega}^*)$ also increase since many of these states are informationally equivalent insofar as their relation to the environmental state is concerned. If the number of signals is increased by increasing temporal redundancy there will, in general, be a longer train of signal patterns in the system and therefore

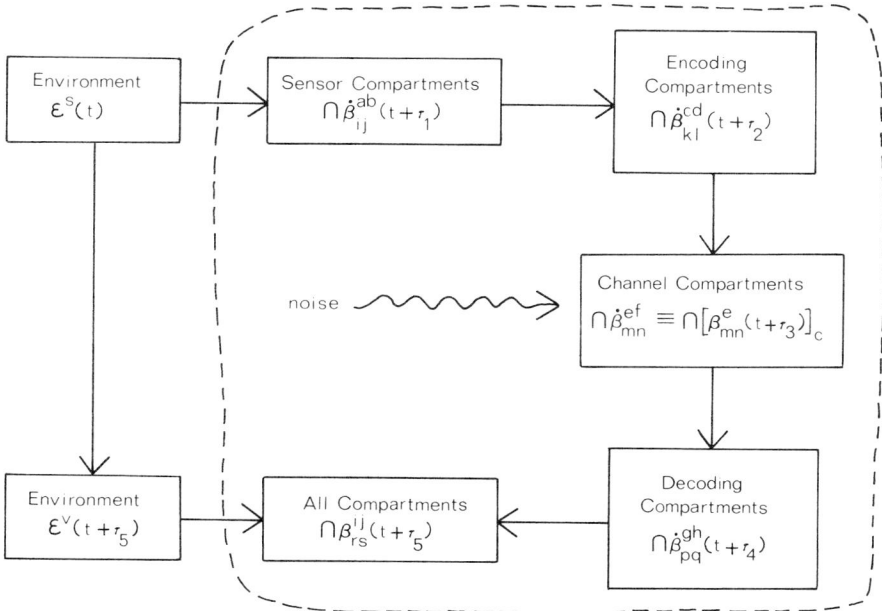

FIGURE 9.1. Sequential view of information transmission in a biological system. The biological system may be a compartment or a complete community. $\cap \dot{\beta}_{ij}^{ab}$ $(t + \tau_1)$ is to be interpreted as the product of the states of all sensor compartments for some informationally distinct partial state of each of these compartments. Similarly, the other products are taken over index sets for the encoding, channel, and decoding compartments respectively. The product $\cap \dot{\beta}_{rs}^{ij}(t + \tau_5)$ is taken over all compartments, and thus may include sensor, coding, and channel compartments (since these may also have to adapt to the environment). The product $\cap[\dot{\beta}_{mn}^{e}$ $(t + \tau_3)]_c$ is an equivalence class of states [as defined in Section 4.11]. The different types of compartments are not necessarily exclusive, but are generally specialized and exclusive with respect to handling information about any component of the environment. The diagram could represent processes in a single organism or in an entire community. The extent to which the states of compartments at time $t + \tau_5$ can be suitable to the state of the environment at this time is limited by the correlation between $\varepsilon^s(t)$ and $\varepsilon^v(t + \tau_5)$ and by noise. However, the effect of the latter can be reduced by increasing the size of the $[\dot{\beta}_{mn}^e(t + \tau_3)]_c$ (see text). If $\tau_1 = \tau_2 = \tau_3 = \tau_4 = \tau_5$, the diagram would represent a view of the states of the sensors, coders, channels, and other compartments at a single instant of time. In this case, of course, the states no longer represent the same signal or responses to the same signal.

more possible "specious present" combinations of signal train patterns. The terms $H(\underline{\hat{\omega}})$ and $H(\underline{\hat{\omega}}|\hat{\omega}^*)$ again increase, just as in the case of spatial redundancy. The only situation in which reliability can increase without an increase in such terms is if it is achieved by decreasing the noise in the channel (i.e., by using components which are physically more reliable). A longer signal train subject to the same amount of noise will exhibit more variability than a shorter train even though both would have the same variability in the noiseless case (because of redundancy in the longer train).

As a simple example of a code which provides error-correcting redundancy, suppose that the signals represented by 0 and 1 are transformed by the encoder to a form which can be represented by 000 and 111. This code (called a Hamming code) is error-correcting if the decoder always converts sequences with two or three zeros into a zero and two or three ones into a one. Under these circumstances, the probability of error goes from p to p^2 (since two mutations of the signals are now the condition for error rather than one). However, the total number of mutations is three times higher in the longer train. The most obvious example of coding in biological systems is in the transmission of nerve impulses. Barlow (1961) has in fact argued that the transmission of sensory information involves a compromise between redundancy and number of impulses. Unary codes may also be used, especially for transmitting analog information. The redundancy of the genetic code (on the average three triplets for an amino acid) makes it error-correcting, though the extent to which this is utilized for repair of damaged base sequences is unknown. This redundancy also confers adaptability by influencing the likelihood that an amino acid will mutate to a more or less similar amino acid.

To more fully illustrate ways in which the information flow scheme depicted in Figure 9.1 can be realized in biological systems, it is interesting to consider a concrete example of vertical information flow between compartments at different levels other than that involving the translation of genes. Cyclic AMP is a messenger molecule which carries information from sensors (receptors) on the cell membrane to receptor proteins associated with subcompartments of the cell, such as the genome, processors in the cytoplasm, or other locations on the membrane. Cyclic GMP is another such messenger molecule, largely produced in the interior of the cell. It is now known that these messenger molecules control nerve impulse activity. The details will be discussed in the next chapter. The important point here is that molecular computation processes involving word processing at the level of DNA and RNA, biochemical reactions, or even subtle conformation changes in individual molecules can be brought to the higher, macroscopic level of nerve impulses through these intracellular messengers and subsequently to the even more macroscopic level of organism action. Conversely, the sensory experiences of the organism can be carried through the nervous system by means of the relatively macroscopic switching behavior of neurons, processed at this level, or brought down to the molecular level by messenger molecules for a finer level of microscopic processing before being brought back to the macroscopic level (Conrad and Liberman, 1982). Later in this chapter some biochemical redundancy mechanisms will be considered which can enhance the reliability of transmission and processing at the molecular level.

9.2. ESSENTIALS OF THE PROOF

A number of proofs and sharpenings of the Shannon theorem have been advanced (cf. Khinchin, 1957; Feinstein, 1958), but none is simpler that the original (Shannon, 1948; cf. also Arbib, 1964). The basic idea is to show that it is always possible to make a code with the property that the sets of output sequences resulting from the subjection of the input sequences to noise do not overlap, thus that it is always possible to reattribute these outputs to the proper input sequences in the decoding operation. This is in fact the convergence in state space idea discussed in Chapter 3 and one way of implementing it is through the Hamming code technique discussed above.

Here we indicate enough of the proof to make clear the formal connection to adaptability theory and to clarify the conditions under which the remarks in the preceding section are applicable. To do this it is first convenient to bring our statement of the theorem into alignment with a more usual statement by defining the capacity as the maximum amount of information per second (determined by varying over all possible sources). (This definition has the same form as the regular capacity discussed in Chapter 4, but with the all-important difference that there we were not dealing with a symbolic system.) The theorem is: *if the entropy generated by the source per second is less than the capacity, it is possible to make a coding system such that the output of the source can be transmitted over the channel with arbitrarily small error, and if the entropy is greater than the capacity (with the difference equal to some constant K), it is possible to make a code such that the output can be transmitted with a frequency of error arbitrarily close to but never less than K.* If the frequency of error is reduced, the uncertainty of the sent message given the received message is also reduced. This uncertainty is sometimes called the equivocation.

Considering n time intervals and assuming the environment to be ergodic and the channel to be memoryless, there are approximately

1. $2^{nH[\varepsilon^s(t)]}$ high-probability message sequences produced by the environment

2. $2^{n\xi H[\cap \dot{\beta}_{kl}^{cd}(t+\tau_2)]}$ high-probability signal sequences produced by the encoders, where $\xi \leqslant 1$ is the fraction of the entropy of the channel associated with such messages
3. $2^{n\xi H[\cap \dot{\beta}_{kl}^{cd}(t+\tau_2)|\cap \dot{\beta}_{pq}^{gh}(t+\tau_4)]}$ high-probability encoder-produced signal sequences which could produce any decoder-received sequence.

Consider the case in which

$$H[\varepsilon^s(t)] < C \tag{9.4}$$

and

$$\xi H\left[\cap \dot{\beta}_{kl}^{cd}(t+\tau_2)\right] - \xi H\left[\cap \dot{\beta}_{kl}^{cd}(t+\tau_2)|\cap \dot{\beta}_{pq}^{gh}(t+\tau_4)\right] = C \tag{9.5}$$

where C denotes the capacity. For simplicity we assume that all of the message sequences produced by the environment are transduced by the sensors (since loss in transduction cannot be corrected in any case). The probability of a message not being represented by a particular encoded signal sequence is

$$p = 1 - 2^{n\langle H[\varepsilon^s(t)] - \xi H[\cap \dot{\beta}_{kl}^{cd}(t+\tau_2)]\rangle} \tag{9.6}$$

The probability of error is the probability that none of the encoder-produced signal sequences which could produce any decoder-received sequence actually does so (apart from the originating sequence). It is given by p taken to the number of such sequences, i.e.,

$$P \approx p^{2^{n\xi H[\cap \dot{\beta}_{kl}^{cd}(t+\tau_2)|\cap \dot{\beta}_{pq}^{gh}(t+\tau_4)]}} \tag{9.7}$$

As $n \to \infty$ this approaches

$$P \approx 1 - \left(2^{n\langle H[\varepsilon^s(t)] - \xi H[\cap \dot{\beta}_{kl}^{cd}(t+\tau_2)]\rangle}\right)\left(2^{n\xi H[\cap \dot{\beta}_{kl}^{cd}(t+\tau_2)|\cap \dot{\beta}_{pq}^{gh}(t+\tau_4)]}\right) \tag{9.8}$$

or, using equation (9.5),

$$P \approx 1 - 2^{-n\langle C - H[\varepsilon^s(t)]\rangle} \tag{9.9}$$

Thus the probability of error decreases as n increases and as the difference between the entropy produced by the source and the capacity increases. As $n \to \infty$ $p \to 0$, which means that at least one and actually most of the codes do the job.

Observe that, as C increases, the possibilities for spatial and temporal redundancy increase, thus leading to the kind of state multiplication described in the preceding section. If n is not large, it is more important for $\xi H[\cap \dot{\beta}_{kl}^{cd}(t + \tau_2)|\cap \dot{\beta}_{pq}^{gh}(t + \tau_4)]$ to be large, but with C remaining large. $H[\cap \dot{\beta}_{kl}^{cd}(t + \tau_2)]$ must be large as well in this case. Thus in practice C must often be relatively large in comparison to $H[\varepsilon^s(t)]$.

Since encoding and decoding operations require either time or high parallelism, it is clear that the use of error-correcting codes to achieve reliability entails certain inescapable costs.

9.3. MORE GENERAL SITUATIONS AND QUALIFYING COMMENTS

The critical assumption in the proof outlined above is that the coding and decoding processes are noiseless. In effect, reliable transmission is achieved by the use of precise systems for processing information. Thus the theory is quite incomplete from the standpoint of biology, since the processing of information is the more ubiquitous situation. Nevertheless, the idea of an error-correcting code and the central-limit-theorem-type convergence which really lies behind the proof make it quite robust. One may reasonably expect it to apply in some fashion to the more general situation of information processing.

The extension has been developed by Winograd and Cowan (1963). I do not want to discuss this treatment in detail here since it involves a number of specializing assumptions, in particular formulation in terms of a particular kind of component (formal neurons). However, the general idea is important and one can adopt the positive point of view that the result holds even under such specializing assumptions.

Imagine a network of components, each of which is capable of processing information. Suppose that each component is replaced by a number of components. The network is now redundant and, if the redundancy is built into the structure of the network according to an error-correcting code and the errors are corrected as soon as possible after they occur, it is possible to carry the Shannon result over to the information processing case (assuming that some technical conventions for defining capacity are incorporated). The increase in the complexity of interconnection, however, means that the original components must be altered to cope with very much more complex inputs, so that the functions of these original components are essentially smeared over the new components. In other words, it is possible to achieve reliable information processing in the presence of noise, but at the expense of greatly increased connectivity among and multifunctionality of the components.

The assumption in the Winograd–Cowan scheme is clearly that the reliability of the components does not decrease as their complexity decreases. This is comparable to the assumption of reliable encoding and decoding devices in the communication case. Many biological components (such as neurons or enzymes) are in fact probably amenable to some diversification in function without a significant decrease in reliability. However, this can only be true to the extent that these functions are based on processes which are inherently reliable, that is, on processes employing either the statistical or the quantum mechanical mechanisms such as those discussed in Section 6.2.

It has often been stated that, in view of the necessity of making assumptions about noiseless coding or independence of reliability from functional complexity, the problem of reliability is unsolved (since clearly there are reliable systems). However, the argument can also be made that the reliability of systems in nature is ultimately based on reliable components and that error-correcting codes are best thought of as strategies for minimizing the cost of reliability. Precise components are costly, but redundancy of components or signals (as well as delay) is also costly. In the broader context of a cost problem, what is happening is that it is sometimes more efficient to have many sloppy components at the expense of an increase in either the number of components or the time and energy expended on passing information around in the system. However, this is not always the case. In modern computers the reliability problem has largely been solved by precision of components, though issues of fault tolerance have been revived by the use of computer-dependent control systems in connection with tasks (such as air travel) in which errors are serious. Enzymes such as peroxidase have an extremely low error rate and the degree to which neurons are unreliable is open to question. It is known that single neurons can reliably respond to one or two quanta of light (Barlow, 1974). It might also be pointed out that Dal Cin and Dilger have shown that the reliability can be obtained without much in the way of redundancy if the components are adaptable in the sense of being able to learn (Dilger, 1974). The information required to adapt must of course itself be processed reliably, requiring either redundancy at a lower level or physical precision.

Another assumption is that it is possible to recognize error. This is easy in the case of information transmission. Any deviation from a one-to-one correspondence between sent and received messages is an error. In the more general case of information processing, there is no such simple criterion. The basic difficulty is that information processing necessarily entails loss of information about the past. For example, the simple logical "or" operation always loses information about the original values of its arguments (since $T = T \vee T, T \vee F, F \vee T$). More generally, any information processing device which does more than count (a cyclic operation) must in general include resetting processes (Krohn and Rhodes, 1963), which of course are irreversible operations. However, noise is also a source of loss of information about the past. This is the difficulty, for there is no *a priori* way to distinguish loss of information resulting from noise (i.e., error) from loss of information associated with information processing. The only way of making the distinction is in terms of its functional significance for the system, which is not known *a priori*. This is, in fact, the same as the problem of distinguishing selective from nonselective indifference, except that in information processing the indifference is restricted to the past and to the finer states of the information transfer picture. In short, information processing is selective dissipation and error is nonselective dissipation.

Does this consideration invalidate the interpretation of the multiplicity of finer states as subserving an error-correcting function? An argument can be made that it does not. In man-made systems it is of course possible to specify in advance what the system ought to do, and therefore to distinguish the selective from the nonselective loss of information about the past. In effect, what one must do is specify a reference system with proper function.* Nature gives us no such reference system. However, it is conceptually legitimate to view any variation in the information processing behavior of a system which decreases its chance of contributing genes to future generations as being an error. In principle one could determine whether a system is more or less subject to errors and one could determine whether a particular variation is an error if many repeat experiments are possible. The justification for distinguishing between the selective and nonselective components of indifference is based on this same consideration.†

---

*This is in fact how capacity is defined in the Winograd–Cowan theory. Once a reference system is specified, one can represent any noisy information processing system by concatenating this reference system with a noisy communication network.

†In emphasizing the dissipative feature of information processing I have disregarded the possibility that information about all intermediate states might be stored on a gigantic memory device. Bennett (1973) has shown that with such a device it would be possible, in principle, to construct a thermodynamically reversible computer. Such a computer would be highly error-prone, and it is evident that all the errors could not be stored in memory for the purposes of achieving reversibility without introducing more errors. In this highly idealized construction the addition of dissipation would be necessary for error correction and may be necessary for speed, but it is not logically necessary for computation. The selective component of dissipation does not enter until the irrelevant information stored in the memory is discarded, as it must be both on the grounds of physical limitation of space and on the grounds of biological efficiency. But of course Bennett's point is not that dissipation-free computation is possible, but that much less dissipation is required for computation than previously thought.

9.4. BIOCHEMICAL PROOFREADING

It is now known that it is possible to enhance the accuracy of gene transcription and translation through proofreading mechanisms (Hopfield, 1974; Ninio, 1975). These proofreading mechanisms all involve the use of error-correcting redundancies. It is therefore interesting to consider the points of similarity and difference between these molecular schemes and the technical ones already considered.

The simplest possible scheme is illustrated in Figure 9.2. In this scheme the frequency of error is reduced by introducing a second enzyme (E_{AC}) which converts an erroneous product (C) back to the original substrate (A). This is the proofreading enzyme, or the proofreading specificity which is added to the original enzyme. A free energy expenditure is required to prevent the proofreading enzyme from driving the reaction in the wrong direction. More generally, the expenditure of free energy allows the insertion of an extra barrier in the reaction pathway to increase the probability that an improper substrate will fall backwards before it scales the final barrier. Infinite energy dissipation is required for optimal proofreading in these kinetic proofreading schemes, though Bennett (1979) has shown that considerable proofreading can in principle be obtained with a dissipation of only 0.1–1 kT/step. There is now substantial evidence that a portion of the energy expended in cellular metabolism is expended on proofreading. According to Savageau and Freter (1979) proofreading of tRNA amino-acylations may account for 2% of the energy required to synthesize a bacterial cell.

Schemes are also possible in which an energized form of an enzyme is responsible for proofreading on a main pathway and an uncharged form of the same enzyme is responsible for converting proofread molecules to product along a second pathway. This is the energy relay scheme proposed

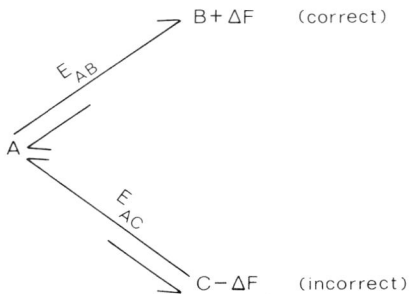

FIGURE 9.2. Simplest proofreading scheme. A is substrate, B is correct product, and C is incorrect product. E_{AB} is supposed to produce B but sometimes fails and produces C. E_{AC} is the proofreading enzyme.

by Hopfield (1980). Savageau (1981) has shown that proofreading is in principle possible at zero energy cost in schemes of this type, aside from the energy required to charge the enzyme. But it is evident that proofreading slows down the reaction and that the enzyme is being asked to use its specificity at least twice and in at least two different ways.

All these proofreading mechanisms are based on the introduction of redundant steps into biochemical reaction schemes. This requires either more types of enzymes or more complicated enzymes. The kinetics of proofreading schemes involves either extra steps in parallel or extra steps in series as well as parallel. The energy-relay mechanisms involve extra steps in series as well as parallel. The greater redundancy means a greater number of finer, functionally equivalent states. By introducing these redundancies, the accuracy of information processes at the molecular level can be enhanced, just as the accuracy of macroscopic computation and communication processes can be enhanced by redundancy. As with the components in the Winograd–Cowan scheme, the reliability of the enzyme must not degrade significantly as it is assigned more functions or as these functions are distributed over a larger number of enzyme species. Unless this can be avoided, inexpensive reliability based on enzyme shape would be sacrificed for expensive reliability based on the multiplicity of enzymes. Enzymes appear to fulfill this requirement much more naturally than macroscopic components.*

Another difference is that enzymes are evolutionarily adaptable. As a result biological systems are able to build the variety of required specificities through the evolution process. Machines clearly do not have this capability at the present time. They depend on humans to build any required new types of components for them. Once the components are given, the limits to attainable reliability are set. But living systems can enhance their reliability

---

*Quantum mechanical models of enzymes can be constructed which have the interesting property that up to a certain point specificity can be increased without sacrificing speed. The idea is that the formation of the complex dynamically opens up decomposition pathways which are distinct from the formation pathways. The low free energy of the complex allows active recognition, therefore enhances the specificity without compromising the speed of complex formation and without compromising the rate of decomposition (because of the dynamic opening of decomposition pathways). The opening of the decomposition pathways is associated with the conformation change of the enzyme during catalysis. The model is called an energy loan model since the energy for conformation change is based on an elastic exchange of potential energy between electronic and nuclear coordinates. The energy loan model is completely consistent with macroscopic thermodynamics, provided that the barriers to complex decomposition are shortened equally in the forward and reverse directions, a feature which follows from the fact that the heights of these barriers do not depend on whether a target covalent bond in the complexed substrate is formed or broken (Conrad, 1979).

through their adaptability. The remarkable reliability of living systems is ultimately based on this adaptability and on the physical features of enzymes which allows their precision to be somewhat independent of their functional complexity. It seems to me to be most interesting that it is the unnaturalness of component adaptability and complexity-independent precision in present-day machines which were the conceptually problematic points in classical theories of reliable information processing.

9.5. INTERDEPENDENCE OF RELIABILITY AND ADAPTABILITY

The basic conclusion is that reliability and therefore small error are possible even in the presence of noise. In general this requires redundancy, either spatial or temporal. Since this redundancy means multiplication of the number of states, the reduction in the indifference term must in general be at the expense of an expansion in the absolute values of the behavioral entropy and the decorrelation entropy. This state multiplication can be thought of as a splitting of finer states into the type described in the information transfer picture. If the redundancy is spatial, many more components will be seen in the system. All reliability schemes based on system design ultimately rest on the assumption that some kind of noise-impervious component plays a role, or that increasing demands on some component does not decrease its reliability. The designs allow effective use of such precise components. They also rest on the assumption that the multiplication of components and states is not too costly in terms of mass and energy. This assumption can always be satisfied if it is signals which are to be proliferated. It is not met if it is functionally distinct states which are to be proliferated.

The embedding of a communication network within the state description of a biological system provided the means for establishing that it is possible in principle to complete the evolutionary tendency without increasing the nonselective component of indifference unduly. The increase in the magnitudes of the behavioral and decorrelation entropies is a price which must be paid for this. This is the content of the proposition about the tendency of the magnitudes in Section 9.2. I have called this a proposition because it has only been proved for embedded communication systems. The arguments put forth for more general situations (involving information processing) are informal and based on the Winograd–Cowan extension of the Shannon theory.

It is important to recognize that in this chapter the principal concern has been the reliability of information processes concomitant to anticipation of the environment. This is the aspect of reliability which is pertinent to the self-consistency of the evolutionary tendency. There are many other aspects of function for which reliability is important. Duplication or multiplication of components is a common phenomenon in biology, ranging from the double presence of kidneys to the use of biochemical proofreading to enhance the reliability of transcription and translation of the genome. The relation to anticipation is indirect in these cases. However, if the system fails for any reason, its ability to anticipate the environment will be compromised. In this sense the connection between the reliability of information processes associated with anticipation and the reliability of information processes associated with communication between components of the system is highly circular, as is the connection between these reliabilities and adaptability. If information transmission and processing are unreliable, adaptability will be ineffective. If adaptability is ineffective the system will suffer damage and information processing will become ineffective. Error will increase, leading to an increase in the nonselective component of indifference. This is also true for internal communication processes. The same reliability principles apply to these processes as apply to processes associated with the transduction and utilization of environmental information. They apply especially naturally at the molecular level due to the natural physical reliability and evolutionary adaptability of molecular components. It is this evolutionary adaptability, analyzed in detail in the next chapter, which is the source of all other forms of reliability and adaptability.

REFERENCES

Arbib, M. A. (1964) *Brains, Machines, and Mathematics*. McGraw-Hill, New York.

Barlow, H. B. (1961) "The Coding of Sensory Messages," pp. 331–360 in *Current Problems in Animal Behaviour*, ed. by W. H. Thorpe and O. L. Zangwill. Cambridge University Press, Cambridge, England.

Barlow, H. B. (1974) "Visual Sensitivity," pp. 198–210 in *Physics and Mathematics of the Nervous System*, ed. by M. Conrad, W. Güttinger, and M. Dal Cin. Springer Lecture Notes in Biomathematics, No. 4, Springer-Verlag, Heidelberg.

Bennett, C. H. (1973) "Logical Reversibility of Computation," *IBM J. Res. Dev. 17*, 525–532.

Bennett, C. H. (1979) "Dissipation–Error Tradeoff in Proofreading," *BioSystems 11*, 85–92.

Conrad, M. (1979) "Unstable Electron Pairing and the Energy Loan Model of Enzyme Catalysis," *J. Theor. Biol. 79*, 137–156.

Conrad, M., and E. A. Liberman (1982) "Molecular Computing as a Link between Biological and Physical Theory," *J. Theor. Biol. 98*, 239–252.

Dilger, E. (1974) "Structural and Dynamical Redundancy," pp. 431–441 in *Physics and Mathematics of the Nervous System*, ed. by M. Conrad, W. Güttinger, and M. Dal Cin. Springer Lecture Notes in Biomathematics, No. 4, Springer-Verlag, Heidelberg.

Feinstein, A. (1958) *Foundations of Information Theory*. McGraw-Hill, New York.

Hopfield, J. J. (1974) "Kinetic Proofreading: A New Mechanism for Reducing Errors in Biosynthetic Processes Requiring High Specificity," *Proc. Natl. Acad. Sci. USA 71*(10), 4135–4139.

Hopfield, J. J. (1980) "The Energy Relay: A Proofreading Scheme Based on Dynamic Cooperativity and Lacking All Characteristic Symptoms of Kinetic Proofreading in DNA Replication and Protein Synthesis," *Proc. Natl. Acad. Sci. USA 77*(9), 5248–5252.

Khinchin, A. I. (1957) *Mathematical Foundations of Information Theory.* Dover, New York.

Krohn, K. B., and J. L. Rhodes (1963) "Algebraic Theory of Machines," pp. 341–384 in *Proceedings of the Symposium on Mathematical Theory of Automata*, ed. by J. Fox. Polytechnic Press, Brooklyn.

Ninio, J. (1975) "Kinetic Amplification of Enzyme Discrimination," *Biochimie 57*, 587–595.

Savageau, M. (1981) "Accuracy of Proofreading with Zero Energy Cost," *J. Theor. Biol. 93*, 179–195.

Savageau, M., and R. Freter (1979) "On the Evolution of Accuracy and the Cost of Proofreading tRNA Aminoacylation," *Proc. Natl. Acad. Sci. USA 76*(9), 4507–4510.

Shannon, C. E. (1948) "The Mathematical Theory of Communication," *Bell System Tech. J. 27*, 379–423; 623–656.

Winograd, S., and J. D. Cowan (1963) *Reliable Computation in the Presence of Noise.* MIT Press, Cambridge, Massachusetts.

Adaptability Theory Analysis of the Genotype– Phenotype Relationship

Reliability is a special case of a biologically more fundamental ability—which I call transformability. The substance of the analysis of reliability was: reliable information processing, insofar as it is based on redundancy, requires an increase in the actual magnitudes of the entropies $H(\underline{\omega})$ and $H(\underline{\omega}|\omega^*)$. It also increases the effectiveness of adaptability by increasing the difference between these two entropies. One of the objectives of this chapter is to show that the gradual transformability of biological structure and function is also based on redundancies which increase the actual magnitudes. The most important example is the transformability of the phenotype in response to genetic variation. This is clearly the *sine qua non* for effective evolutionary adaptation.

A convenient way of picturing transformability is in terms of the structure of an adaptive surface. An example is a plot of efficiency (or fitness) against genetic structure. The adaptive landscape of Sewell Wright and other classical geneticists is such a surface (Wright, 1932). Here what will be of interest is a more detailed, molecular version of the adaptive landscape. It is possible to deduce, using the apparatus of adaptability theory, that gradual change in the function and fitness of a single protein requires the structure of the protein to embody redundancies which buffer the expression of mutation. Mechanistic considerations establish that the degree of buffering is adjustable and rate of evolution considerations establish that it is effectively subject to natural selection and thus adjusted in the course of evolution. The argument extends to multigene systems, except that the mechanisms of buffering are different and another class of mechanisms becomes possible. All such buffering involves a cost in terms of efficiency, implying that increased evolutionary adaptability requires an expenditure of free energy. In effect, biological systems can expend free energy to increase the "climbability" of the adaptive landscape. More precisely, the species is more transformable if it occupies a region of the landscape with more traversible pathways.

Another phenomenon of great importance is purely phenotypic buffering. By introducing buffer compartments whose major function is to decrease the interdependencies among functionally specialized phenotypic compartments, it is possible to greatly decrease the cost of adaptability. Buffer-structuring the phenotype also increases transformability, but not at a net efficiency cost to the individual.

10.1. THE MUTATION–ABSORPTION PARADIGM

To make matters definite and simple and to isolate the essential points it is useful to begin by considering a single gene coding for a single protein. From the informational point of view there are two distinct types of processes, translation and folding (cf. Figure 3.1). The translation process is a straightforward coding of sequences of codons (base triplets) in DNA into sequences of amino acids. In the abstract it basically falls into the class of coding processes which take place in a technical communication system. This is not true for the folding of the sequence of amino acids (primary structure) into a three-dimensional shape (tertiary structure). The latter is an energy process in which higher levels of structural organization emerge from the weak interactions among the constituent amino acids (e.g., van der Waals interactions, hydrophobic interactions, hydrogen bonding, disulfide bonds). The usual statement is that the shape (aside from allosteric shape changes) is determined uniquely by the sequence of amino acids on the basis of at least a local minimization of free energy, with the particular minimum which the system finds possibly determined in part by the way the protein rolls off the ribosome. The function of the protein (what reactions it catalyzes, what quaternary molecular structures it self-assembles into) is generally stated to be determined by this emergent molecular geometry, with the relation between structure and function captured in the famous lock–key metaphor of Ehrlich, or in the more dynamic hand–glove (or induced fit) metaphor of Koshland (1963).

The energy-dependent relation of shape and amino acid sequence has a fundamental but insufficiently appreciated significance. It opens up the possibility for graduated change of shape with graduated changes in amino acid sequence, hence graduated changes in function in response to slight changes in the fine structure of the gene. This does not mean that changes in amino acid sequences always or even almost always give rise to a protein with a useful function, or even that changes in critical amino acids do not often give rise to catastrophically useless proteins. However, it is possible for the protein to be so organized that the number of acceptable single changes in sequence (e.g., point mutations) is significant. This will happen if the interactions which determine the protein's shape overdetermine features of this shape critical for function. Since the degree of overdetermination is something which can be increased or decreased by changing the primary structure, the average degree of shape change in response to a change in primary structure can also be increased or decreased by changing this structure. In short, the evolutionary transformability of proteins and other macromolecules is itself subject to variation and natural selection.

Protein enzymes and other macromolecules may be described in terms of specialized functional capabilities, for example, recognition, control, binding, or catalytic action *per se*. These specialized capabilities can to some extent be associated with particular features of their shape, somewhat in analogy to the attribution of specialized organismic functions to particular organs. In both cases one is of course dealing with a highly coordinated system in which specialized capabilities can only be attributed to a part in the limited sense that it is most affected by perturbations to this part or in the less limited sense that one can understand why and how it is affected by these perturbations. In this sense the specialized functional capabilities of proteins are often described as being associated with particular sites on the molecule (see, for example, Perutz, 1962). According to the argument outlined above, another function of a specifically evolutionary nature can be added to this list. The buffer system enables the protein to absorb mutation and express this gracefully in terms of slight variations of those features of its shape critical for specific physiological functions.

The *mutation–absorption paradigm* is illustrated schematically in Figure 10.1 and in terms of a balls and springs analogy in Figure 10.2. The triangles and different-sized circles represent different amino acids, the solid springs represent strong bonds, and the dashed springs represent weak interactions. As the redundancy of weak interactions increases, the replacement of one amino acid (small circle) by another (large, cross-hatched circle) represents a smaller perturbation to the shape of the active site (represented by the distance and relative orientation of the triangles). Such redundancy can be increased by increasing the number of amino acids, by choosing amino acids in such a way that the redundancy of weak interactions increases, or by utilizing amino acids which have a greater number of close structural analogs. In the latter case the redundancy is in the set of possible differently shaped amino acids in the sequence. The degree of gradualism of shape change can also be increased by utilizing organizational formats which most effectively utilize the various specific mechanisms of redundancy. The most extreme example is the immunoglobins, with their hinged "lobster claws" allowing for flexibility and the looping inside the claws providing a mechanism for gracefully expressing the mutability of the variable and hypervariable amino acids in terms of graduated changes in

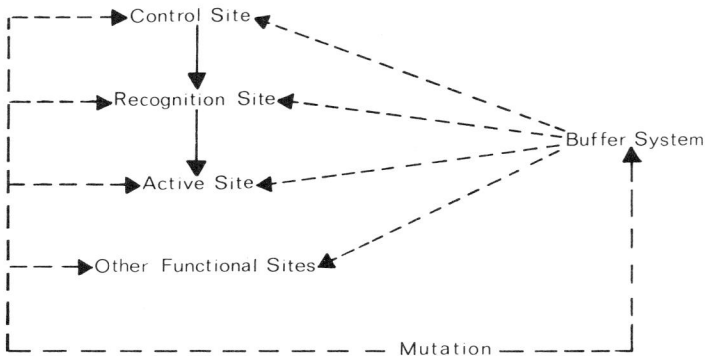

FIGURE 10.1. Functional diagram of an enzyme, including relation of buffering to evolutionary transformability. Solid arrows represent interactions among features of enzyme structure specialized for specific physiological functions. Dotted arrows represent the influence of mutation. The buffer system absorbs some of the effects of mutation, thereby modulating its expression in functionally specific features. This system thus subserves an evolutionary function, implying that enzyme structures should be interpreted as embedding evolutionary as well as physiological functions (from Conrad, 1979b, adapted by permission).

the shape and specificity (see Figure 10.3). Needless to say, a format conferring such effective mutation buffering is to be expected in a macro-molecular species in which transformability of specific affinity to antigen is the *sine qua non* of function.

The occurrence of graduated transformation of shape and function with stepwise change in primary structure is an experimental fact. For example, it is a fact in the case of the immunoglobins. It is also a fact in the case of a number of proteins (such as ferredoxin) that catalytic function remains after the removal of more than half of the amino acids. Another example is provided by the dehydrogenases, in which it is known that tertiary structure is conserved in evolution more than primary structure (Rossman *et al.*, 1975) and in which it is also known (in the case of alcohol dehydrogenase) that point mutations often give rise to slight changes in rate constants (Wills, 1976). A more important class of examples is provided by the large number of variations of given sequences which occur in different species. Again it must be emphasized that this does not mean that all or most point mutations are acceptable or that many sequences, once evolved, are not strictly conserved (the histones, for one, are an example to the contrary). However, it does justify putting the existence of gradualism

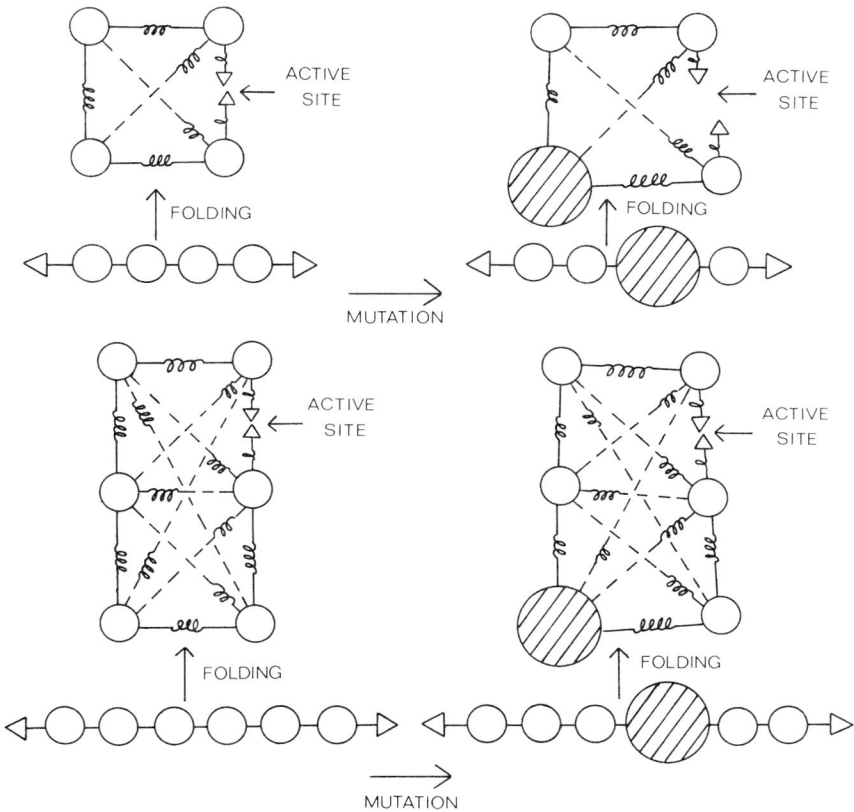

FIGURE 10.2. Balls and springs analogy of mutation–absorption model. Springs on solid lines represent strong bonds and springs on dashed lines represent weak bonds. Small balls, large balls, and triangles represent different types of amino acids. The feature of folded shape critical for function (e.g., catalytic specificity) is the relative position of the triangles. If more amino acids are added to the structure, it is possible for typical but noncritical mutations (involving the replacement of a small ball by a large ball) to produce smaller modifications in the active site (i.e., smaller changes in the distance or angle between the triangles) as well as a larger variety of modifications. The increase in gradualism derives from an increase in the redundancy of weak bonding. However, it could also be based on the utilization of amino acids with a greater number of structurally similar (or redundant) analogs and also on specific organizational formats which allow for effective utilization of redundancy. The balls and springs analogy schematically illustrates mechanisms of mutation buffering and is not intended as a realistic model of protein structure (from Conrad, 1979b, adapted by permission).

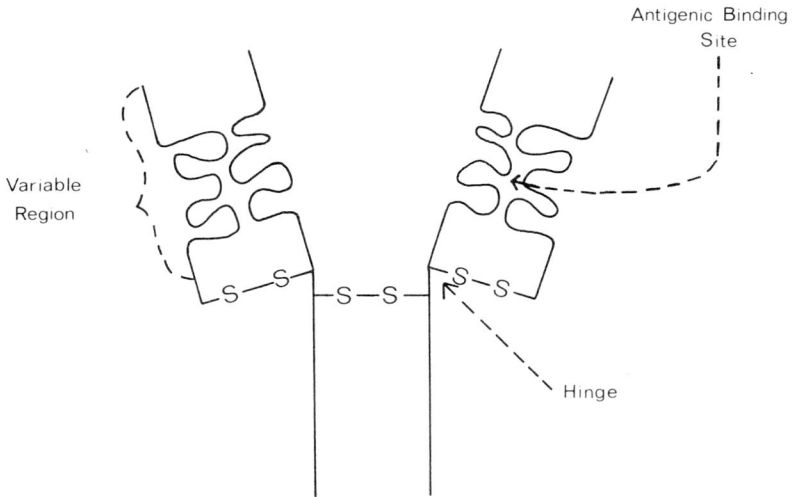

FIGURE 10.3. Schematic illustration of loopings in the "lobster claws" of the immunoglobin molecule (from Conrad, 1979a, adapted by permission; cf. Roitt, 1974).

(along with the uniqueness of folding and the implicitness of function in structure) on a postulatory level. The existence of a buffer system can then be deduced using the methods of adaptability theory. This is done in the next section. Subsequently I establish what will be called the *bootstrap principle of evolutionary adaptability*, viz. that the degree of buffering, hence the evolutionary adaptability of proteins, is adjustable in the course of evolution and adjusted in a way which promotes the evolutionary process. Evolutionary amenability in effect pulls itself up by its own bootstraps by hitchhiking along with the advantageous traits whose emergence it facilitates.

10.2. FORMALIZATION OF THE MUTATION–ABSORPTION MODEL

The basic idea is to treat the mutation process (or, more broadly, the genetic variation process) as a source of uncertainty and the phenotype as the system which must absorb or dissipate this uncertainty. How this is done determines how the genetic uncertainty contributes to adaptability relative to the uncertainty of the external environment.

The assumptions are:

A1. *Unique folding.* Primary structure determines tertiary structure, or, more precisely, it determines a thermal distribution of possible tertiary structures in a given environment and, in some cases, under given initial conditions.   ˎ

A2. *Functional dependence on structure.* Tertiary structure determines function (characterized by parameters such as rate constants, cooperativities, and binding constants).

A3. *Allosteric property.* Enzymes are capable of undergoing shape (and therefore function) changes in response to the substrate or to control chemicals in the environment.

A4. *Functional specialization.* Certain features of the tertiary structure may be specialized for particular aspects of function.

A5. *Gradualism property.* Graduated changes in shape and function are often, but not always, associated with graduated changes in primary structure.

As discussed in the preceding section, A1–A4 are commonly made empirical generalizations about proteins. A5 is also an empirically based statement. However, there is no need to assume that it has the same kind of generality as A1–A4, since the objective is only to show that when A5 does hold, the existence of a buffer system necessarily follows. The allosterism assumption (A3) will not play a direct role in the argument. It is useful for interpretive purposes, since allosteric shifts are part of the variability exhibited by macromolecules, albeit their contribution is to physiological rather than evolutionary adaptability. It should also be noted that the significance of shape for function is to some degree metaphorical, deriving from the lock–key analogy for specificity. In reality the detailed electronic structure and dynamical properties of the protein must play a decisive role. However, as these properties are also presumably determined by primary structure, the only significance of such blurring would be to merge A1 and A2 into a single statement. The argument could be phrased in terms of such a merger, but I believe it will be more transparent if the traditional structure–function distinction is maintained.

The first step is to formalize A1–A4. This cannot be done in terms of the notational conventions in the particular reference structure introduced in Chapter 5 (Figure 5.3), since both the genome and cellular phenome were treated as bottom-level compartments. However, it is perfectly legitimate to describe these in terms of subcompartments, for example individual genes in the case of the genome and various subcellular components, such as proteins, in the case of the phenome. The properties of the protein are primary structure, shape, and function (rate constants). From the mathematical point of view it makes no difference whether we regard the protein as a single compartment whose state is characterized by these three basic types of properties or whether we regard it as two compartments, one for primary structure and the other for shape and rate constants.* The two-compartment choice has the advantage that it makes it possible to treat translation and folding separately. This is because the primary structure is the translated genome and thus, to study the effect of folding in relation to genetic variability independently from the role of translation, it is only necessary to consider the primary structure as in effect the genome.

Consider a single protein species with primary structure G_i and also the set of possible protein structures $G = \{G_1, \ldots, G_n\}$ derivable from G_i by genetic variation. The set of tertiary properties of this species may be denoted by P. However, as discussed earlier, I assume for the sake of exposition that the structural and behavioral properties are separable and therefore that $P = T \times K$, where $T = \{T_1, \ldots, T_n\}$ is the set of tertiary structures and $K = \{K_1, \ldots, K_n\}$ is the set of functional properties. Each K_i is itself a tuple of parameters $K_i = (k_{i1}, \ldots, k_{ih})$, where the k_{ij} are rate constants, binding constants, and so forth. For mnemonic convenience, denote the transition scheme entropy of the primary structure by $H(\omega_G)$ and the entropies for the tertiary structure and functional properties by $H(\omega_T)$ and $H(\omega_K)$.[†] The transition schemes are given by

$$\omega_G = \left\{ p\left(G_j(t + \tau) | G_i(t), \varepsilon^s(t) \right) \right\} \tag{10.1a}$$

$$\omega_T = \left\{ p\left(T_j(t + \tau) | T_i(t), \varepsilon^s(t) \right) \right\} \tag{10.1b}$$

$$\omega_K = \left\{ p\left(K_j(t + \tau) | K_i(t), \varepsilon^s(t) \right) \right\} \tag{10.1c}$$

*The example of a protein illustrates in an extreme but particularly clear way a relationship between properties at two levels. If primary and tertiary levels are treated as separate compartments, the properties of the tertiary level are largely emergent properties which are implicit in the primary level (according to A1). The dependence of the properties of one compartment on the properties of another is rarely so strong.

[†] I am here adhering, as in the preceding chapters, to the simplifying assumption that all sets of states can be treated as being discrete and finite. This avoids technical difficulties associated with a choice of measure (i.e., avoids the necessity of a formalism which handles both the discrete and continuous cases at once) and therefore eliminates some complications which have no material bearing on the conclusion. The set of possible primary structures is of course in principle finite, though perhaps extremely large. An argument might also be made that discreteness is in fact the realistic physical assumption in the case of the tertiary structural properties, insofar as these are really most properly regarded as referring to the possible quantum states of the system. The assumption would seem less realistic with respect to the rate constants; however, the particular choice of measure for discretizing these cannot affect statements which would be true regardless of this choice.

where τ is most naturally taken as the generation time. $H(\omega_G)$ is thus the uncertainty as to primary structure after the reproduction process is complete, given the initial primary structure coded by the same gene locus and also given the initial environment. The source of uncertainty includes all mechanisms of genetic variation at the gene level. Similar interpretations can be given for $H(\omega_T)$ and $H(\omega_K)$ except that the sources of uncertainty include both thermal and allosteric processes in the case of the former and allosteric processes in the case of the latter (since rate constants are by definition average values). The environment scheme will, as usual, be denoted by ω^* and the biota, other than the protein species, by ω_c (i.e., $\omega = \omega_c \omega_K \omega_T \omega_G$).

As usual it is also possible to define conditional entropies which express the correlations between primary and tertiary uncertainty. Reexpressing A1–A4 in terms of restrictions on these conditioned terms:

A1. *Unique folding.* According to the folding property, the primary structure determines the possible tertiary structures, therefore ω_T necessarily provides information about ω_G. Formally,

$$H(\omega_T) - H(\omega_T|\omega_G) > 0 \qquad (10.2a)$$

where $H(\omega_T|\omega_G)$ is the shape uncertainty arising from allosteric or other environmentally induced transitions and also from thermal configuration changes. If the environment is specified it is possible to write

$$H(\omega_T|\omega_G\omega^*\omega_c) \leqslant H(\omega_T|\omega_G) \qquad (10.2b)$$

where $H(\omega_T|\omega_G\omega^*\omega_c)$ is the thermal configurational uncertainty *per se*. The uniqueness of folding expresses the idea that it should be possible to set up a dictionary which associates each protein shape, if specified in sufficient detail, with a single primary structure. Formally,

$$H(\omega_G|\omega_T) = 0 \qquad (10.3a)$$

which implies

$$H(\omega_G|\omega_K\omega_T) = H(\omega_G|\omega_K\omega_T\omega^*\omega_c) = 0 \qquad (10.3b)$$

Note that this does not involve any assumption about the computability of the folded shape from the primary structure; it would be sufficient to establish the associations *a posteriori*. Because of the degeneracy of the genetic code, condition (10.3b) would have to be weakened if extended to the DNA sequence which codes for the primary structure. Condition (10.3b) will be called the nondegeneracy of the genotype–phenotype relation at the protein (or, more precisely, at the polypeptide) level.

A2. *Functional dependence on structure.* The association between rate constants and tertiary structure implies

$$H(\omega_K|\omega_T) = H(\omega_K|\omega_T\omega^*\omega_c) = 0 \qquad (10.4)$$

Furthermore, since the rate constants are average properties connected with the thermal distribution of tertiary structures,

$$H(\omega_K|\omega_G) = H(\omega_K|\omega_G\omega^*\omega_c) = 0 \qquad (10.5)$$

The consideration here is that, if the averaging assumption is good, the association between primary structure and rate constants is always definite, even if the particular tertiary configuration can only be specified probabilistically. The specification of the environment provides no extra information since K represents the complete set of functional parameters, not the subset relevant to any particular environment.

A3. *Allosteric property.* According to (3b) the allosteric uncertainty is less than or equal to

$$H(\omega_T|\omega_G) - H(\omega_T|\omega_G\omega^*\omega_c) \qquad (10.6)$$

A4. *Functional specialization.* The assumption of functional specialization requires that the protein be describable in terms of at least two conceptually distinguishable subsystems, P^I and P^{II}, or, equivalently, in terms of two sets of features of the tertiary structure, T^I and T^{II}, along with sets of functional parameters, K^I and K^{II}, implicit in these features.

The gradualism property (A5) imposes a condition on the magnitude of the entropies, as formulated in the following lemma:

A5. *Gradualism lemma.* A small increase in $H(\omega_K)$ in response to an increase in $H(\omega_G)$ is a necessary condition for a high degree of gradualism of function change in response to genetic change.

Since the only source of variation of interest is genetic, it may be supposed, without loss of generality, that this is the only source of variation. In this case $H(\omega_K) = 0$ when $H(\omega_G) = 0$ and it is sufficient to show that a small value of $H(\omega_K)$ is a necessary condition for a high degree of gradualism.

To establish this, first observe from the defining equation for the entropy, equation (4.6), that in order for $H(\omega_K)$ to increase, the number, n, of possible values of the K_i must increase or the probabilities, $p[K_j(t+\tau)|K_i(t),\ \varepsilon^s(t)]$, must become more equal. In the former case the number of possible distinguishable tuples of rate constants, $(k_{i1},\ldots,k_{ih})$, must increase and in the latter case (which can only account for a delimited entropy increase) the number of tuples whose frequency of occurrence is significant increases. Define the range of values of K as the maximum distance between any two tuples in the ensemble of tuples. A convenient distance function (metric) is given by

$$d\left(K_i, K_j\right) = \max\{|K_{i1} - K_{j1}|,\ldots,|K_{ih} - K_{jh}|\} \tag{10.7}$$

which picks out the largest change in rate parameter as the measure of function change. Other choices for $d(K_i, K_j)$ are also possible. Since it can reasonably be assumed that the number of rate constants required to specify K (i.e., the dimensionality of the tuple) does not increase, it follows that the range of either possible or significantly occurring values of K can only be small if $H(\omega_K)$ is small, with the minimum possible range growing larger as $H(\omega_K)$ grows larger. This means that a small value of $H(\omega_K)$ is a necessary condition for a high degree of gradualism, as stated in (A5).

Note that if the ensemble of possible tuples is small, $H(K)$ will be small whether or not the range of values is small, whereas if $H(K)$ is large the range of values cannot be small. Thus (A5) expresses a necessary but not sufficient condition for gradualism. However, from the standpoint of the argument to be developed, a necessary condition is more powerful since any conclusion derived from it will apply to all proteins. Also note that if the dimensionality, h, of the tuple is large, the range can be smaller for a given value of $H(K)$ and that therefore the lemma would clearly fail if the dimensionality of the tuple were allowed to increase indefinitely (an unrealistic possibility).

THEOREM (mutation buffering).   (1) In the absence of functional specialization, no modulation of the degree of function change is possible.

2. In the presence of functional specialization, gradualism is possible if the protein consists of two subsystems (I and II), where the properties of II are irrelevant to any specific physiological function.

3. This result remains valid when the environment is considered explicitly, except that the total adaptability of the protein includes a number of forms other than mutation buffering.

To prove (1) it is sufficient to show that in the absence of functional specialization the relation between functional and genetic uncertainty is expressed directly by

$$H(\omega_K) = H(\omega_G) \qquad (10.8)$$

Expanding $H(\omega_K \omega_T \omega_G \omega^* \omega_c)$,

$$H(\omega_K \omega_T) + H(\omega^* \omega_c | \omega_K \omega_T) - H(\omega_K \omega_T | \omega_G) - H(\omega^* \omega_c | \omega_K \omega_T \omega_c)$$

$$+ H(\omega_G | \omega_K \omega_T \omega^* \omega_c) = H(\omega_G) \qquad (10.9)$$

Noting that equation (10.3b) implies that

$$H(\omega^* \omega_c | \omega_K \omega_T) - H(\omega^* \omega_c | \omega_K \omega_T \omega_G) = 0 \qquad (10.10)$$

equation (10.9) simplifies to

$$H(\omega_K \omega_T) - H(\omega_K \omega_T | \omega_G) + H(\omega_G | \omega_K \omega_T \omega^* \omega_c) = H(\omega_G) \quad (10.11)$$

This says that uncertainty in the primary structure must be absorbed either in phenotypic uncertainty or in phenotype–genotype degeneracy, $H(\omega_G | \omega_K \omega_T \omega^* \omega_c)$. For a single protein this term is zero [again according to equation (10.3b)]. Eliminating it and expanding (10.9) in a way which isolates structural and functional uncertainty,

$$\{H(\omega_K) + H(\omega_T | \omega_K)\} - \{H(\omega_K | \omega_G) + H(\omega_T | \omega_G \omega_K)\} = H(\omega_G)$$

$$(10.12)$$

Utilizing the functional dependence on structure expressed in equation (10.6), this may be further simplified to give

$$H(\omega_K) + \{H(\omega_T | \omega_K) - H(\omega_T | \omega_G \omega_K)\} = H(\omega_G) \qquad (10.13)$$

The bracketed terms are the only terms which can modulate the effects of primary genetic change, with the unreliability type term, $H(\omega_T | \omega_G \omega_K)$, determining the actual magnitude of $H(\omega_T | \omega_K)$. To interpret the latter, consider the identity

$$H(\omega_K) + H(\omega_T | \omega_K) = H(\omega_T) + H(\omega_K | \omega_T) \qquad (10.14)$$

$H(\omega_K|\omega_T)$ should be set to zero according to equation (10.5), giving

$$H(\omega_K) = H(\omega_T) - H(\omega_T|\omega_K) \qquad (10.15)$$

This means that function change can be made gradual [i.e., $H(\omega_K)$ made small] only to the extent that specification of function provides no information about tertiary structure. To the extent that this is the situation, the structure–function relation is degenerate.

The connection between gradualism and structure–function degeneracy gives a significant hint about the basic requirements for a mutation-buffering system. Certainly structure–function degeneracy, taken in the sense that many structures can perform the same function, is quite generally characteristic of biological and other function-performing systems—for example, any number of different organizations of the keyboard of this typewriter would allow for my typing this sentence. For a single protein, however, equation (10.15) is not physically realistic since two different species of protein cannot reasonably be expected to be identical in all their rate constants. This means that an *a posteriori* detailed specification of all rate constants is just as good for specifying an equivalence class of tertiary structures as specification of primary structures would be, and therefore that

$$H(\omega_T|\omega_K) = H(\omega_T|\omega_G\omega_K) \geqslant 0 \qquad (10.16)$$

In combination with equation (10.13) this gives equation (10.8), thus establishing part 1 of the theorem, viz. that modulation of the effect of primary variation on function is not possible under assumptions A1–A3. The conclusion does not extend to assumption A4 since this did not enter the argument.

The proof of part (2) of the theorem shows how it is possible to escape the above difficulty by using functional specialization (A4) to effect a physically acceptable form of structure–function degeneracy (to be called specialization degeneracy). In the presence of functional specialization $H(\omega_K\omega_T\omega_G\omega^*\omega_c)$ is replaced by the equivalent two-subsystem form, $H(\omega_{K^{\mathrm{I}}}\omega_{T^{\mathrm{I}}}\omega_{K^{\mathrm{II}}}\omega_{T^{\mathrm{II}}}\omega_G\omega^*\omega_c)$, where $H(\omega_{K^{\mathrm{I}}}\omega_{T^{\mathrm{I}}}\omega_{K^{\mathrm{II}}}\omega_{T^{\mathrm{II}}}) = H(\omega_K\omega_T)$. This means that equation (10.11) can now be written

$$H(\omega_{K^{\mathrm{I}}}\omega_{T^{\mathrm{I}}}\omega_{K^{\mathrm{II}}}\omega_{T^{\mathrm{II}}}) - H(\omega_{K^{\mathrm{I}}}\omega_{T^{\mathrm{I}}}\omega_{K^{\mathrm{II}}}\omega_{T^{\mathrm{II}}}|\omega_G) = H(\omega_G) \qquad (10.17)$$

where the genotype–phenotype degeneracy has again been set to zero.

Expanding to the two-subsystem analog of equation (10.12),

$$\{ H(\omega_{K^{\mathrm{I}}}) + H(\omega_{T^{\mathrm{I}}}|\omega_{K^{\mathrm{I}}}) + H(\omega_{T^{\mathrm{II}}}|\omega_{K^{\mathrm{II}}}\omega_{T^{\mathrm{I}}}) + H(\omega_{K^{\mathrm{II}}}|\omega_{K^{\mathrm{I}}}\omega_{T^{\mathrm{I}}}\omega_{T^{\mathrm{II}}}) \}$$

$$- \{ H(\omega_{K^{\mathrm{I}}}|\omega_G) + H(\omega_{T^{\mathrm{I}}}|\omega_G\omega_{K^{\mathrm{I}}}) + H(\omega_{T^{\mathrm{II}}}|\omega_G, \omega_{K^{\mathrm{I}}}\omega_{T^{\mathrm{I}}})$$

$$+ H(\omega_{K^{\mathrm{II}}}|\omega_G\omega_{K^{\mathrm{I}}}\omega_{T^{\mathrm{I}}}\omega_{T^{\mathrm{II}}}) \} = H(\omega_G) \qquad (10.18)$$

Simplifying with restrictions associated with functional dependence on structure [of the form used in equations (10.4) and (10.5)] and also with the structure–function degeneracy restriction expressed by equation (10.16),

$$H(\omega_{K^{\mathrm{I}}}) + H(\omega_{T^{\mathrm{II}}}|\omega_{K^{\mathrm{I}}}\omega_{T^{\mathrm{I}}}) - H(\omega_{T^{\mathrm{II}}}|\omega_G\omega_{K^{\mathrm{I}}}\omega_{T^{\mathrm{I}}}) = H(\omega_G) \quad (10.19)$$

The condition that $H(\omega_{K^{\mathrm{I}}})$ be small is now

$$H(\omega_{T^{\mathrm{II}}}|\omega_{K^{\mathrm{I}}}\omega_{T^{\mathrm{I}}}) > H(\omega_{T^{\mathrm{II}}}|\omega_G\omega_{K^{\mathrm{I}}}\omega_{T^{\mathrm{I}}}) \qquad (10.20)$$

This inequality does not contradict any assumption made so far. Furthermore, there is no physical reason why it should not be possible since K^{I} is the set of rate constants associated with T^{I} and therefore it would not be expected to give complete, or even very much, information about T^{II}, whereas G specifies T^{II} up to thermal noise. Thus part (2) of the theorem is established since the term $\{H(\omega_{T^{\mathrm{II}}}|\omega_{K^{\mathrm{I}}}\omega_{T^{\mathrm{I}}}) - H(\omega_{T^{\mathrm{II}}}|\omega_G\omega_{K^{\mathrm{I}}}\omega_{T^{\mathrm{I}}})\}$ can be used to regulate $H(\omega_{K^{\mathrm{I}}})$.

The term $H(\omega_{T^{\mathrm{II}}}|\omega_{K^{\mathrm{I}}}\omega_{T^{\mathrm{I}}})$ represents a structure–function degeneracy, but in the particular form which is physically allowable. What equation (10.19) says is: the degree to which the rate constants implicit in one substructure of the protein can change gradually is limited by the extent to which specification of this substructure and its functional attributes fails to give information about some second substructure. This second substructure is thus the buffer system, since it can absorb primary variations in a way which modulates their expression in the first substructure. However, in order for the function change of the protein as a whole to be gradual, the attributes implicit in the second substructure must be irrelevant to any specific function. The conclusion is thus that buffering is necessary for gradualism and that in order for buffering to be possible there must be features of the structure irrelevant to any specific function (as in the balls and springs analogy in Figure 10.2).

The purpose of part (3) of the theorem is to look at the buffer system in the context of the various possible forms of protein adaptability. To do this it is necessary to shift the point of view, considering the environment as the source of uncertainty with which the protein must cope and the primary structural uncertainty as providing one of the mechanisms for coping. To do this, expand $H(\omega_{K^{\mathrm{I}}}\omega_{K^{\mathrm{II}}}\omega_{T^{\mathrm{I}}}\omega_{T^{\mathrm{II}}}\omega_G\omega^*\omega_c)$ to give

$$H(\omega_{K^{\mathrm{I}}}) + \{H(\omega_{T^{\mathrm{I}}}|\omega_{K^{\mathrm{I}}}) - H(\omega_{T^{\mathrm{I}}}|\omega_G\omega^*\omega_c\omega_{K^{\mathrm{I}}})\} + \{H(\omega_{T^{\mathrm{II}}}|\omega_{K^{\mathrm{I}}}\omega_{T^{\mathrm{I}}})$$

$$- H(\omega_{T^{\mathrm{II}}}|\omega_G\omega^*\omega_c\omega_{K^{\mathrm{I}}}\omega_{T^{\mathrm{I}}})\} + H(\omega^*\omega_c|\omega_{K^{\mathrm{I}}}\omega_{T^{\mathrm{I}}}\omega_{K^{\mathrm{II}}}\omega_{T^{\mathrm{II}}})$$

$$\geq H(\omega^*) + H(\omega_c|\omega^*) + H(\omega_G|\omega^*\omega_c) \qquad (10.21)$$

where the genotype–phenotype degeneracy term,

$$H(\omega_G|\omega_{K^{\mathrm{I}}}\omega_{K^{\mathrm{II}}}\omega_{T^{\mathrm{I}}}\omega_{T^{\mathrm{II}}}\omega_c\omega^*)$$

has as usual been set to zero. The terms

$$H(\omega_{K^{\mathrm{II}}}|\omega_{K^{\mathrm{I}}}\omega_{T^{\mathrm{I}}}\omega_{T^{\mathrm{II}}}), \; H(\omega_{K^{\mathrm{I}}}|\omega_G\omega_c\omega^*), \quad \text{and} \quad H(\omega_{K^{\mathrm{II}}}|\omega_G\omega_c\omega^*\omega_{K^{\mathrm{I}}}\omega_{T^{\mathrm{I}}}\omega_{T^{\mathrm{II}}})$$

have also been set to zero, in accordance with equations (10.4) and (10.5). But the structure–function degeneracy restriction, equation (10.16), cannot in general be used to eliminate the terms in the leftmost brackets [as was the case in equation (10.18)]. This is because specification of all the relevant rate parameters is not sufficient to specify the particular allosteric form, whereas specification of the primary structure and environment is. Thus the terms in the leftmost curly brackets represent the amount of environmental uncertainty that is absorbed in allosteric transitions. It may also include a contribution from externally induced shape changes of a nonfunctional nature, such as shape changes concomitant to denaturation. The second set of curly brackets represents buffer system adaptability, although the magnitude of the individual terms of course also reflects thermal noise and environmentally induced shape changes. $H(\omega^*\omega_c|\omega_{K^{\mathrm{I}}}\omega_{T^{\mathrm{I}}}\omega_{K^{\mathrm{II}}}\omega_{T^{\mathrm{II}}})$ represents the degree to which both the structure and function of the enzyme forget (or are indifferent) to the environment. High indifference means that the protein is capable of performing the same function in a variety of environments and without allosteric shape change. The term $H(\omega_c|\omega^*)$ represents the extent to which the behavior of the internal milieu external to the protein correlates to the behavior of the environment. If correlation is poor, the protein in effect sees a more uncertain surround. The term $H(\omega_G|\omega^*\omega_c)$ is the uncertainty in the primary sequence as reduced by the environment of the protein (including both the environment of the biota and the internal milieu external to the protein). But since selection (including the restricting effect of changes at other gene loci) occurs after buffering, the time scale is such that $H(\omega_G|\omega^*\omega_c) \approx H(\omega_G)$ in this case.

If $H(\omega^*)$ is taken as $H(\hat{\omega}^*)$ the entire left-hand side of equation (10.21) becomes the adaptability of the protein species. This splits into a phenotypic part and an evolutionary (genotypic) part. The phenotypic part consists of allosteric adaptability and indifference. If these are not sufficient to absorb short-term changes in the environment, these changes must then be absorbed into functional uncertainty which is concomitant to denaturation or other diminution of function. The evolutionary component of adaptability is expressed in $H(\omega_{K^{\mathrm{I}}})$ and in the buffer system adaptability which modulates this. The buffer system terms might also represent processes which buffer the protein from the environment as well as buffering the genetic variation. If the protein is well structured for mutation absorption, it should also be well structured for absorbing external perturbations to its shape. But the effectiveness of the genetic buffering would be reduced if features of T^{II} assume a specific function, even a specific function connected to adaptability, such as allosteric adaptability. The more specific T^{II} becomes as a perturbation absorber, the less effective it can be as a mutation absorber, since mutation becomes increasingly likely to destroy the specificity. The implication is seemingly paradoxical, for if the effectiveness of genetic buffering depends on the extent to which the difference between $H(\omega_{T^{\mathrm{II}}}|\omega_{K^{\mathrm{I}}}\omega_{T^{\mathrm{I}}})$ and $H(\omega_{T^{\mathrm{II}}}|\omega_G\omega^*\omega_c\omega_{K^{\mathrm{I}}}\omega_{T^{\mathrm{I}}})$ makes no contribution to adaptability relative to the environment, it follows that it would in principle be impossible for adaptability to decrease to its minimal allowable value. If it did, the uncertainty of the environment would then be greater than the effective adaptability of the system. The paradox is only apparent, however, since the effectiveness of the buffering increases as the magnitudes of the individual terms increase. The magnitude of the difference between these terms is not so important. It might be noted that if a long-term change in the environment is met by a change in T^{II} rather than by a modulated change in T^{I}, the initially physiologically functionless buffer system would assume a specific physiological function. This is of course evolution by transfer of function, except that the transfer is from an evolutionary function (viz. the buffer system function) to a physiological function rather than from one physiological function to another.

Equation (10.21) represents a particular choice of correlated statistical processes in the system, hence a particular choice of point of view for describing the various forms of adaptability. There is another way of establishing part (3) of the theorem, directly and without any arbitrariness of viewpoint. Recall the equation for the evolutionary tendency expressed in terms of effective entropies, equation (6.5). Each of the terms in equation

(10.21) could be made to correspond to terms in the effective entropy form of the equation, provided of course that they are expressed in terms of normal rather than mnemonic indices and that each is assigned a normalizing coefficient. Thus the effective entropy form of the equation includes as terms $pH(\omega_{T^{II}}|\omega_{K^{I}}\omega_{T^{I}}) - qH(\omega_{T^{II}}|\omega_{G}\omega^{*}\omega_{c}\omega_{K^{I}}\omega_{T^{I}})$, where p and q are normalizing coefficients. Parts (1) and (2) of the theorem establish that this is a buffer system. Its occurrence in the effective entropy equation establishes part (3) of the theorem.

10.3. THE BOOTSTRAP PRINCIPLE OF EVOLUTIONARY ADAPTABILITY

The essence of the mutation buffering concept is that two versions of a protein, M and M', may have the same physiological function, but nevertheless one may be more amenable to evolution than the other. To produce such a protein, all one would have to do is to add appropriate redundancies, either of weak bonding or of number and type of amino acids. The argument so far has been based on the empirical observation that graduated shape and structure change does in some cases occur. However, I have not yet shown that natural selection could add these redundancies. This is the work of the present section. It is important, not only because it establishes that the buffering capability is an evolving (rather than a fixed) property of proteins, but because it provides an alternative argument for mutation buffering which is completely distinct from the adaptability theoretic argument of the previous section. Furthermore, this alternative argument is completely self-contained. It does not even depend (as does the argument of the previous section) on the empirical fact that graduated change does occur in some cases. It depends only on the geometrical and functional traits of the protein (assumptions A1–A3), on the recognition that gradualism is a trait subject to the same assumptions (since the various redundancies determining gradualism are determined by the primary structure), and on the theory of evolution by variation and natural selection.

The conclusion to which this argument leads is summed up in what I call the bootstrap principle of evolutionary adaptability (Conrad, 1977, 1979a). This is: *the degree of gradualism with which protein function changes with stepwise changes in primary structure is both a condition for and a consequence of evolution by variation and natural selection.* In principle the argument is quite simple. Evolution proceeds most rapidly when it proceeds through single genetic changes, where each genetic form in the sequence has selective value, or at least is viable. It is increasingly slow as the number of simultaneous steps required to make the jump from one viable form to another increases. The reason of course is that the probability of such a jump is the product of the probabilities of the individual changes, assuming that these are independent. (Here a single change could be a point mutation, duplication, recombination, or any other indecomposable genetic process.) This means that practically all evolution of new, useful traits will occur in those protein species with the most nearly optimal gradualism. Thus the gradualism trait will hitchhike along with the advantageous traits whose probability of occurrence it increases, despite the fact that it may be an energetic disadvantage to the organism (because it means larger proteins, more types of amino acids, more complicated amino acid formats).

A simple model is sufficient to illustrate how fast rate of evolution decreases with decreasing gradualism (Conrad, 1978). The model is not intended to be realistic, but rather to be simplified in a way which is less favorable to the conclusion than a more realistic model would be. Again we consider the protein species coded by a single gene locus. G_u is a sequence of bases $b_1 \ldots b_{(3n)}$, where each of the b_i is one of the four types of nucleotides in DNA. G_u thus codes for some sequence of n amino acids, $a_1 \ldots a_n$, where each of the a_i is one of the approximately twenty types of amino acid found in protein. Suppose first that G_0 has fitness W_0, or more precisely, that the organism which carries G_0 has fitness W_0. Suppose also that G_m, with fitness W_m, is the nearest protein with a higher fitness level and that to reach G_m requires at minimum m alterations in the amino acid sequence of G_0. The only way to jump between these two sequences (or the points representing them in the landscape) is to do so by m simultaneous changes in the appropriate amino acids, subject to the condition that no changes occur in any other amino acids. The probability for such an occurrence is

$$P\left(G_0 \xrightarrow{m} G_m\right) = \frac{N_i}{N_0} = \frac{P^m(1-p)^{n-m}}{20^m} \tag{10.22}$$

where N_0 is the number of organisms with gene G_0 at time t, N_1 is the expected number of organisms with gene G_m after one generation, and p is the mutation probability. The probability of an appropriate amino acid change, however, is approximately $p/20$, since the change could be to any of the 19 or 20 other amino acids. The average number of generations required for the appearance of a protein G_m is thus

$$\bar{\tau}_{0m}^{(m)} = \frac{1}{N_0 P\left(G_0 \xrightarrow{m} G_m\right)} = \frac{20^m}{N_0 p^m(1-p)^{n-m}} \tag{10.23}$$

where (m) indicates the number of required simultaneous genetic events. If more than one G_m is expected to appear after one generation (i.e., if $N_1 > 1$) the $\bar{\tau}_{0m}^{(m)}$ would be smaller than unity. Note that $p/20$ is actually an average probability because the redundancy structure of the genetic code is such that not all transitions between amino acids are equally probable. Also note that we make the reasonable assumption that the probabilities are independent. In fact this is a least favorable assumption, since any other assumption would in general increase the number of expected generations required for the appearance of an arbitrarily chosen sequence.

Now consider the extreme alternative situation in which G_0 can change into G_m by m single changes in amino acid sequence with the property that each W_i in the sequence is at least slightly greater than its predecessor. Formally,

$$G_0 \xrightarrow{1} G_1 \xrightarrow{1} \cdots \xrightarrow{1} G_i \xrightarrow{1} G_{i+1} \xrightarrow{1} \cdots \xrightarrow{1} G_{m-1} \xrightarrow{1} G_m \qquad (10.24)$$

such that $W_i < W_{i+1}$ for all i from 0 to m. The probability for the appearance of G_m now depends on the probability of m appropriate single-step mutations. There are $m!$ possible ways in which this can happen in m steps. The worst case assumption, however, is that only one way is possible. Thus in the worst case the expected number of generations required for a single step in this chain is just a special case of equation (10.23),

$$\bar{\tau}_{i(i+1)}^{(1)} = \frac{20}{N_i(t)\,p\,(1-p)^{n-1}} \qquad (10.25)$$

except that N_i is now a function of time since the number of organisms carrying gene G_i begins to grow as soon as the first one appears. The average number of generations required for the appearance of G_m is the sum of all the individual evolution times:

$$\bar{\tau}_{0m}^{(1)} = \sum_{i=1}^{m} \tau_{i(i+1)}^{(1)} = \sum_{i=1}^{m} \frac{20}{N_i(t)\,p\,(1-p)^{n-1}} \qquad (10.26)$$

The expected evolution times can thus be determined if the $N_i(t)$ are known, i.e., if it is known how fast each population in the sequence grows. This is clearly complicated since the population size changes in a way which depends on the increase in fitness following the appearance of G_i, on the appearance of alternate alleles or inevitable changes at other loci (both of which are ignored), and on the interaction with other populations. To avoid these complications, we again make the most unfavorable assumption, viz.

that mutation and all other mechanisms of genetic change are turned off until the new population effectively grows to the same size as the old population (i.e., until N_i grows to N_0) and that there is no further growth. Then equation (10.26) becomes

$$\bar{\tau}_{0m}^{(1)} = \frac{20m}{N_0 p (1-p)^{n-1}} + (m-1)D \qquad m \geqslant 1 \qquad (10.27)$$

where D is the number of generations (delay time) before we allow mutation to be turned on and it is not necessary to consider delay following the first appearance of G_m. In general D would be different for each step in the series, but it can always be taken sufficiently large to ensure that $\tau_{0m}^{(1)}$ is an underestimate (aside from the underestimation inherent in assuming no mutation until N_i reaches N_0).

Relative evolution times for the stepwise and simultaneous occurrence of m changes are given by the ratio

$$F(m) = \frac{\bar{\tau}_{0m}^{(1)}}{\bar{\tau}_{0m}^{(m)}} = P^{m-1} \left[\frac{20m + (m-1)DN_0(1-p)^{n-1}}{20^m (1-p)^{m-1}} \right] \qquad (10.28)$$

A number of illustrative values are listed in Table 10.1. For a mutation rate of $p = 10^{-8}$ and a step length of two, the stepwise mode of evolution is 10^9 faster than the simultaneous mode, assuming a population size of one million, a delay time of one thousand generations, and a protein with three hundred amino acids. For a step length of three, the stepwise mode is 10^{18} faster and for a step length of eight it is 10^{64} times faster. The speed advantage of the single-step mode is enormous even if only two amino acid changes are required and rapidly becomes astronomically enormous as the number of required amino acid changes increases. Any reasonable choice of the mutation rate would give basically the same result (as indicated in Table 10.1). Population size and differential growth rates (expressed in the values of D) could vary widely. However, values of $F(m)$ are quite insensitive to changes in this parameter and also quite insensitive to changes in the number of amino acids in the protein. Indeed, for any reasonable choice of generation time the expected waiting time for the appearance of a particular sequence through, say, simultaneous changes in five amino acids would be longer than the known age of the earth for even the most generous choice of parameters. Furthermore, as m gets larger, the rate advantage of a decrease in step length becomes greater [since $1/F(m) - 1/F(m+1) \approx 1/F(m)$]. This is important, for it means that the selective advantage of an increase in the degree of gradualism increases as the degree of gradualism departs from the

TABLE 10.1
Order of Magnitude Values of F(m) for Selected Values of the Step
Length (m), Mutation Rate (p), Population Size (N_0), Delay Time (D), and
Protein Size (n)

m	p	N_0	D	n	$F(m)$	
1	10^{-8}	10^6	10^3	3×10^2	1	Increase of $F(m)$
2	10^{-8}	10^6	10^3	3×10^2	10^{-9}	with step length
3	10^{-8}	10^6	10^3	3×10^2	10^{-18}	
4	10^{-8}	10^6	10^3	3×10^2	10^{-28}	
8	10^{-8}	10^6	10^3	3×10^2	10^{-64}	
2	10^{-6}	10^6	10^3	3×10^2	10^{-6}	Variation of $F(m)$
2	10^{-10}	10^6	10^3	3×10^2	10^{-11}	with mutation rate
2	10^{-8}	10^6	10^3	10	10^{-9}	Insensitivity of $F(m)$
2	10^{-8}	10^6	10^3	1110	10^{-9}	to protein length
2	10^{-8}	10	10^3	3×10^2	10^{-9}	Insensitivity of $F(m)$
2	10^{-8}	10^9	10^3	3×10^2	10^{-9}	to population size
2	10^{-8}	10^8	1	3×10^2	10^{-9}	Variation of $F(m)$
2	10^{-8}	10^6	10^5	3×10^2	10^{-8}	with delay time

one-step case (all other things being constant). In short, the one-step strategy is not only optimal, but a global attractor in the sense that the evolutionary advantage (in terms of expected evolution times) for approaching it is strong when close and even stronger when less close. After the one-step situation is reached, further enhancement of gradualism produces a less dramatic, constant factor speedup in the rate of evolution. Two single-step pathways are better than one, but not by as much as one single-step pathway is better than even a larger number of double-step pathways.

The conclusion is quite robust and hardly surprising. The potential rate of evolution is so much faster in the step-by-step mode than in the simultaneous mode that virtually all evolution of new traits must occur in populations which vary in the direction of more nearly optimal gradualism. However, this means that selection for practically all protein properties which confer an individual selective advantage will also be tied to selection for the gradualism as well. In effect, the degree of gradualism is both a condition for and a consequence of evolution by variation and natural selection, from which it follows that species which predominate in the course of evolution are inevitably species whose complement of protein molecules is well adjusted for evolution. This, however, is equivalent to our original statement that degree of gradualism is adjusted for efficient evolutionary behavior in the course of evolution—and adjusted through a bootstrap process.

It is important that the rate of evolution argument requires the fitness at each step to increase, or at least not to decrease significantly. By itself this does not imply a requirement for gradualism. Rather, the requirement derives from the unlikelihood that a significantly altered protein will have functional value. There are an enormous number of possible proteins (at least on the order of 20^{300}) and it can hardly be expected that the majority of these would have functional value. This means that the number of functional proteins is extremely sparse in the total number of possible proteins and therefore it is unlikely that primary structures can be selected in such a way that single changes in primary structure produce proteins with very different but nevertheless functionally useful properties. However, they can be selected (by selecting for the buffering mechanisms described in the previous section) in such a way that single changes in sequence are not unlikely to be associated with slight changes in function, hence slight changes in fitness. The degree of gradualism can always be altered gradually since it is always possible to add or delete redundancy in a step-by-step fashion.*

It is also important that the argument does not make the claim that gradualism is by itself an advantage to the individual organism. In general an increase in gradualism means an increase in the number of amino acids in the protein, the number used by the protein, or the introduction of formats for increasing the redundancy of weak bonding which would otherwise be unnecessary. In each case the free energy expended by the organism in constructing its proteins must increase. In order for gradualism to evolve, this energetic disadvantage would have to be outweighed by the advantage to the species. Superficially this might seem to imply that the evolutionary bootstrapping of gradualism depends on group selection in the sense that the individual organisms in the population are sacrificing efficiency to increase the efficiency of the population. The argument does not involve any such altruistic component. The offspring of the organisms which vary in the direction of gradualism are more likely to be gifted with

---

*Furthermore the numbers in Table 10.1 imply that a series of many such evolution-enhancing mutations can always occur faster than an increase in fitness which requires two simultaneous mutations. If the falloff in population size is not considered, one fitness-increasing double mutation is equivalent to about 10^9 evolution-enhancing mutations. Suppose the evolution of a protein reaches a bottleneck, that is, a point for which no single point mutation can increase fitness. Further evolution could always proceed more rapidly by drifting through a sufficient number of evolution-enhancing mutations, each of which is slightly fitness-reducing, than by waiting for the improbable fitness-increasing double mutation (see Conrad, 1982). It is interesting that P. M. Allen and W. Ebeling have calculated that a slightly fitness-reducing mutation will persist in a simple predator–prey system for a number of generations which is quite adequate to provide an alternative justification for this conclusion (Allen and Ebeling, 1983, and personal communication).

an increase in efficiency over the offspring of parents whose genetic constitution makes them more efficient but less amenable to gradualism. Hence the gradualism property hitchhikes along with the desirable traits to which it gives rise.* If all evolution of such carrier traits could somehow be terminated, the gradualism property might decline—but of course such termination is impossible and thus gradualism is, so to speak, hanging onto the coattails of the evolution it makes possible.

The optimal adjustment of the gradualism property would clearly mean that single amino acid changes should be accompanied by a maximum possible increase in fitness. I want to emphasize that the bootstrapping of evolutionary adaptability does *not* imply this type of optimization. Evolution may tenably be regarded as optimizing to the extent that a type arising as a result of variation and favored by selection will continue to be favored as long as the conditions of selection do not change. If the population is not initially at an optimum which is at least local, it should evolve toward this, assuming that the optimum is accessible. This consideration, however, cannot be extended to a property (such as the gradualism property) which facilitates further evolution, for the conditions of selection necessarily change. Rather the evolved degree of gradualism is a condition for past evolution of the species and therefore tends in the direction of optimality only to the extent that the degree of gradualism required for past success can be extrapolated into the future.† This consideration suggests that a number of possible strategies of buffering may develop. If the specialization

---

*The term *hitchhiking* was orginially used in the context of selection for high mutation rates (Cox and Gibson, 1974) and subsequently in the context of selection for high recombination rates (Strobeck *et al.*, 1976). In these cases a gene-enhancing recombination or mutation hitchhikes along with a fitness-increasing gene belonging to the same linkage group. In the case of a single protein the structural alteration which increases amenability to evolution is on the same gene. A more critical difference is that the hitchhiking effect is stronger for amenability than for recombination and mutation. This is because of the more pronounced effect which amenability has on the rate of evolution. For a computational model illustrating that evolutionary amenability bootstraps even under assumptions unfavorable to it, see Conrad and Rizki (1980).

†Adaptability of any type is of course a product of evolution relative to the past environment and therefore not necessarily suitable to the future. However, it is in principle (in an experimental situation) possible to hold the uncertainty of the external environment constant and to define adaptability operationally in terms of this fixed uncertainty (Section 6.11). If the structure of environmental uncertainty stays the same the system's adaptability structure should remain suitable. Amenability to evolution (transformability) is different in that the appearance of a new trait may alter the constellation of factors which determine the probability for the appearance of still another useful trait. Extrapolation into the future depends on the fact that the redundancy mechanisms described in this chapter have a general facilitating effect. Note that even if this were not the case, it would not lead to any contradiction in the general theory. This is because the gradualism property is represented in the absolute magnitude of the entropies.

of the protein increases, yet further evolutionary development is possible, the degree of buffering will increase. However, if the traits of the protein are well tuned and the frequency of fitness-increasing modifications decreases, an advantage will accrue to species in which it varies in the direction of decreasing gradualism. This possibility has an amusing though highly speculative consequence, for it would mean that the gradualism property would have been most important early in the development of life, during the age of basic biochemical evolution and adaptive radiation of enzyme types. One might thus imagine that many present-day proteins, even proteins whose sequences are locked into a very narrow range, had larger, more highly buffered forebears.

10.4. ISSUE OF NEUTRALISM AND SIGNIFICANCE OF THE MAGNITUDES

The discovery of unexpectedly large numbers of variants of proteins, either allozymes (coded at the same loci) or isoenzymes (coded at different loci), has raised questions about the role of selection in evolution (Kimura, 1968; King and Jukes, 1969; Lewontin, 1974; Nei, 1975; Bargiello and Grossfield, 1979). In many cases genetic distance between proteins with a common ancestor serves as an evolutionary clock, indicating that the rate of evolution depends more on rate of mutation than on selection. This type of evolution is sometimes called non-Darwinian, an unfortunate name since it is really just the special case of Darwinian evolution in which selection is neutral.

The adaptability theory analysis in the present chapter suggests the following interpretation. Since amenability to evolution requires that slight changes in primary structure are not unlikely to be associated with slight changes in function, it is inevitable that numerous variations of any protein which has not lost its evolutionary adaptability will make either the same or a similar contribution to fitness. This does not mean that selection is unimportant, but on the contrary that some degree of neutralism or near-neutralism is the inevitable concomitant to the conditions under which selection can effectively act. In this sense the phenomena which support the hypothesis of neutral selection are not only consistent with but implied by the classical selectionist theory.

This interpretation generalizes to the multigene situation. The more amenable such systems are to evolution, the larger the number of variations on the phenotype which have similar selective value. This means that in both the single- and multigene case high development of evolutionary transformability means heightened genetic and phenotypic variability, therefore an increase in the actual magnitudes of $H(\underline{\omega})$ and $H(\underline{\omega}|\omega^*)$. These magnitudes also increase because the greater size of the genome in general leads to a larger ensemble of genome and phenome states. Increase in the magnitude does not mean that the adaptability of the system increases, since neutral variation would increase both terms by the same amount. However, while graceful transformability means more variations, hence larger magnitudes, it also means that fewer of these variations are nonfunctional. Thus the increase in the magnitudes in principle makes it possible for the evolutionary tendency equation, equation (9.1), to approach an equality without the indifference term becoming excessively large. If the magnitudes became as small as possible, the indifference term would have to increase since that state-to-state behavior of nonfunctional forms could never be expected to reflect accurately the state-to-state behavior of the environment. Note that no claim is made that the indifference term could disappear altogether, as it might in principle if the issue were one of reliability of function rather than its transformation. This is clearly not possible in a random search process. The only claim being made is that by organizing the system appropriately the structure of the search problem becomes more amenable to step-by-step search and hence less wasteful.

The above consideration answers the question raised at the beginning of this chapter concerning the relation between transformability and the magnitudes. The answer is: in order for the adaptability to approach a minimum value compatible with the uncertainty of the environment and without an undue increase in indifference the magnitudes of the transition and anticipation entropies must increase. The increase is due to an increase in the ensemble of functionally distinct states, although these functionally distinct states may become so similar that they grade into functionally equivalent states (i.e., into neutralism). The significance of the magnitudes has been discussed in the context of genotype–phenotype relations, where, because of its decisiveness for organic evolution, it assumes it greatest importance. However, the same considerations apply to any possible physiological processes which utilize evolutionary mechanisms, such as Burnet's clonal selection theory of antibody formation.

I have recently found that sequence data analyzed by Professor M. Volkenstein (1979) provide empirical support for the bootstrapping model and therefore for the interpretation described here. Volkenstein defined the replaceability of amino acids in terms of their functional similarity to other amino acids and showed that codons coding for the same amino acid may have different replaceabilities. This is possible since the functional similarities of the amino acids arising from point mutations of different codons which code for the same amino acids may be different. Thus the situation is that silent mutations—mutations with absolutely no effect on protein structure and function—can lead to changes in amino acid replaceability, hence to changes in amenability to evolution. A careful examination of Volkenstein's analysis of RNA sequence data for hemoglobin indicates a significant bias in favor of the more replaceable codons. Biases in codon frequency could allow for different degrees of gene expression in milieus with different tRNA concentrations. But the preference for more replaceable amino acids cannot be due to selection on the individual, suggesting that the codon frequencies in this evolutionarily plastic protein have to some extent been controlled by the hitchhiking mechanism. [For a discussion of these data and a formulation of the bootstrapping principle in terms of measures of replaceability, see Conrad and Volkenstein (1981).]

10.5. ADAPTABILITY OF THE ADAPTIVE LANDSCAPE

A very general principle begins to emerge from the considerations of the previous section. One can imagine an enzyme or an organism much smaller and energetically more efficient than any present-day organism, essentially a smaller version of a man, energetically more efficient than the author or any anticipated reader, but capable of no less as far as the specifics of man's life is concerned. Yet such a man could never arise through evolution, nor would a race of such men have any evolutionary future—for the toll of such absolute optimality would be the most extreme inamenability to evolution.

The entire idea can be fruitfully summed up in the framework of a molecular version of the classical adaptive landscape. For the present purposes the classical fixed environment landscape can be thought of as a plot of fitness against gene frequencies (see Figure 10.4a). The landscape may also be enlarged to allow for environmental change by adding axes for each of the environmental parameters. In this picture of evolution (developed by S. Wright and other geneticists) one sees the evolution process in terms of populations climbing the multidimensional hills and peaks in this landscape. I call the landscape classical because genotypes are being described in terms of gene frequencies and evolution in terms of changes in gene frequencies. For the present purpose (and indeed for many purposes) it is necessary to consider the finer level of detailed sequence of nucleotides in the gene (or of translated amino acids if consideration is restricted to a single protein). The most convenient way to introduce this refinement is to replace the gene frequency axis by a set of axes for specifying nucleotide sequence, with a distinct axis for each possible nucleotide type at each possible position. The construction, in projected form, is illustrated for a fixed environment landscape in Figure 10.4b,c and for a global landscape in Figure 10.4d. The presence of a nucleotide of a particular type is represented by a one on the appropriate axis and its absence by a zero. The fixed environment landscape is thus a multidimensional cloud of points in the interval zero to one on both the nucleotide and fitness axes (where fitness is taken as a number between zero and one). The environment-parameterized landscape is a hypercloud of lines whose projection on the fitness and nucleotide axes is between zero and one. If all the points in this hypercloud are connected by straight lines, the envelope of these lines becomes a fitness hypersurface, with multidimensional peaks and valleys.

The fitness hypersurface has a global structure (the topography of the adaptive peaks and valleys) and a local structure which determines the ease with which particular peaks may be climbed by an evolving population. Thus, a particular pathway, either up or between peaks, may be easily traversible in the sense that it can be negotiated through a series of single-step changes in base sequence which do not involve any significant decreases in fitness, or it may be nontraversible in the sense that multistep changes in primary structure are necessary to jump crevices in fitness.

To make the concept of traversibility precise, define the genetic distance between two genes to be the minimal number of changes of zeros and ones on the nucleotide axes required to make them identical. Thus the genetic distance between two genes is the Hamming distance between their adaptive landscape representations. The negotiability of any neighborhood of a gene represented on the fitness surface is the minimum genetic distance which must be traversed to reach a gene with a higher level of fitness. If this minimum distance is equal for two different neighborhoods, their traversibility is the same, though one may be steeper in the sense that the fitness rises more rapidly. A gene coding for a protein whose function changes more gradually with single-step changes in primary structure is clearly likely to have a more negotiable neighborhood than a gene coding for a protein function which changes radically—since gradual function changes will be accompanied by slight changes in fitness, while radical function changes are much more likely to be lethal than advantageous. The concept of traversibility of the landscape can also be extended to higher levels of genetic organization by defining genetic distances in terms of single-step changes involving genetic changes which affect blocks of nucleotide axes in a uniform way.

According to the bootstrap principle of evolutionary adaptability, populations will tend to move to regions of the landscape which are easily traversible, either because the pathways between peaks are smoother, because the peaks themselves are more climbable, or because there are peaks

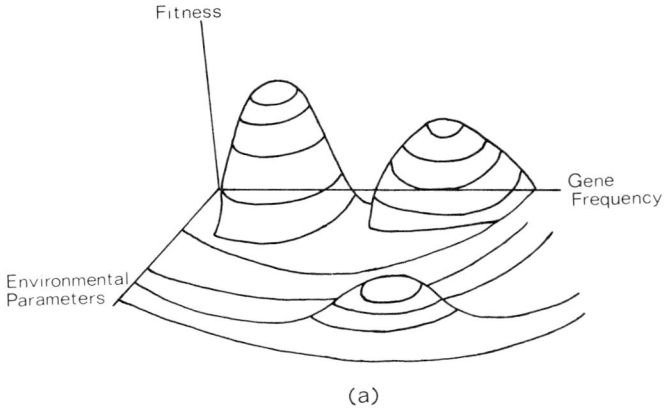

(a)

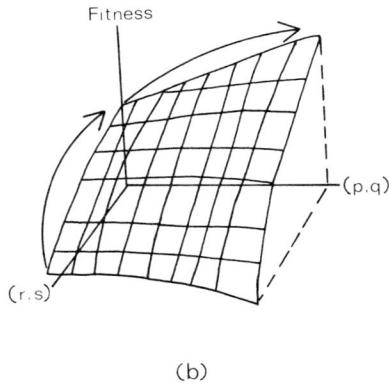

(b)

FIGURE 10.4. Classical adaptive landscape and projections of molecular adaptive landscapes. (a) A classical environmentally parameterized landscape. Any plane parallel to the gene frequency–fitness plane is an environmentally fixed landscape. (b) Projection of a molecular adaptive landscape onto two axes (nucleotide of type p at position q and nucleotide of type r at position s). The most fit situation is one in which p is present at q and r is absent at s and the least fit the one in which p is present at q and r is present at s. The arrows indicate that it is possible to go from (r, s) to the highest fitness point in two single steps. (c) A two-axis projection in which it is not possible, in the projection, to go from (r, s) to (p, q) in a series of single steps, since the intermediate steps have zero fitness. A two-step jump is necessary. (d) The projection of an environmentally parameterized landscape along a single nucleotide axis and along one environmental axis (from Conrad, 1979a, adapted by permission).

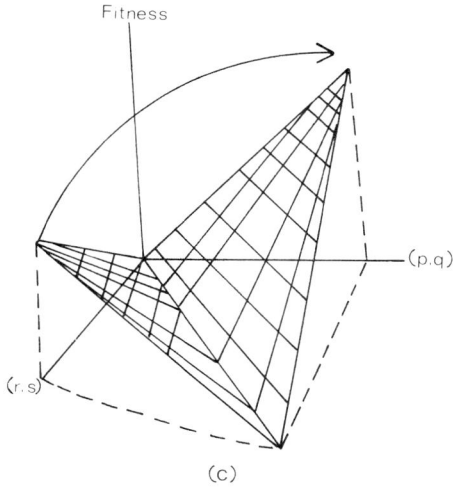

(c)

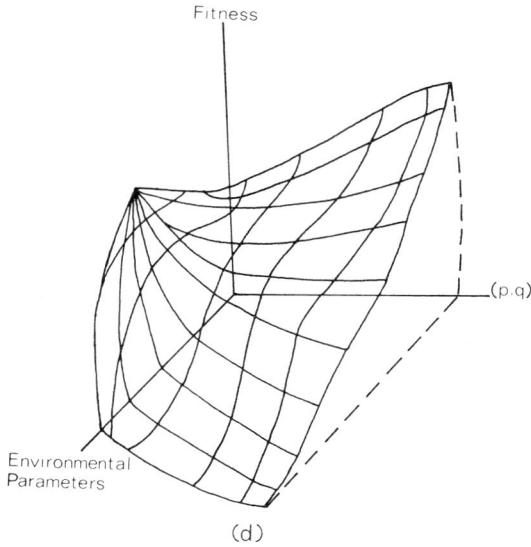

(d)

FIGURE 10.4. (*continued*)

in the region. This is always possible because pathways between less- and more-traversible regions are always traversible—since gradualism can always be added in a gradual way. If a genetic variation moves a population to a region of the landscape which is more traversible, this variation will be drawn along with the evolutionary movement allowed by this increase in traversibility. Since any such variation is a cost in terms of efficiency, a population must expend free energy to occupy a region of the landscape which is more amenable to evolutionary hill climbing. In effect, then, biological systems use free energy to increase the traversibility of the adaptive surface on which they are evolving.

It is important that the argument does not imply that primitive biological systems occupied a region of the landscape which is not easily traversible. On the contrary, it is much more natural to suppose that the lineage of present-day systems traces back to those varieties of prebiological polymers which came into the world with significant evolutionary adaptability. Indeed, in some cases the traversibility of the landscape, either for the species or for some part of its genome, may decrease. This is a possibility for a highly adapted population which can hold its position without any further requirement for evolutionary adaptability. There are other, more flexible mechanisms of evolutionary conservatism—such as position on the chromosome, inversion, dominance, and recessiveness.

So far the picture is of a single population moving on a fitness surface which is static. In reality such a surface is occupied by many populations simultaneously climbing either the same or different peaks. Each such population is part of the environment of every other population. In principle this can be incorporated into the adaptive surface by introducing into the set of environmental axes a numbers axis for each possible genome sequence. Thus, as the numbers associated with each population change, each point on the surface will move, just as it would move as any environmental parameter changes. It might be pushed either higher or lower on a given peak, off a peak, or onto another peak. Since it is hardly reasonable that arbitrary shifting of a point along any of the environmental axes would push it uphill as often as it would push it downhill, a condition for the fitness of the various species represented on the surface to be nondecreasing is that the hills do not move relative to them faster than they can be climbed. This is an important point, for it means that selection for rate of evolution is not based solely on the advantages accruing to those species which can optimize more rapidly, but also on the disadvantage of de-optimization in those species which cannot keep pace with the relative motion of the landscape.*

---

*This is reminiscent of Van Valen's red queen hypothesis, according to which all species have to keep evolving in order to stay in place on peaks moved by the evolution of other species (Van Valen, 1973).

10.6. EXTENSION TO MULTIGENE SYSTEMS

Now I want to justify the claim that the bootstrapping argument and the concept of smoothing the adaptive landscape apply to the multigene complex as well as to the single gene. The generalized principle is: *the degree of gradualism with which the phenotype changes with independent changes in the genotype is adjustable and adjusted for effective evolutionary behavior in the course of evolution.*

The rate of evolution part of the argument is in principle the same as for the individual gene, except much more complicated. I do not want to develop it in detail, but would like to indicate why the same type of dramatic difference in the relative rates of stepwise and simultaneous pathways of evolution is to be expected. The extra complication comes from the fact that a number of processes other than mutation have to be considered (recombination, crossing over, gene duplication, development of polyploidy, and others). The probabilities are not the same for these processes and not the same for particular instances of them. However, to the extent that the processes are independent, particular combinations of two, three, or more are unlikely to occur simultaneously, and to the extent that they are not independent, most combinations are unlikely to occur at all.

To put the situation into a framework analogous to the single-gene case, define the genetic distance between any two genomes as the number of independent processes which would be required to transform one into another. If two or more simultaneous crossovers are required, the probability of traversing the distance becomes small, just as described by equation (10.23), except that the probabilities for each contributing crossover would be different. If recombination is allowed, the probability of a favorable outcome becomes rapidly smaller as more particular chromosomes are required to pair in the formation of the zygote. In the single-step situation acceptable genotypes arise with reasonable frequency from single cross-over events or single recombination processes. The importance of these processes is not that they make it possible to bypass the step-by-step mode of evolution at the single-gene level, but rather that they make it possible to bring together independently evolved species of the gene or of the genome in single steps—in short they make it possible to utilize hierarchy to take full advantage of the step-by-step mode.

The rate argument is clear enough. However, the bootstrap principle depends not only on this, but also on the possibility of variable genetic and developmental mechanisms which allow multigene gradualism to hitchhike along with the traits whose evolution it facilitates. Clearly gradualism cannot, at this level, depend on folding, for it is individual proteins which fold or which interact with other individual proteins to form quaternary structures. The relationship between genotype and phenotype is thus not subject to the highly restrictive conditions (A1–A4 in Section 10.2) which apply to the relationship between primary structure and tertiary properties. This opens up a number of possibilities. These may be classified, with the help of adaptability theory, by expanding $H(\omega_{P^{\mathrm{I}}}, \omega_{P^{\mathrm{II}}}, \omega_G, \omega^*, \omega_c)$ to give

$$\{H(\omega_{P^{\mathrm{I}}}) + H(\omega_{P^{\mathrm{II}}}|\omega_{P^{\mathrm{I}}})\} - \{H(\omega_{P^{\mathrm{I}}}|\omega_G) + H(\omega_{P^{\mathrm{II}}}|\omega_{P^{\mathrm{I}}}\omega_G)\}$$

$$+ H(\omega_G|\omega_{P^{\mathrm{I}}}\omega_{P^{\mathrm{II}}}\omega^*\omega_c) = H(\omega_G) \tag{10.29}$$

The mnemonic conventions used elsewhere in this chapter are here retained, except that ω_G is the transition scheme entropy of the untranslated genome rather than the translated primary structure for a single gene, $\omega_{P^{\mathrm{I}}}$ and $\omega_{P^{\mathrm{II}}}$ are the transition schemes for phenotypic compartments I and II (with no attempt to separate structural and functional properties), and ω_c is the transition scheme of all the biota components other than the organism.* Equation (10.29) is otherwise the same as equation (10.18) except that the genotype–phenotype degeneracy has not been eliminated. Since the base sequence of the genome provides no extra information about the environment, terms of the type in equation (10.10) have been eliminated, as before.

According to the gradualism lemma (A5, Section 10.2), the increase in $H(\omega_{P^{\mathrm{I}}})$ in response to an increase in $H(\omega_G)$ must be small in order for P^{I} to change gradually in response to genetic variation. This is possible in one or a combination of the following circumstances:

1. $H(\omega_{P^{\mathrm{II}}}|\omega_{P^{\mathrm{I}}}) - H(\omega_{P^{\mathrm{II}}}|\omega_{P^{\mathrm{I}}}\omega_G)$ is large. This is the same as the specialization degeneracy term [cf. equation (10.20)] which made mutation buffering possible in the case of a single protein, except that P has not been split into T and K. Specialization degeneracy is possible to the extent that the subsystems of the organism are independently alterable. The genome must be broken into sets of genes, at least one of which has a strong influence on P^{I} but not on P^{II}. This limits the ramification of genetic change, although at the cost of integration of the system. The mechanism of specialization

---

*For ω_G the correspondence to the reference structure conventions in Figure 5.3 is: $\omega_G = \omega_{i0}\omega_{(i+2)0}\cdots\omega_{(i+n-2)0}\omega_{(2+n)0}$, where i can equal any of the values allowed by the reference structure. Adopting the convention that $\Pi\omega_{ij}/\Pi\omega_{rs}$ is the product of all transition schemes running over index sets $i \in I$, $j \in J$, excluding those schemes running over index sets $r \in R \subset I$, $s \in S \subset J$, the product scheme corresponding to $\omega_{K^{\mathrm{I}}}\omega_{T^{\mathrm{I}}}\omega_{K^{\mathrm{II}}}\omega_{T^{\mathrm{II}}}$ is ω_{u3}/ω_G for the value of i corresponding to u. The product scheme corresponding to ω_c is ω_{15}/ω_{u3}. Note that if all cellular genome schemes are the same, $H(\omega_{i0}|\omega_{(i+2)0}) = 0$.

degeneracy is clearly different at the organism and the polypeptide levels, for in the latter case there is only one gene. If the protein is a single polypeptide, the compartments correspond to features of the shape, with features less relevant to specific function subserving buffering. However, in a protein which self-assembles from two polypeptides, the possibility for buffering is increased, since each polypeptide could be modified by modifying the gene which codes for it without any effect on the other (until they interact). Clearly the possibilities for this type of channeling of genetic change very much increase in multigene systems.

2. $H(\omega_G|\omega_{P^{\mathrm{I}}}\omega_{P^{\mathrm{II}}}\omega^*\omega_c) > 0$. This is the genotype–phenotype degeneracy term. It was zero when G was the primary structure of a single protein, or, more accurately, a single polypeptide. Because of the degeneracy of the code, it would be greater than zero for a single gene. Degeneracy may allow for more reliable coding in this case, but it can only contribute to the transformability of the protein by biasing the probabilities for a normal buffer system mechanism (viz. amino acid replacements). The situation is entirely different at the whole-genome level. At this level genotype–phenotype degeneracy becomes the most important vehicle for gradual transformability. There are three major known mechanisms. First, the genotype–phenotype relation is degenerate in the sense that many alternative orderings of genes are for all practical purposes completely equivalent from the standpoint of development. This is what makes possible higher-level genetic shuffling processes and is therefore the basis for hierarchies of stepwise variations of genes and blocks of genes. The second mechanism is quantitative action of genes. The genotype–phenotype relation is degenerate because alternative genotypes—genotypes with a few genes coding for a potent enzyme or with many genes coding for an attenuated enzyme—are entirely equivalent from the standpoint of development. However, they are not equivalent from the standpoint of evolution since a trait arising from the additive effect of attenuated enzymes coded by redundant genes is clearly more gradually transformable. The redundancy of genes is mechanistically analogous to the redundancy of weak bonding in an enzyme, except that the gradualism is now based on change in enzyme concentration rather than change in enzyme shape (with both affecting the rates of processes).

The third mechanism is regulation. This is not so well understood, especially in eukaryotes. However, there is evidence that it may be the most important mechanism of gradualism and indeed some authors have suggested that many species are principally differentiated by changes in regulatory properties (e.g., Britten and Davidson, 1971). Gradual transformability through alteration of regulatory genes falls into the genotype–phenotype degeneracy category because any particular phenotype could in principle be

coded by alternative genomes without easily transformable regulatory properties, just as a radio could always be built (and at decreased cost) to receive
only one station. A tunable and a nontunable radio could be equivalent as
far as any frequency is concerned, just as a regulatory system transformable
through genetic change and one not so transformable could be equivalent as
far as the particularities of physiological function are concerned.

3. $H(\omega_{P^1}|\omega_G) < H(\omega_{P^1})$. This is a reliability condition, for it means that
P^1 (or the range of plastic variation of P^1) has a highly determinate relation
to the genetic structure. In biological parlance the development of the
system is said to be canalized. By itself an increase in canalization does not
imply an increase in gradualism—since reliability is the special case of
transformability in which function is constant rather than gradually varying
in response to perturbation. There is, however, an indirect connection,
proposed by C. H. Waddington (1957). According to Waddington's theory,
traits suitable to a particular environment can be gradually assimilated into
the genome (without any Lamarckian inheritance of acquired characteristics) by selecting for organisms which most easily acquire these traits by a
phenotypic adaptation process. Such selection will also select for those
organisms which are most prone to develop the trait in the absence of the
environment, hence to develop it constitutively. One can imagine primitive
versions of a trait being transformed into highly sophisticated forms in this
way. The process is based on transforming the pattern of canalization, and
therefore on mechanisms of the type described in (1) and (2) above. In
principle, such Darwinian assimilation of environmentally induced traits
could also occur in a single gene, except that the mediating mechanisms
would be the buffer system mechanisms associated with folding.

The conclusion is that for multigene systems two classes of mechanisms
become available for buffering the effects of genetic variation, whereas only
one class is available for single proteins. It perhaps says something about
biological nature that at least one known mechanism can be identified
which belongs to each class and, if Waddington's theory is correct, degree of
canalization (the reliability term) can be used in an indirect way. The
buffering mechanisms are not the same as the mechanisms which play a role
in a single protein, where the problem is to utilize the free energy of folding
to increase the internal redundancies in a single macromolecule. However,
as with the protein, each of the gradualism-increasing mechanisms involves
a cost in terms of free energy. Reducing the ramifications of genetic change
means duplication of genes and synthetic activities, allowing alternative
orderings of genes complicates the problems of readout and regulation,
quantitative inheritance means more genes and more allocation of resources

to less potent forms of an enzyme, and tunable regulatory systems mean more complicated regulatory apparatus than would otherwise be necessary.

Multigene bootstrapping is important for evolution. But single-gene bootstrapping is more important, for protein catalysts are the ultimate determinants of the structure and behavior of biological systems. Their evolution provides the pieces whose arrangement into larger complexes makes higher levels of genetic variation possible. Even if the evolution of the gene as a piece is basically complete, its amenability to evolution allows it to be more particularly molded to other genes with which it must cofunction and to which it must be coadapted.

10.7. BOOTSTRAPPING OF THE CYCLIC NUCLEOTIDE SYSTEM

F. H. Kirkpatrick (1979) has suggested that one of the baffling phenomena in current work on intracellular regulation may be explained on the basis of bootstrapping mechanisms operating at both the single-gene and multigene levels. The phenomenon is the ubiquity and multiplicity of signal systems involving cAMP, cGMP, calcium, and protein phosphorylation. These substances serve as intracellular messengers which activate cAMP- and cGMP-dependent protein kinases. In turn the kinases activate other proteins by catalyzing their phosphorylation. Calcium plays a number of roles, but a principal function is to activate phosphodiesterases which break the cyclic nucleotides down into the inactive mononucleotides. According to Kirkpatrick:

> ...at the moment, it is possible to explain any one of a variety of forms of regulation by one form or another of these general mechanisms; but it is difficult to envisage how all of them can be operating simultaneously. A cell will typically have dozens of phosphorylated proteins, when carefully analyzed, and most of these reactions appear to be regulated via cAMP, cGMP and/or intracellular calcium. It is not clear why cyclic pyrimidines and cyclic deoxynucleotides have not been employed to provide more precise control. ... The "bootstrapping" principle helps explain why the existing modes of regulation have been selected. Phosphorylation has been selected as a reversible post-translation modifier of conformation, which can be applied to a variety of proteins without requiring modification of their primary sequence. Instead, the sequence of the protein kinase (or of its regulators) can be varied. It seems clear that a cell with this capability is more amenable to [evolution] than one without it. Such a cell will rapidly develop a variety of protein kinases and phosphatases with variable sensitivity to regulation.

The enhancement of evolutionary amenability in the cyclic nucleotide system is based on specialization degeneracy at both the single-gene and multigene levels. In the usual single-gene case the active site is practically a constant structure. Its protein surround contains the highly replaceable amino acids and serves as the mutation-buffering system. This should be the case for the kinases or its regulators. But the cyclic nucleotide system exaggerates this segregation of active site and mutation buffer to the point where the mutation buffer (the kinase) is completely separated from the principal active entity, namely the phosphorylated protein. This is just the type of channeling of genetic change which belongs to the multigene specialization degeneracy term (Section 10.6). A similar phenomenon also belongs to this category is posttranslation modification through methylation.

Cyclic AMP is sometimes called a second messenger since it is produced by adenylate cyclase principally in response to receptor activation by a hormone or mediator. The separation of first messenger (hormone or mediator) and second messenger is another complicating feature of this system which enhances evolutionary amenability. The extra interface allows for independent changes in the way different cells interpret the same external signals. By increasing the number of signals (hormones, mediators) in the system of signals it is also possible to create a situation in which the first messages from cell to cell can be changed independently. These are both examples of specialization degeneracy. They enhance amenability to evolution by increasing the chance that single genetic events will lead to small changes in structure and function, just as the use of bootstrapped kinases does at the intracellular level. But there is a noteworthy difference. The interface between a first and second messenger allows for independence between purely phenotypic compartments as well as between phenotype and genotype. The multiplication of first messengers has the same effect. In both cases the membrane serves as a buffer which dissipates the effects of hormones and mediators other than those which are recognized by receptors. The resulting increase in independence decreases the amount of modifiability required for purely phenotypic adaptabilities. Thus this is the type of organization which is an advantage to the individual as well as to the lineage.

This form of organization—to be described in the next section as buffer structuring of the phenotype—is possible whenever the process being considered involves a direct interaction among subcompartments of the phenotype in addition to the dependency of these phenotypic compartments on the genotype. Whether the genome is composed of more or less replaceable codons has no visible effect on the phenotype, only on its evolutionary transformability. Thus hitchhiking provides the pathway of evolution. But the organization of the phenome—whether compartments are more or less independent—affects both individual adaptability and evolutionary transformability. Direct selection as well as hitchhiking selection can therefore contribute to its evolution.

Cyclic 3',5'-AMP and -GMP are now known to control nerve impulse activity in certain central nervous system neurons (Liberman *et al.*, 1975; Bloom, 1975; Greengard, 1976). In large neurons of the snail *Helix lucorum* cAMP has a depolarizing effect and cGMP has a predominately hyperpolarizing effect. This has been established by Liberman, Minina, and Shklovsky-Kordy in an elegant series of microinjection experiments. The data fit well to a reaction–diffusion model (Liberman *et al.*, 1982; Conrad and Liberman, 1982b), which I believe supports the extension of "adaptive landscaping" to brain processes. According to the reaction–diffusion hypothesis the pattern of presynaptic input to a neuron initiates a pattern of cAMP production. Mediators released by the presynaptic fibers activate receptors in the postsynaptic membrane, which in turn activate the enzyme (adenylate cyclase) which catalyzes the production of cAMP. The cAMP must diffuse from its point of origin to the kinase which recognizes it. The basic idea is that different patterns of presynaptic input give rise to different diffusion patterns of cyclic nucleotide. How these different diffusion patterns are interpreted depends on the distribution of the kinases on the membrane or at the axon hillock. The kinases are believed to activate gating proteins which control the movement of ions across the membrane. By altering the concentration or locations of the kinases it is possible to alter the way the neuron interprets a pattern of a presynaptic input. The interpretation can also be altered by varying either the phosphatase or binding proteins which control the rate at which cAMP or cGMP diffuses. The distance between points of cyclic nucleotide production and point of action on kinases controls the timing of the response. The concentration of kinases controls the frequency, hence the intensity, of the response. The intriguing point is that this would allow the neuron to serve as a gradually variable analog for an effector cell (such as a muscle cell) which it controls. It is only necessary to bootstrap evolutionarily amenable kinases, phosphatases, or binding proteins. But there is a difference. The same amenability can be harnessed for graceful learning of pattern recognition and motor control tasks. So in this case evolution-enhancing features of the cyclic nucleotide system which initially could only evolve through the hitchhiking mechanism acquire phenotypic value for the individual, totally apart from the value connected with buffer structuring. The situation is similar to that of the immune system, in which the evolution-enhancing features of the immunoglobins confer a selective advantage on the individual. In both these cases an adaptive surface is smoothed, allowing for more effective optimiza-

tion of molecular pattern recognition in the immune system and for more
effective optimization of, say, spatial pattern recognition and motor control
in the (proposed) neuronal example.*

   Another especially interesting, but speculative, example of multigenic
bootstrapping has been brought to my attention by T. Bargiello and J.
Grossfield. They observed that some current theories for control of DNA
readout in eukaryotes involve the geometry of the DNA molecule or of the
DNA–protein complex. By selecting geometries with superfluous features, it
would be possible to buffer the effect of mutations on features of shape
critical for regulation, hence to enhance the evolvability of the regulatory
properties.

10.8. THE BUFFER STRUCTURE PRINCIPLE OF
PHENOTYPIC ORGANIZATION

   Now I want to look more closely at the phenomenon of buffer
structuring. By introducing inexpensive buffers phenotypic compartments
whose adaptabilities are costly but necessary can be made more independent of one another. Their contribution to adaptability is then more
efficient. The introduction of such buffers increases the absolute magnitudes
of the behavioral and anticipation terms equally, thus does not increase the
information-processing power. But it does increase the transformability of
the phenotype by ameliorating the interdependencies among functionally
important compartments.

   In pure buffer structuring the repertoire of the buffer systems does not
include responses specifically suited to absorbing environmental disturbances. Instead it serves solely to dissipate the effects of modifications in
the compartments which do respond specifically, therefore to prevent the
ramification of disturbances in the system. Compartment I handles disturbances in class I. Compartment II handles disturbances in class II. But
the adaptive behavior of I is also a disturbance to II and conversely. To
make I and II as independent as possible, the buffer system b is introduced.
In the case of the genotype–phenotype relation b smoothed out the effects
of genotypic changes on crucial phenotypic compartments. Here b is
smoothing out the effects of changes either of phenotypic compartment I on
phenotypic compartment II or of phenotypic compartment II on I, or both.

   The situation is illustrated in Figure 10.5. Adaptability is highest for a
given total modifiability of the costly compartments when the two modifia-
bility terms, $H(\hat{\omega}_I)$ and $H(\hat{\omega}_{II})$, are large and the two-way independence

*I identify the kinases and gating proteins with the controlling proteins in the evolutionary
learning scheme mentioned in Section 6.2 (see Conrad, 1981).

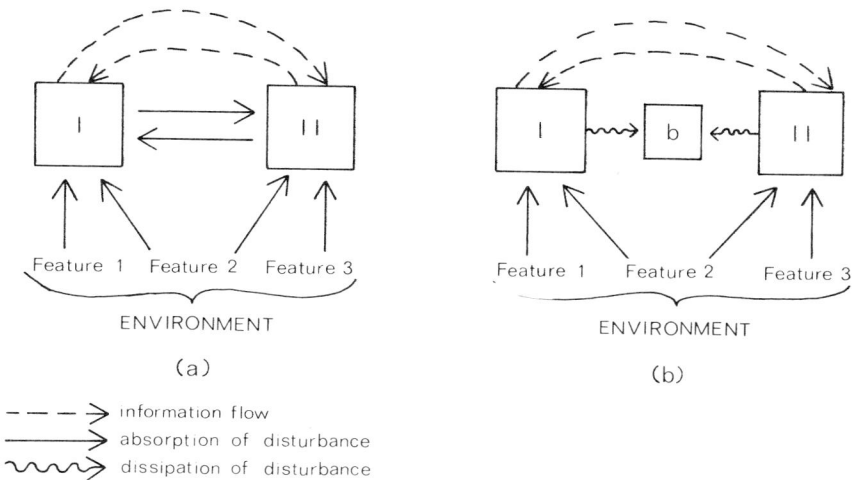

FIGURE 10.5. Buffer structure principle. The insertion of the buffer b between compartments I and II eliminates unnecessary interdependencies between them. As the number of compartments increases, either the number of ways of coupling to the buffer must increase or the number of buffers must increase. The latter situation is the usual one in energy metabolism (see text).

terms are small. This is not possible in Figure 10.5a, since I and II interact directly. The introduction of the buffer system in Figure 10.5b introduces the new modifiability term, $H(\hat{\omega}_b)$, and the new independence terms $H(\hat{\omega}_b|\hat{\omega}_I)$, $H(\hat{\omega}_b|\hat{\omega}_{II})$, and $H(\hat{\omega}_b|\hat{\omega}_I\hat{\omega}_{II})$. All these terms can be high since the organization need not fulfill special requirements imposed by the environment. These independence terms may be small at no great cost, since $H(\hat{\omega}_b)$ is not costly. But the terms $H(\hat{\omega}_I|-)$ and $H(\hat{\omega}_{II}|-)$ can now be high. This is because I and II are isolated and there are no specific functional requirements on b other than that it confer this independence. The terms $H(\hat{\omega}_b)$ and $H(\hat{\omega}_b|-)$ can only make a contribution to adaptability if they exceed the anticipation terms, $H(\hat{\omega}_b|\omega^*)$ and $H(\hat{\omega}_b|\omega^*-)$. In fact these differences should be small, else the buffer system would become another costly form of adaptability. In that case the magnitudes of both $H(\hat{\omega})$ and $H(\hat{\omega}|\omega^*)$ increase. The increase lessens the cost of adaptability by increasing independence.

Note that in the diagram compartment I has responsibility for feature 1 of the environment, compartment II has responsibility for feature 3, and both have responsibility for feature 2. Their modifiabilities and anticipation entropies should be independent as regards features 1 and 3, but should be correlated as regards feature 2. However, the lack of independence is necessary in this case. I and II might also be correlated by information flow between them which coordinates their activities, but again the resulting lack of independence is then desirable.

E. E. Sel'kov (1975, 1979) has developed a model of cellular energy metabolism which serves to illustrate the connection among buffer structuring, transformability, and adaptability. The basic idea is that an optimal scheme for energy metabolism should allow for maximum flexibility in the energy demands of different energy-consuming reactions and allow for this flexibility despite an uncertain distribution of substrates in the external environment. The linchpin element is the universal energy carrier ATP. This is the major product of the degradative reactions and the major energy source for all the consuming reactions. To maintain the independence of these consuming reactions, it is necessary to organize the metabolic structure in such a way that ATP (or energy charge, $ATP + ADP/2$) is as stable as possible. Actually it appears to be maintained so that rhythmic changes in ATP are stabilized. The principal feature which makes this stabilization possible is the existence of pools in which high-energy intermediates can be reversibly deposited. These pools are connected to autocatalytic reactions involving ATP. ATP initiates the reactions leading to the intermediates and is produced by further reaction of these intermediates. By funneling these intermediates into the buffering pool, it is possible to make the reaction quite independent of the variation in the substrate concentration and of the requirements of the consumers. By using many such units to control ATP concentration it is possible to enhance greatly the independence of the consumers and to shift completely to different substrates.

It is important that the cyclic reactions involving ATP always involve a net expenditure of ATP. This is the cost of independence. Some other controls are present which reduce this cost by inhibiting the occurrence of these cyclic reactions when they are not needed. But the independence is conferred by the uncertainty in the size of the buffering pools and it requires energy to dissipate these variations in pool size.

Sel'kov and co-workers have shown that these organizational features are characteristic of a number of basic energy processes, including glycolysis, glycogenolysis, fructolysis, glycerolysis, and oxidative phosphorylation. If the independence of different intracellular energy-consuming processes were not reduced in this way, the independence of different features of compartments of the cell would be compromised. There are many substances other than ATP which are critical for a variety of processes. The same type of buffer pool mechanisms can be used to enhance the independence of these processes, hence of different compartments with which they are associated. The compartments may be at different levels, including levels above the cell. An example is cell size and structure, features associated with the organism level. Changes in cell size and cell number are important mechanisms of growth adaptation in plants and in animal tissues as well. Both are energy-consuming processes. But to the extent that they can be kept independent the amount of adaptability which can be obtained from a given degree of tissue modifiability is increased. A second cross-level example is the genotype–phenotype relation. Buffer structuring of the phenotype increases its gradual transformability in response to genetic variation by increasing the independence of different processes and compartments. Under these circumstances allometric relations become much less severe as constraints on evolutionary adaptation. As with the genetic buffering mechanisms discussed earlier, the structural stability of the developmental process is increased, in general the situation when independence is increased. But a crucial point of difference is that purely genotypic buffering does not decrease the cost of adaptability. It is possible to build the same phenotype, including the same phenotypic adaptabilities, with genes or genetic systems which are more or less amenable to evolution.

The Sel'kov model is interesting since it can be used to show how a single organizational feature can contribute to adaptability, reliability, and transformability, as well as to buffer structuring of different modes of adaptability *per se*. The buffer structuring of different modes of adaptability (such as cell size and number) arises from the independence created by buffering pools. But the buffering pools also contribute to adaptability since they help the cell to dissipate variations in degradative or consuming processes resulting from internal failures. This buffering contributes to transformability since a genetic change often changes the requirements of a consuming reaction. To the extent that different features of a system can be varied independently, the possibility for gradualistic evolution increases.

Processes other than energy processes are also buffer structured. I have already considered the important cross-level example involving first messengers (hormones and mediators) at the organism level and second messengers (cyclic nucleotides) at the cellular level. Here the buffer structuring involves the signal system of the organism. The multiplication of first messengers and the separation of these from the second messengers by the membrane both enhance independence. They buffer structure different modes of adaptability by segregating the signaling systems involved in these modes. This segregating also increases transformability. But there is no contribution to any particular mode of adaptability and no contribution to reliability. Obviously much less dissipation is required for the membrane to filter out the undesired signals than for the buffering pools to absorb variations in the energy input or in the demands of consuming reactions. But the extra chemical complication required is a major energetic cost.

The examples considered—development of immunity, cyclic nucleotide control systems, the interface between first and second messenger systems, the role of the cyclic nucleotide system in learning, and the organization of cellular energy metabolism—all point to a fundamental difference between physiological control mechanisms and control mechanisms in technical systems. A furnace controlled by a thermostat allows a house to be protected from changes in the external temperature. This is a simple feedback system in which the uncertainty of the state of the furnace controls the state of the house. The state of the house is monitored by the thermostat, which sends controlling signals to the furnace. It is usual to think of this type of feedback control as paradigmatic of physiological control. But many more features of biological systems are modifiable. The basic properties of the furnace and house do not change. If they were biological tissues the situation would be quite different. Each component of tissue can adapt to stresses imposed on it. The number of adaptive features is qualitatively different. This is because the basic level of construction—the biochemical level—is also a level at which control is occurring. Even the components at this level can adapt, though through the evolution process (also in the development of immunity and possibly in some forms of CNS learning). As a consequence the buffer structuring of the phenotype is much more important in living systems than would be concluded on the basis of our experience with much less effectively adaptable technical systems. The modifiability of so many features in biological systems would in fact be extremely disadvantageous if it were not for the buffer structuring of the phenotype. Without it new adaptations in any one component would ramify throughout the whole system, in general causing maladaptive changes in other components and disturbing the structure of control. In our present-day technical systems this problem is avoided by simply fixing certain features over a great range of conditions. The components of the furnace and the house are fixed features. By using components which cannot adapt, great independence is achieved and the problem of the components adapting to each other is entirely avoided, but at the expense of greatly restricted adaptability. In the evolution of biological systems and humans the trend has been just the opposite. This is why the final adaptability is vested in the human who constructs and repairs the house.

10.9. BOOTSTRAPPING IS POSSIBLE ONLY IN NONDECOMPOSABLE SYSTEMS

The possibility for the immensely rich direction taken by biological systems is ulimately based on the possibility of bootstrapping of evolutionary amenability. In the opinion of the author this process is the major *link* between living and nonliving nature, yet the possibility for its occurrence is the major *distinction* between living systems and present-day technical systems. The reason for this difference is subtle. One way of appreciating it is to consider why present-day technical systems cannot bootstrap.

Systems cannot bootstrap if they are constructed from standard components, where the number of possible types of components is limited. Technical information-processing systems such as present-day digital computers are of this character. They are constructed from a limited number of types of standard switching elements. Unlike the proteins of biological systems—the real switching components in biology—these standard components cannot be gradually transformed. In fact they cannot be transformed at all once the system is constructed. They are the *given* primitives. The only way a system constructed from such primitives might conceivably be made more amenable to evolution is by organizing the interconnections among them differently. However, systems of standard building blocks have a powerful feature which is incompatible with amenability to evolution. The powerful feature is *structural programmability*: it is possible to effectively design them so that they realize any rule which can be expressed in terms of a computer program.* More precisely, it is possible to find an algorithmic process (A_1) which can be used to hardwire any algorithmic process (A_2) into such systems. In general, however, this means that it is not possible to predict in advance how much the wired-in algorithm A_2 will change if the interconnections among the components and their initial states are changed. If this were possible, it would be possible to answer the question: will the algorithm executed by the original network and the algorithm executed by the modified network both reach similar states when presented with the same input? To reach similar states it is here sufficient to say that they reach any states in arbitrary subclasses of their state sets. However, if this problem could be solved it would be possible to solve the famous halting problem for

---

*A concrete example is the construction of networks of McCulloch–Pitts formal neurons to realize finite-state automata. Suppose that the network is made solely out of formal neurons which fire when the sum of the inputs to them is equal to or greater than 2, that each input line is in one of two states (0 or 1), and that input x_j to the finite automaton is coded by the firing of external input line j, and that all neurons in row j receive inputs from line j. The program of the automaton can be represented by a set of triples $\langle q_i x_j q_k \rangle$, where the transition is from state q_i to state q_k. For simplicity taking the automaton as a semiautomaton (state equals output), the canonical construction algorithm is: for each triple $q_i x_j q_k$ draw a connection between all neurons in column i and the neuron in row j of column k. The state q_i is coded by the firing of a neuron in column i and q_k is coded by the firing of a neuron in column k. This construction algorithm was suggested by a somewhat different one described by Minsky (1967), which should be consulted for details. For further discussion of the notion of structural programmability see Conrad (1974a,b).

Turing machines, which is unsolvable. To establish this it is only necessary to associate the subclasses of states with halt states, that is, states reached by a system executing a defined computation. To decide whether A_2^{orig} (the algorithm executed by the original network) and A_2^{mod} (the algorithm executed by the modified network) ever reach these subclasses it is necessary to have another algorithm A_3 which reaches a state specified as a halt state only if A_2^{orig} and A_2^{mod} do not. Then we will certainly receive a definite answer to the question posed in a finite amount of time, for either A_2^{orig} and A_2^{mod} will reach one of the states specified or A_3 will. Is A_3 a possible algorithm? Suppose that A_3 is applied to itself. Then if A_3 reaches one of the specified states it does not reach one of the specified states, and if it does not reach one of the specified states it does. This is an obvious contradiction, therefore the assumption of the existence of A_3 has been reduced to an absurdity. Since an algorithm A_1 connects the structure of the system to the algorithm it executes, changing this structure will in general change the algorithm it executes; and since it is in general impossible to specify in advance how different the behavior of two algorithms will be it must in general be impossible to specify in advance how different the behavior of two different structurally programmable networks will be.*

Strictly speaking, the halting problem argument only applies to systems with a potentially infinite number of states. Any real system is of course finite. However, what is unsolvable with no finite restriction is in general extremely difficult to solve even when such a restriction is imposed. In fact this corresponds to the common experience with computers. Programs (whether or not hardwired) are very sensitive to slight alterations (other than alterations in parameter). In general it is impossible to put a metric on the expected degree of function change in such systems, therefore impossible to define a concept of gradual transformability. This is why such systems cannot bootstrap amenability to evolution.

---

*For a general discussion of unsolvability see Davis (1958). The unsolvability property discussed here has societal implications when combined with adaptability theory. The replacement of humans by automatic (structurally programmable) information-processing systems decreases $H(\hat{\omega}|\hat{\omega}^*)$ insofar as it enhances prediction and integration. But at the same time it shrinks $H(\hat{\omega}^*)$ since the likelihood that failures or alterations will give rise to functionally acceptable configurations is small. Biological systems ultimately depend on evolution, therefore on $H(\hat{\omega})$, for achieving adaptability. Improvements in information processing ultimately arise through the evolution process or through physiological processes which utilize the same underlying physiochemical principles. Some modern societies are reversing this by attempting to base adaptability primarily on structurally nonprogrammable computing systems. Dramatic reductions in $H(\hat{\omega}|\hat{\omega}^*)$ may be achieved, but these will be counterproductive if $H(\hat{\omega}^*)$ decreases as well.

The distinction between structurally programmable and structurally nonprogrammable systems can be illustrated by returning to the house analogy. A contractor wants to build a variety of different houses, each suitable to a different environment. There are two strategies. First we may give him a limited set of possible components and ask him to develop a different blueprint for constructing each different type of house. This is the way things are actually done in our world of human industry and construction. Allowing the different blueprints to evolve by variation and natural selection is out of the question. The blueprint is really a rule (or program) for building the house. One or two changes in it will nearly always result in the construction of a house which is not well formed—just as one or two changes in a computer program will nearly always lead to its crashing. In general many changes are necessary, implying a dependence on the intelligence of the programmer or designer. The second strategy avoids this. We restrict the contractor to a single basic type of blueprint and allow him an indefinite number of types of components. We allow no changes in this blueprint other than changes in the specifications of the components. The sequence of steps for building the different houses is always the same. If a continuous gradation of component types can be specified, it will be possible to generate a continuous gradation of different realizations of the house.

The second strategy—the one which does not depend on the existence of an intelligence competent to design—is the one used by biological systems. Recall from our discussion of cellular self-reproduction (Section 3.1) that DNA can be thought of as a blueprint for the cell. Subsequences of this DNA (the genes) code for the switching elements (enzymes). There are for all practical purposes an indefinite number of possible switching components which could be specified—at least 20^{300}. If the genes and the genetic systems are organized in such a way that they are gradually transformable, it is possible to change the cell in significant ways without in any basic way altering the sequence of steps required for its construction. What is changed is not the construction rule, but the realization of this construction rule.

The above considerations can be summarized in terms of a trade-off principle: structural programmability is incompatible with amenability to evolution. The sacrifice of decomposability opens up the possibility of efficiently generating a family of gradated realizations of an organism by an evolution process. If the sequence of bases in DNA were the program of the organism, this would not be possible. It is not properly thought of as such a program—at least if one uses the notion of *program* in its proper sense, as a formal expression of the transition function of a system. By changing the sequence of bases in DNA we may change the "program" of an organism, but this does not mean that the DNA sequence is a formal expression of this program. If it were we could not reasonably expect to transform the realizations gradually, for we would again be faced with the self-computability contradictions described earlier. The contradiction is avoided because of the indefiniteness in the number of possible genes. Unlike computer programs, the sequence of amino acids coded by the sequence of bases assumes its function through an energy-dependent folding process. In this respect the difference between biological systems and present-day computers is enormous. In living systems one is working with an indefinite number of emergent primitives, whereas in present-day computing systems one is not. When the number of primitives is allowed to become indefinite, one loses the ability to communicate algorithms to the system effectively, but one gains the possibility of transforming the system gracefully through variation and selection. This is the subtle key to the success of evolutionary search in producing so remarkable a variety of functional life forms.

10.10. RELATION TO OTHER DISCUSSIONS OF EVOLUTION AND DEVELOPMENT

In concluding this chapter I would like to observe that a gradualistic relation between genome and phenome is not only a fact of protein biology, but also well established at morphological levels of organization, for example from Simpson's studies of the evolution of horses (Simpson, 1951). It is probably fair to say that it is a central element in the so-called modern synthetic theory of evolution (cf. Mayr, 1963). The idea is also expressed in the famous topological transformation of D'Arcy Wentworth Thompson, who has shown how different types of fish can be produced through the smooth distortion of a coordinate grid (Thompson, 1917). Alternative, saltationist conceptions of evolution have also been put forth, in particular by Goldschmidt (1940). But it is clear from probabilistic considerations that genomic saltations are unsupportable as contributors to the course of evolution. Qualitative phenotypic changes in response to continuous genotypic change cannot be excluded on these grounds. Genes either control rates of reactions (including assembly processes) or control other genes which control rates of reactions. A continuous change in these rate parameters may of course give rise to a discontinuous change in the state-to-state behavior of the system. In recent times this idea has been discussed by Thom (1970) in the context of catastrophe theory. The presumption is that qualitatively distinct, archetypal forms arise through such continuous genetic evolution. However, even if this is so, most evolution involves quantitative radiation of given types and the gradual changeability of the rate constants is still assumed.

Some paleontologists have recently reemphasized the fact that the fossil record often appears to exhibit stretches in which change is minor and other periods in which changes are relatively fast, giving the appearance of a punctuated equilibrium (Gould, 1977). But this does not contradict the requirement that increases in fitness arise through small changes in genotype. This requirement is still most likely to be met when the relationship between genotype and phenotype allows for gradualism insofar as possible. It is important to recognize that gradualism in these paleontological discussions refers phenomenologically to rate of evolution, not to the degree of phenotype change in response to genotype change. But the fossil record is most likely to exhibit nongradual aspects (rapid rates of evolution) when the relationship between genotype and phenotype has a gradual aspect. According to adaptability theory the periods of rapid change in the fossil record are the result of crises which are inevitably concomitant to the minimal tendency of adaptability, with the capacity to respond rapidly to these crises being in large part based on the gradual aspect of the genotype–phenotype relation.

The central element in the present argument is that the essential gradualistic relation of genotype and phenotype at both the protein and organism levels is mediated by certain definite forms of organization which are themselves the products of selection. This selection is not secondary selection in the weak sense that selection for optimum mutation rate would be secondary selection. A system with a nonoptimum mutation rate will evolve at a rate which is less than it could otherwise be, whereas a system without explicit mechanisms for modulating gene variation will not evolve at all. Such mechanisms are not free. They must be paid for in terms of an organism and a developmental process which is more wasteful and more complex than it might otherwise be, therefore in terms of an expenditure of free energy and an increase in the difficulty of predicting the particular effects of any genetic change.

REFERENCES

Allen, P. M., and W. Ebeling (1983) "On the Stochastic Description of a Predator–Prey Ecology," *BioSystems* (in press).

Bargiello, T., and J. Grossfield (1979) "Biochemical Polymorphisms: The Unit of Selection and the Hypothesis of Conditional Neutrality," *BioSystems 11*, 183–192.

Bloom, F. E. (1975) "The Role of Cyclic Nucleotides in Central Synaptic Function," *Rev. Physiol. Biochem. Pharmacol. 74*, 1–103.

Britten, R. J., and E. H. Davidson (1971) "Repetitive and Non-repetitive DNA Sequences and a Speculation on the Origins of Evolutionary Novelty," *Q. Rev. Biol. 46*(2), 111–138.

Conrad, M. (1974a) "The Limits of Biological Simulation," *J. Theoret. Biol. 45*, 585–590.

Conrad, M. (1974b) "Molecular Automata," pp. 419–430 in *Physics and Mathematics of the Nervous System*, ed. by M. Conrad, W. Güttinger, and M. Dal Cin. Springer-Verlag, Heidelberg.

Conrad, M. (1977) "Evolutionary Adaptability of Biological Macromolecules," *J. Mol. Evol. 10*, 87–91.

Conrad, M. (1978) "Evolution of the Adaptive Landscape," pp. 147–169 in *Theoretical Approaches to Complex Systems*, ed. by R. Heim and G. Palm. Springer-Verlag, Heidelberg.

Conrad, M. (1979a) "Bootstrapping on the Adaptive Landscape," *BioSystems 11*, 167–182.

Conrad, M. (1979b) "Mutation–Absorption Model of the Enzyme," *Bull. Math. Biol. 41*, 387–405.

Conrad, M. (1981) "Algorithmic Specification as a Tool for Computing with Informal Biological Models," *BioSystems 13*, 303–320.

Conrad, M. (1982) "Natural Selection and the Evolution of Neutralism," *BioSystems 15*, 83–85.

Conrad, M., and E. A. Liberman (1982) "Molecular Computing as a Link between Biological and Physical Theory," *J. Theor. Biol. 98*, 239–252.

Conrad, M., and M. M. Rizki (1980) "Computational Illustration of the Bootstrap Effect," *BioSystems 13*, 57–64.

Conrad, M., and M. Volkenstein (1981) "Replaceability of Amino Acids and the Self-Facilitation of Evolution," *J. Theor. Biol. 92*, 293–299.

Cox, E. C., and T. C. Gibson (1974) "Selection for High Mutation Rates in Chemostats," *Genetics 77*, 169–184.

Davis, M. (1958) *Computability and Unsolvability.* McGraw-Hill, New York.

Goldschmidt, R. (1940) *The Material Basis of Evolution.* Yale University Press, New Haven, Connecticut.

Gould, S. J. (1977) *Ontogeny and Phylogeny.* Belknap Press of Harvard University Press, Cambridge, Massachusetts.

Greengard, P. (1976) "Possible Role for Cyclic Nucleotides and Phosphorylated Membrane Proteins in Postsynaptic Actions of Neurotransmitters," *Nature 260*(5547), 101–108.

Kimura, M. (1968) "Evolutionary Rate at the Molecular Level," *Nature 217*, 624–626.

King, J. L., and T. H. Jukes (1969) "Non-Darwinian Evolution," *Science 164*, 788–798.

Kirkpatrick, F. H. (1979) "Commentary," *BioSystems 11*, 181–182.

Koshland, D. E. (1963) "The Role of Flexibility in Enzyme Action," *Cold Spring Harbor Symp. Quant. Biol. 28*, 473–482.

Lewontin, R. C. (1974) *The Genetic Basis of Evolutionary Change.* Columbia University Press, New York.

Liberman, E. A., S. V. Minina, and K. V. Golubtsov (1975) "The Study of the Metabolic Synapse. I. Effect of Intracellular Microinjection of 3′, 5′-AMP," *Biofizika 20*, 451–456.

Liberman, E. A., S. V. Minina, N. E. Shklovsky-Kordy, and M. Conrad (1982) "Microinjection of Cyclic Nucleotides Provides Evidence for a Diffusional Mechanism of Intraneuronal Control," *BioSystems 15*, 127–132.

Mayr, E. (1963) *Animal Species and Evolution.* Harvard University Press, Cambridge, Massachusetts.

Minsky, M. (1967) *Computation: Finite and Infinite Machines.* Prentice-Hall, Englewood Cliffs, New Jersey.

Nei, M. (1975) *Molecular Population Genetics and Evolution.* North Holland/American Elsevier, New York.

Perutz, M. F. (1962) *Proteins and Nucleic Acids.* Elsevier, Amsterdam.

Roitt, I. (1974) *Essential Immunology*, 2nd. ed. Blackwell Scientific Publications, Oxford, England.

Rossman, M. G., A. Liljas, C. Branden, and L. J. Banaszak (1975) "Evolutionary and Structural Relationships among Dehydrogenases," pp. 63–102 in *The Enzymes*, Vol. 11, 3rd ed., ed. by P. D. Boyer. Academic Press, New York.

Sel'kov, E. E. (1975) "Stabilization of Energy Charge, Generation of Oscillations and Multiple Steady States in Energy Metabolism as a Result of Purely Stoichiometric Regulation," *Eur. J. Biochem. 59*, 151–157.

Sel'kov, E. E. (1979) "The Oscillatory Basis of Cell Energy Metabolism," pp. 166–174 in *Pattern Formation by Dynamic Systems and Pattern Recognition*, ed. by H. Haken. Springer-Verlag, Berlin.

Simpson, G. G. (1951) *Horses*. Oxford University Press, Oxford, England.

Strobeck, C., J. Maynard Smith, and B. Charlesworth (1976) "The Effects of Hitchhiking on a Gene for Recombination," *Genetics 82*, 547–558.

Thom, R. (1970) "Topological Models in Biology," pp. 89–116 in *Towards a Theoretical Biology*, Vol. 3, ed. by C. H. Waddington. Edinburgh University Press, Edinburgh.

Thompson, D'A. W. (1917) *On Growth and Form*. Cambridge University Press, Cambridge, England.

Van Valen, L. (1973) "A New Evolutionary Law," *Evol. Theor. 1*, 1–30.

Volkenstein, M. (1979) "Mutations and the Value of Information," *J. Theor. Biol. 80*, 155–169.

Waddington, C. H. (1957) *The Strategy of Genes*. George Allen and Unwin, London.

Wills, C. (1976) "Production of Yeast Alcohol Dehydrogenase Isoenzymes by Selection," *Nature 261*, 26–29.

Wright, S. (1932) "The Roles of Mutation, Inbreeding, Cross-Breeding, and Selection in Evolution," *Proc. Sixth Int. Cong. Genet. 1*, 356–366.

11

Compensation in Organisms and Populations

The antecedent chapters justify a generalization about compensating changes in biological adaptability. This generalization, which I call the principle of biological compensation, is: *changes in the adaptability of one level or unit of organization tend to be compensated by opposite changes in the adaptability of some other level or unit.* The changes in adaptability may be due to changes in the modifiability of units (e.g., culturability, developmental plasticity, gene pool diversity), to changes in the independence of these different modes of modifiability, to changes in the ability to anticipate the environment, or to changes in the indifference to the environment. The latter involve changes in the degree to which regions of the environment or features of them are avoided. Changes in the ability to anticipate the environment may require increases in the absolute magnitudes of the modifiability, independence, and anticipation terms. The total adaptability will tend to increase or decrease if the uncertainty of the environment changes. While the adaptability structure of a biological system is the product of its entire past evolution, the tendency itself is independent of the time scale used to determine the uncertainty of the environment. According to the efficiency arguments, there should always be a tendency for the total adaptability to move in the same direction as the uncertainty of the environment.

The statement is a generalization about compensating changes in the adaptability structure of biological systems. By itself it does not specify the detailed changes which will take place in the course of evolution. It does provide a criterion for specifying which patterns of change are possible and which are not. To know the actual patterns something must be said about the costs and advantages of different types of adaptability given the characteristics of the environment and the anatomical and physiological characteristics of the biological system.

This chapter is devoted to exploring and exemplifying the principle of compensation insofar as it applies to organisms and populations. The starting point is the concept of homeostasis. The formalism of adaptability theory allows for a precise definition of homeostasis as well as for a precise statement of the conditions required to achieve it. A second starting point is in the environment. The adaptability theory formalism also allows for analogs of the classical definitions of the niche. Once these definitions are constructed, it is possible to make useful statements about the role of niche breadth in compensation. The real heart of the chapter is the discussion of patterns of adaptability in organisms and populations. The basic question is, what factors influence the allocation of homeostatic and other adaptability mechanisms in organisms and populations, given characteristics of the niche and characteristics of their organization? Fundamental organizational constraints of biological systems (along with the general connection between organizational complexity and adaptability) do suggest generalizations about these patterns. Section 11.6 reviews adaptability structures exhibited by different basic types of life plans (protista, plants, animals) in the light of these generalizations. Constraints on allowable patterns of adaptability also have a reciprocal and in the long run decisive influence on the evolution of organizational constraints. The most dramatic example, and the one most pursued, is the evolution of the brain. Section 11.9 sums up the situation as regards patterns of adaptability in terms of a simple pictorial representation which I call a vector diagram.

11.1. HOMEOSTASIS

That the conditions for life fall within a restricted range has been most fruitfully expressed in the famous dictum of Claude Bernard (cf. 1949) that the maintenance of the stability of the internal milieu is the condition for a free life.* The "wisdom" in Cannon's classic study, *The Wisdom of the Body* (1939), refers to this capacity of organisms to regulate their internal milieu.

---

*The formulation has an interesting implication for the structure of biological theory. There are two extremes of emphasis. One emphasizes organizational mechanisms (such as negative feedback) and seeks to explain life phenomena in terms of this organization. The second emphasizes the physiochemical conditions and seeks explanations in terms of these. The two views have different practical consequences, particularly for analyses of the etiology of disease. The linguistically sensitive reader will note that Bernard's statement necessarily incorporates both views. The significance of the special physiochemical conditions is that it allows the special organization which maintains them to persist; the significance of the special organization is that it allows the special physiochemical conditions which maintain it to persist. A conceptually satisfactory understanding of biological nature cannot be based on a separate consideration of informational and physiochemical organization.

The mechanism of such wisdom largely involves intricate error-correcting feedbacks, an idea pursued in Wiener's *Cybernetics* (1961). However, the feedback loops need not be mechanistically explicit and the maintained internal condition need not be a steady state. It might be a dynamic condition such as heartbeat or breathing, or the sequence of events underlying the cell cycle. Waddington (1957) has shown that developmental systems exhibit a generalized form of homeostasis, which he calls homeorhesis. What is resistant to disturbance is not the state of the internal milieu but rather the trajectory of development. Bernard's generalization still holds: too much deviation from a peculiar class of states or sequences of states is incompatible with the peculiar freedoms exercised by living systems.

The idea of homeostasis—both the stability of the milieu and the wisdom required to maintain it—is expressed precisely and quantitatively by the evolutionary tendency equation. Suppose that we focus our attention on a particular compartment, say compartment c_{rs}. In this case equation (6.5) may be written

$$H(\hat{\omega}_{rs}) + \left[\sum_{\substack{i \neq r \\ j \neq s}} H(\hat{\omega}_{ij}) - \sum_{i,j} H_e\left(\hat{\omega}_{ij} | \prod_h \hat{\omega}_{h0}^* \right) + \sum_h H_e\left(\hat{\omega}_{h0}^* | \prod_{i,j} \hat{\omega}_{ij} \right) \right]$$

$$\rightarrow \sum_h H_e(\omega_{h0}^*) \qquad (11.1)$$

The behavioral uncertainty of compartment c_{rs} can be made arbitrarily small by increasing the magnitude of the expression in the brackets. Making $H(\hat{\omega}_{rs})$ small corresponds to increasing the degree to which the trajectory of c_{rs} is controlled. Increasing the magnitude of the expression in the brackets corresponds to increasing the potential wisdom of the controllers. This means either increasing the behavioral uncertainty of the other compartments, increasing the ability to anticipate, or increasing the indifference. These are the possible forms of "wisdom."

As a concrete example, suppose that we are dealing with the behavior of the organism in terms of its genome and phenome. In this case $\hat{\omega}_{rs} = \hat{\omega}_{r1}$. $H(\hat{\omega}_{r1})$ small means that the pattern of gene action is maintained constant despite environmental disturbance. This is the dynamic form of homeostasis which Waddington calls homeorhesis. There are undoubtedly numerous detailed mechanisms, ranging from enzyme activation and inhibition at the biochemical level, through cellular and physiological regulatory mechanisms analogous to those operating in the mature organism, to regulation or protection provided by a parent or an encasing shell. These mechanisms could be specifically expressed in equation (11.1) by specifying the value of each important term. Such specification would automatically specify all the control interdependencies within the system. What is really more interesting, however, is to recognize general patterns of interdependency and compensation.

I believe that the first person to recognize with real clarity the importance of behavioral uncertainty in regulation was W. Ross Ashby (1956). The idea is of course present in Shannon's fundamental identity, but Ashby recognized explicitly that regulation of one component of a system meant that variety was necessary elsewhere in the system. He called this the *Principle of Requisite Variety*. Equation (11.1) expresses this principle in the context of the hierarchical and compartmental structure of biological systems. Considerations of efficiency suggest that the required uncertainties are bounded not only from below, but from above as well. Information processing considerations suggest that the functionally distinct states which contribute to the requisite variety should be distinguished from the informationally distinct states which allow them to be effectively used for control. The structure of the equation exposes the importance of indifference and most especially the importance of the dependency structure of the system. Homeostasis is a relative concept. Different compartments may mutually regulate one another relative to different types of disturbances, or some features of a compartment may be controlling relative to outside compartments while other features may be controlled. In the next sections we use the adaptability theory formalism to form a picture of biological control unbiased by this favored compartment point of view.

11.2. THE NICHE AND ITS ROLE IN COMPENSATION

DEFINITION. The potential niche of compartment c_{rs} (usually an organism or population) is the set of compartments $\{c_{mn}|(m, n) \neq (r, s)\}$ such that $H(\hat{\omega}_{mn}) > H(\hat{\omega}_{mn}|\hat{\omega}_{rs})$. The realized niche is the set of compartments $\{c_{mn}|(m, n) \neq (r, s)\}$ such that $H(\omega_{mn}) > H(\omega_{mn}|\omega_{rs})$. (The c_{mn} can include biotic compartments and regions of the environment. The biotic compartments cannot include c_{rs} or its subcompartments. If c_{mn} is a region of the environment $\hat{\omega}_{mn} = \hat{\omega}_{h0}^*$ for some h.)

The niche of an organism or population is thus the set of biotic and abiotic compartments in its environment to which it is not indifferent. The set of compartments to which it is actually sensitive is the realized niche. The largest set to which it is potentially sensitive is the potential niche. Thus if the organism (or type of organism) is potentially sensitive to events in more regions of space in which it can persist indefinitely, it has a potentially broader niche. [If it could not continue to persist $H(\hat{\omega}_{mn}|\hat{\omega}_{rs})$ could certainly not continue to be less than $H(\hat{\omega}_{mn})$.]

It should be recalled that the potential entropies are defined by varying over all possible environments. The organism might be able to live in a less uncertain environment which has some compartments (organisms, regions) not included in the most uncertain environment. In that case the potential niche is not coextensive with all the possible niches. However, it would be coextensive for a type of organism or a population since the most uncertain environment would always include one region for each tolerable environment (provided that using an arbitrary number of regions is allowed). The definitions could be further refined to distinguish between different features of the compartments, in particular between different macroscopic properties of the abiotic regions. From the formal point of view these different features could be treated as different compartments.

A measure of potential *niche breadth* is given by $1/H(\Pi\hat{\omega}_{mn}|\hat{\omega}_{rs})$. Niche breadth then decreases if the organism or species is indifferent to more regions and biotic compartments. It is also possible to define notions of niche binding and external specialization. The *niche binding* increases as the difference between the potential and realized niche breadths increases. This happens if $\{1/H(\Pi\hat{\omega}_{mn}|\hat{\omega}_{rs})-1/H(\Pi\omega_{mn}|\omega_{rs})\}$ increases. Under these circumstances there are more possibilities for moving the organism or population to new environments. Its restriction to a narrow realized niche is presumably a matter of history. The *external specialization* increases as the breadth of the potential niche decreases. This may be due to highly specific adaptation, in which case it is as advantageous as it can be, or it may be due to lack of adaptation to other potential environments, in which case it is in no way advantageous.

As discussed in Section 5.8, external specialization also increases if the organism requires more nutrients or can use fewer sources of food. Since the c_{mn} include biotic compartments, the magnitude of the niche breadth (but not the niche binding) will reflect the absolute magnitudes of the behavioral uncertainty terms. This can be eliminated by replacing the $\hat{\omega}_{mn}$ by the $\underline{\hat{\omega}}_{mn}$.

We can now state the principle of compensation as it applies to niches: if the behavioral uncertainty and anticipation terms are fixed, but the uncertainty of the environment increases, then the potential niche breadth of the organism or population decreases. The niche binding also decreases. To prove this it is only necessary to inspect the evolutionary tendency equation as written for the organism or population of interest. Thus if c_{rs} is the object of interest

$$H(\hat{\omega}_{rs})+H\left(\hat{\omega}_{rs}|\Pi\hat{\omega}_{h0}^{*}, \prod_{n \neq 0} \hat{\omega}_{mn}\right)+\left\{ \sum H_{e}(\hat{\omega}_{h0}^{*}|\hat{\omega}_{rs})+ \sum_{n \neq 0} H_{e}(\hat{\omega}_{mn}|\hat{\omega}_{rs})\right\}$$

$$\rightarrow \sum H_{e}(\omega_{h0}^{*})+ \sum_{n \neq 0} H_{e}(\omega_{mn}) \tag{11.2}$$

where the abiotic transition schemes, $\hat{\omega}_{h0}$, are explicitly separated from the biotic ones and the sums are taken over all indices except r and s. If the only way of compensating an increase of the terms on the right (the biotic and abiotic environmental uncertainty) is through the indifference term, the terms in the curly brackets must decrease, implying that the system's potential for occupying regions of space decreases or that the potential number of features of the environment to which it can be sensitive decreases. To the extent that the actual uncertainty of the environment can decrease without the actual indifference increasing, the realized niche is smaller and the binding to this niche increases. Note that if equation (11.2) is summed over all values of the indices, it may not be identical to the evolutionary tendency equation for the biota as a whole. The reason is that $\Sigma H(\hat{\omega}^*|\hat{\omega}_{ij})$ may be larger than $H(\hat{\omega}^*|\Pi\hat{\omega}_{ij})$. In this case the behavior of the community would give more information about the behavior of the environment than would the behaviors of all the populations taken individually.

An alternative formulation is: if external specialization increases, the behavioral uncertainty and ability to anticipate can decrease (assuming that the uncertainty of the environment stays the same or decreases). However, it should be born in mind that an increase in external specialization does not mean an increase in adaptedness to the environment. An increase in degree of adaptation could occur without an increase in external specialization. But this would usually require a rather fundamental evolutionary development. Increase in external specialization should *allow* for an increase in degree of adaptation even in the absence of any such development. This is because it should be easier to fit an organism specifically to a narrower than to a broader niche, and by definition increase in external specialization means niche narrowing.

The potential niche corresponds to Hutchinson's fundamental niche, defined in terms of a hyperspace each of whose axes represents either biotic or abiotic environmental variables (Hutchinson, 1957). The fundamental niche is the maximum hypervolume whose projections on all the axes are tolerable. Hutchinson's realized niche can be thought of in terms of the actual volume occupied. This corresponds to the realized niche defined above. To translate the adaptability theory formulation into the hyperspace formulation it would be necessary to construct a space for each compartment (or for features of the partial states of these compartments if this is more natural). But as we have not assumed a continuum of states, the imagery of a hypervolume is in general not entirely accurate.

The adaptability theoretic and hyperspace definitions of the niche differ in one crucial respect. The hyperspace definition is strictly organism-dependent, whereas the adaptability theory definition is only partially so. Because it is defined in terms of components of the evolutionary tendency equation, it has logical connections to other components of adaptability, both within the organism and in its environment. In the next chapter I exploit this key point by considering the environment's adaptability to the influences of the biota as well as the biota's adaptability to the influences of the environment.*

11.3. FACTORS AFFECTING THE ALLOCATION OF THE ENTROPIES

In Chapter 5 a major concern was to use the theory to classify the major forms of adaptability (cf. Tables 5.2 and 4.1). Now I want to use the law of compensation to say something about the actual allocation of these adaptabilities in organisms and populations. This is a matter of costs and advantages. These are determined by the particularities of the organism, by its anatomy and physiology, and by the particularities of the environment. The situation may initially appear hopelessly complex, yet important generalizations can be teased out of this complexity.

Let $\bar{U}$ be the average efficiency (or fitness) of the population. The average is taken over the efficiencies in each possible environmental situation, weighted by the probability of that situation. The ensemble of situations is the same ensemble which enters into the definition of adaptability. We regard efficiency as a primitive concept of biology in the sense discussed in Chapter 7. That is, thermodynamically grounded statements can be made about the connection between changes in efficiency and changes in adaptability, but efficiency itself is a path-dependent quantity, dependent on many factors.

The principal factors are listed in Table 11.1. Each of these can be characterized by some measure or index. The diversities (behavioral diversity, topographic and biotic diversity of the niche, diversity of niche states) can all be characterized in terms of appropriate entropy measures. The internal specialization of a population, organism, or cell increases as the number of different types of components increases and as their relative frequencies become more inequitable (see the end of this section for a formal definition of this measure). The thermodynamic richness of the niche

---

*The adaptability theory definition thus avoids the criticism that it precludes in principle the possibility of making any statements about the niche as an ecosystem property. It might be noted that von Uexkull (1926) used the paradigm of function circles to think about ecosystems. Recently Patten (1978) has proposed a dual model of the niche which formalizes this notion. The niche is defined in terms of the potential input sequences to a system and in terms of its potential output sequences. If I understand Patten correctly, the possible niche structure of an ecosystem should in principle be given by the possible consistent pairs of such sequences. In essence, these would correspond to stable function circles.

TABLE 11.1
Principal Factors Affecting Efficiency

A. Organism-dependent factors
  1. Adaptability
  2. Internal specialization (number of component types and disproportionality of their occurrence)
  3. Behavioral diversity
  4. Biomass turnover per unit volume[a]
B. Niche-dependent factors
  1. Macroscopic parameters of the niche (e.g., temperature, salinity)
  2. Thermodynamic richness of the niche (e.g., free energy, nutritive quality)
  3. Structural (topographic) diversity of the niche
  4. Biotic diversity of the niche
  5. Diversity of niche states
  6. Range and timing of environmental disturbances

[a]Must be considered during periods of population growth or decline.

can be described in terms of the average free energy available to the population occupying this niche and in terms of the diversity of nutritive factors. The range of environmental disturbance can be specified by specifying the range of variation of the environmental parameters. The timing of environmental disturbances could be specified by specifying the distribution of correlation times for the environment (using correlation time in the sense defined in Section 4.8).

The heuristic hypothesis is that the efficiency of a population is, to a reasonable approximation, a function of the factors listed in Table 11.1 (as quantified by the measures indicated above). Symbolically

$$\bar{U} = \bar{U}\left(\mathcal{Q}_0,\ldots,\mathcal{Q}_4, \mathcal{Q}_{-1}, \mathcal{C}_0,\ldots,\mathcal{C}_4, \mathcal{E}_1,\ldots,\mathcal{E}_n\right) + K \qquad (11.3)$$

where K is a constant, included since only changes in efficiency are important. The $\mathcal{Q}_s$ ($s = 0,\ldots,4$) are the behavioral and anticipation contributions to the adaptability from the level s, $\mathcal{Q}_{-1}$ is the indifference component, the $\mathcal{C}_s$ are the internal specializations of a typical compartment at level s in terms of its subcompartments at the next lower level, and the $\mathcal{E}_i$ include niche-dependent parameters and biomass per unit available energy. Level 4 is the population level, 3 the organism level, 2 the organ level, 1 the cell level, and 0 the gene level. The behavioral and anticipation contributions are given by the pair

$$\mathcal{Q}_s = \left\langle \sum_r H_e\left(\hat{\omega}_{rs}\right), \sum_r H_e\left(\hat{\omega}_{rs} \mid \bigcap \hat{\omega}_{h0}^*, \bigcap_{n \neq 0} \hat{\omega}_{mn}\right) \right\rangle \qquad (11.4)$$

The indifference, which must be for the complete system, is given by

$$\mathcal{Q}_{-1} = \sum_h H_e(\hat{\omega}_{h0}^*|\hat{\omega}_{i4}) + \sum_{n \neq 0} H_e(\hat{\omega}_{mn}|\hat{\omega}_{i4}) \qquad (11.5)$$

where i labels the population. Since it is reasonable to assume that the adaptabilities of different compartments and levels are independent so far as costs are concerned, equation (11.3) can also be written as

$$\bar{U} = \sum_i \bar{U}_i(\mathcal{Q}_i, \mathcal{C}_0, \ldots, \mathcal{C}_4, \mathcal{E}_1, \ldots, \mathcal{E}_n) + K \qquad (11.6)$$

where $\bar{U}_i$ are component efficiency functions.

The allocation problem is: given the $\mathcal{C}_s$ and $\mathcal{E}_i$, what distribution of the $\mathcal{Q}_i$ maximizes $\bar{U}$? A reasonable assumption is that internal specialization, as an index of the complexity, changes much more slowly than the adaptability, compatible with these constraints. However, one can also ask what kind of changes might take place in the $\mathcal{C}_s$ given a specification of the $\mathcal{E}_i$ and of the environmental uncertainty. In other words, the short-run question concerns the evolution of patterns of adaptability. The long-run question concerns the evolution of patterns of internal specialization.

The detailed form of the efficiency function is of course not known, nor are gain and loss functions for specific forms of adaptability given the internal specializations of the various levels known. In the absence of a systematic theory of adaptability it really could not be expected that field workers or experimentalists would be motivated to collect such data. However, it is still possible to obtain a useful qualitative understanding of the spectrum of adaptabilities in different major life plans. This is because well-founded statements can be made about the qualitative dependence of $\bar{U}$ on fundamental constraints inherent in biological organization.

MEASURE OF INTERNAL SPECIALIZATION. The diversity of compartment c_{ij} in terms of its subcompartments at level $j-1$ is given by

$$\mathcal{D}_{ij} = \sum_r \rho[c_{k(j-1)}]_r \log \rho[c_{k(j-1)}]_r \qquad (11.7)$$

where $\rho[c_{k(j-1)}]_r$ is the proportion of subcompartments of type r. The maximum value of $\mathcal{D}_{ij}$, to be denoted by $\mathcal{D}_{ij}(\max)$, occurs when all the $\rho[c_{k(j-1)}]_r$ are equal. The disproportionality will be defined as

$$\mathcal{S}_{ij} = \mathcal{D}_{ij}(\max) - \mathcal{D}_{ij} \qquad (11.8)$$

The pair of indices

$$\mathcal{C}_{ij} = \langle \mathcal{D}_{ij}(\max), \mathcal{S}_{ij} \rangle \qquad (11.9)$$

describes the internal specialization. The addition of more types of components (all else constant) always increases the internal specialization. If the number of components is constant, any decrease in the evenness of their occurrence also produces an increase.

11.4. THE CONNECTION BETWEEN ADAPTABILITY AND COMPLEXITY

Before discussing the fundamental constraints on $\overline{U}$ it is worth exploring the general connection between the degree of interdependence of the parts of a system and the cost of adaptability. The connection is quite natural in adaptability theory. But interdependence is much harder to measure directly than the number of components and the frequency of their occurrence. Thus it will also be useful to consider the connection between interdependence and internal specialization.

THEOREM (cost of interdependence). The cost of adaptability in terms of efficiency increases if the interdependence among components increases, provided that the absolute magnitudes of the behavioral and anticipation entropies do not increase.

ARGUMENT. The argument is similar to that used to establish the self-consistency principle of hierarchical adaptability theory (Section 6.9). From the definition of effective entropy (Section 5.5)

$$\sum H_e(\hat{\omega}_{rs}) = f \sum H(\hat{\omega}_{rs}) + \text{independence terms} \qquad (11.10)$$

and

$$\sum H_e(\hat{\omega}_{rs} | \Pi \hat{\omega}_{h0}^*, \Pi \hat{\omega}_{mn}) = f \sum H(\hat{\omega}_{rs} | \Pi \hat{\omega}_{h0}^*, \Pi \hat{\omega}_{mn}) + \text{independence terms}$$

$$(11.11)$$

where f is a normalizing constant, the c_{rs} are all the compartments of interest (say a population and all its subcompartments), and the c_{mn} are all the biotic compartments in the environment of c_{rs}. (If the c_{rs} include all biotic compartments, the $\Pi \hat{\omega}_{mn}$ disappears.) The degree of dependence may be defined as

$$\Theta = \frac{\Sigma H(\hat{\omega}_{rs})}{\Sigma \text{ independence terms}} \qquad (11.12)$$

This increases if one or more of the independence terms decreases.* If Θ increases, but the indifference and environmental uncertainty are constant, the system must compensate by increasing $\Sigma H(\hat{\omega}_{rs})$ or by decreasing $\Sigma H(\hat{\omega}_{rs}|\Pi\hat{\omega}_{h0}^{*}, \Pi\hat{\omega}_{mn})$. In the former case the extent of modification which the system must undergo to support the required adaptability increases; in the latter case anticipation improves. According to the evolutionary tendency argument (Section 6.1) both these types of compensation increase the cost of adaptability in terms of efficiency. There is one exceptional situation. According to the reliability theorem of Section 9.1, an increase in the magnitude of both the behavioral and anticipation terms may be concomitant to an increase in the reliability with which information about the environment is transmitted or processed; it may also be concomitant to an increase in the computational power of the system. In this case the increase in interdependence would allow for the difference

$$\sum H_e(\hat{\omega}_{rs}) - \sum H_e(\hat{\omega}_{rs}|\Pi\hat{\omega}_{h0}^{*}, \Pi\hat{\omega}_{mn}) \tag{11.13}$$

to become greater, in which case the cost in terms of efficiency may go down.

Interdependence is connected to centralization. A system whose components are highly interdependent is centralized. But how is centralization or decentralization to be measured? One way is to perturb the organism or society. If some perturbations localized to particular components have significant global effects on all components it is sure that we are not dealing with a collection of independent subsystems acting in parallel.

Another way to assess interdependence is through structure. The peculiarities of flexibly constrained dynamic systems are too individual for exception-free statements. Nevertheless there are important principles. The first is the *diversification principle*. If the number of component (say cell) types increases, it is possible for each component to be more specialized for particular tasks. Any such specialized type must depend on other types to perform the task which it no longer performs. If the variety of interactions and interdependences in such a system do not increase, it will not be able to collect on the investment required to produce more types. The second principle is the *linchpin principle*. Rare or unique components often play a critical structural or functional role. Thus the addition of such components often indicates more of an increase in interdependence within the system than does the addition of new types of component in equitable proportions.

---

*Note that Θ is basically the opposite of the degree of autonomy [equation (8.1)] used to establish the cross-correlation between adaptability theory and dynamics.

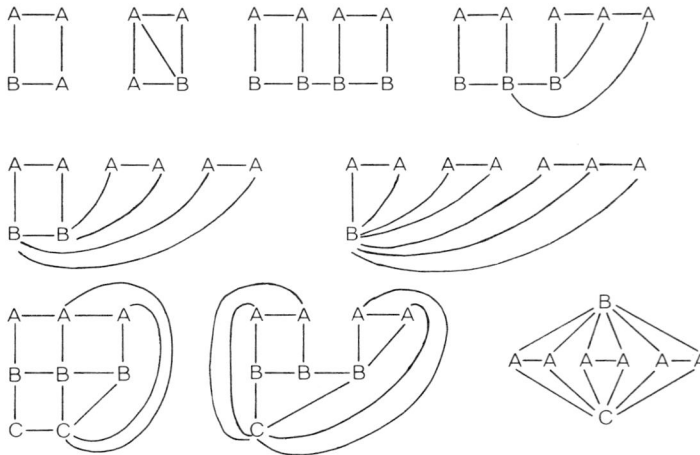

FIGURE 11.1. Interdependence structures. Different letters stand for different types of components and lines between letters represent interactions among components. The assumption is that one of each type of component interacts with one of each other type and, if possible, one of its own type. All the components belonging to the same system must be connected either directly or indirectly.

The point is familiar from experience with machinery. The typewriter on which I am working at the moment consists of components, some of which are used repetitively (such as standard springs and screws), some of which are used repetitively with only slight variation (such as the keys), and some of which are unique (such as the platen). The functioning of any two keys may be independent, but all are dependent on the platen.

It is for this reason that internal specialization provides a better index of degree of interdependence than does diversity standing alone. The situation is illustrated schematically in Figure 11.1. According to the diversification principle at least some components of a given type interact with some components of any other given type and with some components of their own type as well (unless some interaction is precluded by a constraint). The linchpin effect appears automatically as long as not all components interact with each other. Then many components will necessarily interact with any rare component. Diversification and linchpinning both increase the number of interdependences among the component parts of a system. As linchpinning increases, the degree to which the system's organization is centralized increases.

There are three connections which are worth noting. The first is the connection between internal and external specialization. An internally spe-

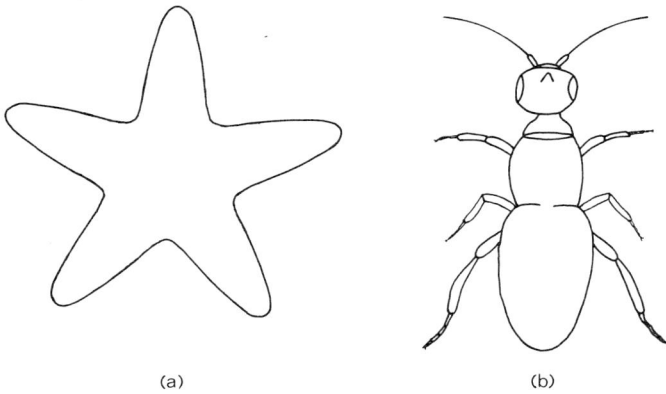

(a)   (b)

FIGURE 11.2. Connection between internal specialization and morphological symmetry. In the less symmetrical situation (b) there are fewer possible choices of small sections which are indistinguishable on the basis of their position relative to distinctive points of reference. Internal specialization increases as long as the increase in geometric distinguishability is correlated to an increase in histological distinguishability.

cialized system may or may not be externally specialized. But its component parts are more externally specialized. One can define a concept of niche for the cells or organs of a multicellular organism which is analogous to the niche concept defined for an organism or population. The only difference is that the environment of the cell or organ includes the body of the organism itself. The advantage of increasing internal specialization is that it allows for niche narrowing of organs or cells, yet without niche narrowing of the organism itself.

The second connection is that between the internal specialization of an organism and its symmetry. As one goes to lesser degrees of symmetry (say from radial to bilateral symmetry), the internal specialization necessarily decreases (Figure 11.2). The distribution of components cannot be as equitable. Thus relationships between adaptability and internal specialization should correlate to relationships between adaptability and morphological symmetry.

The third connection is between internal specialization and amenability to evolution. It may be recalled, from the last chapter, that graceful phenotypic change in response to likely genotypic changes is essential for effective evolution. High interdependence among the parts of a system reduces the chances for such graceful change. The formal reason for this can be seen by referring to Sections 10.6.1 and 10.6.2 of Chapter 10. Increase in interdependence decreases $H(\omega_{p\text{II}}|\omega_{p\text{I}})$, therefore decreases the specialization degeneracy term $[H(\omega_{p\text{II}}|\omega_{p\text{I}}) - H(\omega_{p\text{II}}|\omega_{p\text{I}}\omega_G)]$. The degree to which genetic change can be buffered depends on this term. To achieve the same degree of gradualism, it would be necessary to add extra buffering mechanisms (which is costly) or to increase the genotype degeneracy, thereby increasing the number of possible single-step genetic changes. Diversification and linch-pinning may increase efficiency, but they are also an Achilles heel.

Note on a corroborating dynamical argument. It is significant that the argument can also be phrased in a dynamical framework. In general the stability of a differentiable dynamical system decreases as the number of components increases. Here stability is taken in the asymptotic orbital sense, that is, in the sense that if a disturbance to a system is not too large, it will eventually reassume, with arbitrary closeness, the dynamical behavior it would have had in the absence of this disturbance. The reason is that the stable state (or trajectory) must be located in a valley of the system's phase space. All the eigenvalues must be negative, preferably large and negative. But as the number of components increases, the dimensionality of the space increases, decreasing the likelihood of valleys. The situation has been most thoroughly studied by May (1973) in the context of ecosystem models in which the components are species. The essence of May's analysis is that for a wide variety of dynamical models the probability of stability decreases as the number of species, the connectance, or the interaction strength in-creases, with a sharp transition to instability for large systems (cf. also Gardner and Ashby, 1970). An implication is that segmenting components into hierarchical groups increases the probability of stability, assuming this hierarchization reduces the total number of interactions. Increase in the number of interactions among components would have the opposite effect, since it would impose more restrictions on the possible eigenvalues. Thus the probability of stability should decrease as the frequency of components becomes more inequitable (under the frequency–connectivity assumption of this section). Insofar as stability of a rearrangement is a prerequisite for its acceptability, the dynamical analysis therefore corroborates the conclusion that rearrangements of a complex system are less likely to be acceptable than rearrangements of a simpler system.

11.5. FUNDAMENTAL CONSTRAINTS

The theorem can be crystallized into a biologically more specific form by taking hierarchical structure into account. Major considerations are as follows:

1. The cost of genetic plasticity [the $H_e(\hat{\omega}_{(u-1)0})$ terms] increases as the developmental and morphological complexity increases. If an organism consists of many types of highly interdependent parts, it is less likely that a random change in the relationships among these parts will be functionally useful.
2. The cost of developmental and morphological plasticity [the $H_e(\hat{\omega}_{u3})$, $H_e(\hat{\omega}_{u2})$, and $H_e(\hat{\omega}_{u1})$ terms] increases as the developmental and morphological complexity increases. The reason is related to the reason for the increasing cost of genetic plasticity. If the system consists of many types of highly interdependent parts, the transitions between different functional morphological forms would generally require more total disassembly and reassembly. Developmental plasticity would require the maintenance of disproportionately more genetic information to specify different functionally viable organizations.
3. The cost of culturability [the $H_e(\hat{\omega}_{u4})$ terms] increases as the potential reproduction rate of the organisms decreases and as the energy required to produce an offspring increases.
4. If there are different types of disturbance or if the duration of disturbance varies, the advantage of any particular form of adaptability will in general have diminishing returns ($\partial \overline{U}/\partial \mathcal{Q}_s < 0$).

   In the absence of diminishing returns the adaptability spectrum would always collapse to a single, least costly form of adaptability. This would happen if there were only one type and time scale of disturbance. A more detailed understanding of a system's spectrum of adaptabilities is possible if these are specified. However, the dependence of $\overline{U}$ on the distribution of types and time scales of disturbance may also be implicitly expressed by imposing a suitable requirement on the $\overline{U}_i$.

   Some disturbances have a mortal effect on almost every organism in a population, and others a mortal effect on only a few at any one time. If disturbances which knock out a large fraction of a population occur every few generations, culturability will be important. However, one cannot reason *a priori* about the effects of a disturbance (except for the most extreme and unusual types of disturbance) without considering the entire adaptability structure of a population. A population which opts for culturability may do so because there is no other way of coping with the disturbances to which it is exposed, but the mortal effect of these disturbances may be due to the fact that the population has sacrificed other forms of adaptability. Similarly, the time scale of a disturbance relative to the generation time is a biological as well as an environmental property.

One further factor which should be considered is the behavioral diversity [γ, as defined by equation (4.53)]. If the behavioral pattern of a compartment (say an organism or a society) is diverse, this will in general make the adaptability of that compartment more expensive. This is because variations on more complex behavioral sequences are more likely to be nonfunctional than variations on simpler sequences. It is reasonable to expect behavioral diversity and internal specialization to be correlated—a structurally more diverse and more centralized system can support more complicated behavioral patterns. But this correlation is not a logical necessity.

11.6. PATTERNS OF ADAPTABILITY IN POPULATIONS

These basic constraints are responsible for important trends in evolution. As a life form becomes progressively more complicated, certain forms of adaptability become progressively more costly. Compensation is necessary, with concomitant evolution of structures which support the compensating adaptabilities. The situation is summarized in Table 11.2, which lists organizational features of the major kingdoms and the types of compensations required by these features. Key points are as follows:

1. *Protista.* Protista (bacteria, blue-green algae, protozoa, and other entities of the microbial world) are potentially capable of rapid reproduction and also of assuming long-lived dormant (spore) forms. This combination allows culturability to be a particularly effective form of adaptability. Their relative organizational simplicity also allows them to support high genetic plasticity, including genetic plasticity based on haploidy in combination with either obligatory or facultative asexual reproduction and high mutation rate; conjugation or other forms of genetic interchange involving sexual reproduction, in some cases involving a number of different mating types; and transduction of genetic material by viruses. Switching between different ploidy numbers is possible and can be controlled by the rate of DNA reproduction in comparison to cell reproduction. Adaptations can be maintained by high ploidy numbers and asexuality, by inbreeding, or in fact by extensive gene flow if there is outbreeding. It is when gene flow between the members of a population is incomplete (i.e., under conditions of isolation) that divergence is most likely. All these mechanisms are not necessarily or even generally used. But in principle they are available and there is evidence that they do play as important role in the microbial world

TABLE 11.2
Allowable Patterns of Adaptability

	Protista	Higher plants	Higher animals
Main organizational feature	Unicellular, relatively simple	Multicellular, open growth system	Multicellular, closed growth system
Genetic adaptability	Potentially high	Potentially high	Restricted by closure of growth system
Developmental and physiological adaptability	Potentially high, but restricted by unicellularity	Potentially high	Restricted by closure of growth system
Culturability	Potentially high	Restricted by generation time and cost of development	Restricted by generation time and cost of development
Immunologic adaptability	Not well developed	Not well developed	Highly developed, compensates for restrictions on other modes of adaptability
Behavioral adaptability	Not well developed	Not well developed	Highly developed, compensates for restrictions on other modes of adaptability

as they do in the world of plants (where they have been much more intensively studied). Stebbins (1960) has argued that asexuality in bacteria, as in other forms, is derivative, implying a fundamental role for sex. Ravin (1960) has argued not only that sex is important in bacteria, but that transduction and possibly even transformation under natural circumstances are important as well. Protozoa show a continuous array of breeding systems, ranging from extreme outbreeding through various degrees of out- and inbreeding to extreme inbreeding and finally asexual reproduction (Sonneborn, 1957). Algae appear to be somewhat similar to bacteria with regard to sex and its absence, with the absence more important during periods of rapid population growth. The relative morphological simplicity of protista potentially allows for high developmental and morphological plasticity in addition to culturability and genetic plasticity. Some mechanisms are known, such as induction and repression mechanisms in bacteria and physiological adaptability based on diploidy [important in diatoms (cf. Patrick, 1954)] or on multiple gene loci in protozoa. Other mechanisms are undoubtedly important and morphological plasticity is a fact. The potentiality for high culturability and high genetic and high morphological plasticity extends to other of the so-called "lower" life forms. Fungi have fantastically complicated sexual systems; Raper (1954) describes several types of life cycles and sexuality, with a variety of basic kinds of life histories. Major morphological or physiological saltations are possible under different conditions.

2. *Higher plants.* For higher plants culturability is a potential mode of adaptability only for variations occurring on time scales of a few years. It is important for some colonizers and for plants simple enough to be annuals, but the lengthened generation times inherent in multicellular organization preclude the degree of effectiveness which culturability has in single-celled organisms. However, the relative simplicity of multicellular organization in higher plants does allow for high genetic, developmental, and morphological plasticity. The number of cell and organ types is not large. But, more important, the growth system is open. There is a rule of growth, but no fixed final form in terms of size and number of organs, or even in terms of the metrical aspects of the geometry. Those forms that are broadly classed as higher plants can grow basically indefinitely at the tips (by cell division at the apical meristems) and around the girth (divisions in the cambrial cylinder). Adaptive changes in morphology can thus be mediated through flexible changes in the branching pattern and support structures, and most especially through differential elongation (by water absorption) of new cells at the tips. The latter is the mechanism of morphological plasticity which enables plants to turn toward the light. The openness of the growth system allows for multiple occurrence of each type of organ (branches, leaves, flowers, roots) and therefore for very reduced dependence of different parts of the plant on one another (it is possible to cut a branch off a tree without its bleeding to death).

The same features which allow for high morphological and developmental plasticity allow for high genetic plasticity as well. Hybridization between forms which are very different in appearance and in some cases quite different in chromosomal structure is possible. Heterozygosity and diversity of the gene pool can be maintained through hybridization or apomixis (cf. Stebbins, 1950). Extensive hybrid complexes are found in nature and make possible the persistence of intermediate types with no immediate selective advantage (Grant, 1957). Apomixis, or asexuality, makes possible the maintenance of a variety of pure strains, hence the maintenance of highly specific adaptations. Facultative apomixis, which may in fact be the most usual case (Clausen, 1954) allows for switching between the two strategies. Self-fertilization has the same effect as apomixis (as evidenced by the fact that it either does not occur or only rarely occurs in conjunction with facultative apomixis). Gene flow, the mechanism of enhancing gene pool diversity in hybrid complexes, can also have the opposite effect of maintaining homogeneity of the population, providing that it is very extensive. In this case genetic isolation, the necessary condition for speciation, is prevented. Increase in ploidy number, a mechanism infeasible in the morphologically more complex animal kingdom, is an important mechanism of variation in plants.

   3. *Higher animals.* As with higher plants, the relatively long generation times of higher animals reduce the potential effectiveness of culturability as a mode of adaptability. The increased thermodynamic cost of producing more complex offspring also increases the cost of culturability. This complexity is also incompatible with as high a degree of developmental, morphological, and genetic plasticity as is found in the plant kingdom. There are an order of magnitude more cell types, more types of organs, and much greater interdependence among different parts of the system (the Achilles heel property). The crucial point is that the growth system is closed. The growth rule generates a definite number of bones, with not very much variation from one member of the species to the next. The generated pattern of nerve connectivity is almost as invariant, the musculature somewhat less so, and the arterial connectivity is only slightly more variable than the musculature. The venous system is allowed more variability and the lymphatics still more, but even these assume a specific final pattern in any given organism. Open growth for any of these major organ systems is a disease. This does not mean that the organs are not plastic. Even for the skeletal system there is a well-known histological generality (Wolf's law) that a bone subject to stress will exhibit growth in the direction of that stress. But the topology of the system is fixed and only delimited metrical changes are possible. Skin can darken, fur can change color, exercise can increase the size of muscles, but the high morphological and developmental plasticity available to the plants is clearly precluded by the closure of the growth system and by the excessive amount of genetic information which would be required to support alternative modes of development.
   These features also preclude high genetic plasticity in higher animals. Sexual recombination, crossing over, and mutation are as indispensible for the evolution of animals as they are for that of plants. In general sexuality is even more obligatory in animals than in plants, though parthenogenesis (the equivalent of apomixis) is possible in some species. However, gene flow

between very different forms usually results in nonfunctional offspring and in fact does not occur in nature. The chromosomal structure of the parent organisms must be identical, single step increases in ploidy is in general not feasible, and variations on breeding systems are limited. The reason is that variations on the complex interdependence structure of a multicellular animal are much less likely to be viable than variations on the much simpler interdependence structure of a plant. The greater applicability of the biological species concept in zoology than in botany or microbiology is testimony to this reduction in genetic plasticity. The extensive hybrid complexes possible among plants and protista do not occur in birds, mammals, or reptiles.

The conclusion is that the protista can draw on culturability, and on genetic, developmental, and morphological plasticity; plants are restricted as regards culturability, but can draw on genetic and developmental plasticity, and to a great extent on morphological plasticity; but animals are restricted as regards all these modes of plasticity. Either the niche must narrow or the animal must compensate with alternative modes of adaptability. The most striking alternative modes are behavioral plasticity, topographical plasticity, plasticity of social organization, and the immune system of the body. Each of these requires distinctive organ systems and levels of organization. Behavioral plasticity requires the high development of the central nervous system, in conjunction with organs of perception and control. The flexible constraints of the skeletal and muscular system confer degrees of freedom on animals which are not available to plants. The plastic use of these constraints depends on the high development of learning and memory mechanisms, as in mammals and primates. In the absence of such a high development, these flexible constraints can still be tapped for adaptability, but in the form of topographical rather than behavioral plasticity. The genetic, developmental, and morphological plasticity of birds is extremely limited. Their learning capabilities are also not impressive. But these limitations are compensated by the fact that their position in space may be most uncertain. This is plasticity based on coordination of highly flexible constraints with cerebellar balance mechanisms rather than cerebral learning mechanisms.

The plasticity of social organization is also based on the nervous system, together with sensors and effectors, except that the constraints involve communication and hence are of a symbolic rather than a mechanical nature. The plasticity of such organizations may be coordinated to behavioral plasticity, or it may involve switching between a limited repertoire of stereotyped behaviors. In the latter case the mechanism of social plasticity is analogous to the mechanism of developmental plasticity. In both cases the mechanism involves a new, higher level of constraint. The remaining remarkable system is the defense system of the body (the reticuloendothelial system). While there are many mechanisms of defense, almost certainly including important unknown mechanisms, the most striking is the immune system. According to the clonal selection theory of Burnet (1959) cells with surface antibody which bind sufficient antigen are stimulated to produce more cells carrying that antibody, or to secrete antibody directly into the body fluids. There are sophisticated, only incompletely understood mechanisms for control and allocation of the system's resources, as well as mechanisms of memory which enable the organism to respond more rapidly to a second invasion by the foreign entity. The key point is that the immune system provides a mode of adaptability which enables the system to deal with disturbances which occur on a short time scale, including disturbances arising from the mutability and culturability of microorganisms. The incredible fact is that the immune system fights fire with fire. Its modus operandi is the culturing of defender cells. In the most complex vertebrates, where culturability is most precluded at the organism level, the immunity mechanism reaches its most elaborate form; the loss of culturability at the organism level is compensated by the evolution of culturability within the organism.

The trends in compensation in higher plants and animals exhibit important parallelisms. As one moves to the very most complex forms, one finds an increasing allocation of resources to the embryo. Higher animals maintain the environment of the growing organism, either in an egg or in a placenta or within the mother's tissues. The greater delicacy of the developmental process and the greater energy required for development impose the necessity for greater allocation of control to compartments outside the developing system (e.g., disturbances are dissipated by a shell or absorbed by the mother). They also restrict litter size, further interfering with culturability. To a lesser extent, the higher plants show a similar trend. The germ is packaged even more carefully and the developing embryo provided with more of a head start. Some germline cells (i.e., endosperm) may be produced to feed those which are destined to develop into a plant, thereby reducing the potential "litter" size. As in higher animals, increases in the complexity of the developmental pattern (for example by the addition of new organs such as flowers) interfere with developmental and morphological plasticity. Great morphological plasticity is still possible, but the major developmental and morphological saltations which make it difficult to identify the species become increasingly infrequent. The species concept becomes clear in the higher plants, as it does in the animals, indicating the same kind of reduction in sexual and morphological plasticity. In some cases loss in morphological plasticity is compensated by support structures which increase indifference to the environment, as in large trees. In other cases special reproductive organs (in particular flowers) develop which enable the plant to utilize mobile animals to secure effective outbreeding and also topographical plasticity. Alternatively, topographical plasticity may be achieved through evolution of seed dispersal mechanisms which involve mechanical forces or which depend on air currents or other vagaries of the environment. In either case the location of the seed may become quite uncertain, despite the fact that the location of the individual plant is fixed in space. The compensating evolution of flowers allows plants to tap the behavioral and social adaptability of animals in order to regulate the critical aspects of their reproductive process. Though based on symbiosis, it is a development which parallels the development of behavioral plasticity and social organization in the animal kingdom.*

11.7. EVOLUTION OF THE HIGHER NERVOUS SYSTEM AS COMPENSATION

The issue inevitably arises, what are the causal relations between the evolution of constraints (such as a neuromuscular system and brain) which require the evolution of compensating adaptabilities and the evolution of the constraints (such as a neuromuscular system and brain) which make these compensations possible? Like the famous chicken and egg, the relations are certainly circular.

The issue can be viewed in terms of the advantage of specialization of labor. Such specialization means greater differentiation of the system (more cell types, more organs, more castes). The structural diversity and hence the interdependence among the parts of the system increases. The advantage is that the more specialized system earns a greater energy and resource return for a given energy and resource expenditure. This is balanced by two disadvantages. More energy and resources are required for the development

---

*An assumption in this section is that the genetic variability resulting from sexuality contributes to adaptability, as it must according to adaptability theory. Some recent discussions (e.g., Williams, 1975) have questioned whether the advantage of sex in terms of leaving more viable offspring can outweigh its cost in terms of potential reproduction rate. If not, more subtle evolutionary mechanisms, such as hitchhiking, or more subtle advantages, such as repair or specialization of labor, may play a role. But these alternative mechanisms of origin do not contradict the contribution of sex to adaptability. Furthermore, in evaluating the importance of sex as a mode of adaptability it should not be forgotten that the uncertainty of the environment of a metazoan organism is always substantial due to processes in the microbial world (Bremermann, 1980). Another point is that the disadvantage of sex in terms of reproduction rate should not be overestimated since most populations incorporate regulatory mechanisms which control their growth rate in any case.

of the system in the first place. Once the system is developed, adaptability is more costly. One way of reducing the extra cost in terms of adaptability is to narrow the niche. The problem is that this also reduces the potential advantage in terms of increased energy return. The second possibility is to develop compensating modes of adaptability. The selective advantage of specialization in effect induces a selective advantage for the development of new compartments or levels which allow for these compensations. In turn any such new compartment or level allows for further specialization of labor, thus reiterating the process.

In the above scenario there is a transfer of function from a general adaptability function to some specific physiological function. This is not the necessary course of events. Variations which ultimately lead to a new compartment or level might be selectively favored because of their contribution to the specific functions performed by the organism, because of their contribution to adaptability, or for both reasons. In many cases, however, the requirement for adaptability may be dominant. Furthermore, a variation leading in the direction of a new compartment or level may often more easily provide selective value in terms of adaptability than in terms of specialization of labor. In either case, however, there will be an interplay in terms of selection for adaptability and selection for specialization of labor, with the different factors dominating at different times.

The evolution of the human brain is the most dramatic example. The vertebrates and in particular the mammals are characterized by a high degree of internal differentiation and the development of a high degree of interdependence among the differentiated parts. In some cases their evolution is accompanied by niche narrowing, but in many cases compensating adaptabilities appear. The expansion of the central nervous system provided an important arena for such adaptabilities, both by increasing information processing capabilities and by allowing for increased behavioral plasticity. It also provided an arena for further specialization, in particular the hand–eye coordination of the primates. In turn this opened up more possibilities for adaptability. By itself this would not have led to the evolution of the modern brain. Superfluous intelligence is superfluous adaptability, hence costly. The required extra ingredient is niche expansion. Increasing adaptability makes it possible to increase niche breadth, provided that the species can outcompete its competitors in each part of the expanded niche. The brain, as both a powerful organ of information processing and a powerful organ of adaptability, allowed for this situation in unusually high degree. Rather than adaptability falling in the direction of the uncertainty of the environment, the uncertainty of the environment could increase to match the adaptability. The latter in turn would increase as long as niche expansion were possible. This type of causal interplay (which has the character of positive feedback) would certainly account for the remarkably rapid evolution of human brain size. Furthermore, it has the noteworthy property that only one large-brained species will remain at the end of the process. Niche expansion in combination with competitive exclusion will inevitably eliminate all other large-brained forms (with the possible exception of forms in physically inaccessible environments).

Remark on the comparative aspect of behavior. A debate perennial to science revolves around the extent to which behavior is plastic or specific. This issue should be considered in the light of the principle of compensation. Plasticity of behavior is one means of compensating for restrictions on nonbehavioral modes of adaptability. The existence of organisms with stereotyped behavior does not imply that plasticity of behavior in other organisms (such as humans) is illusory any more than the existence of populations with low genetic diversity implies the nonexistence of genetically diverse populations. The issue is clouded by the fact that even the behavior of the most plastic system depends on constraints which are fixed or which change only very slowly. The existence of such deep constraints is not inconsistent with the plasticity of behavior any more than the fixed structure of most computer programs is incompatible with the possibility of their generating an extremely large repertoire of behaviors. Fixed constraints, such as the genetic code, underlie genetic diversity as well. The proper interpretation is not that such constraints imply nonplasticity, but rather that forms of adaptability such as genetic and behavioral plasticity provide the homeostatic mechanisms which, paraphrasing the words of Bernard, maintain the stability of these constraints. Studies of plasticity versus stereotypy of behavior in particular organisms should be generalized with the same care as statements about the plasticity or fixity of genetic structure, that is, in the light of the spectrum of adaptabilities.

11.8. EXAMPLE OF A DIFFERENT KIND (HOMEOTHERMY VERSUS POIKILOTHERMY)

It should be clear that the rather explosive evolution of the brain does not imply any general trend to greater adaptability and internal differentiation in evolution. It is a product of a peculiar interplay of adaptability, internal differentiation, and niche breadth. Any apparent directedness is only a consequence of the fact that the complex forms necessarily appear later than the simple ones. The niches they occupy were either previously unoccupied, or occupied with forms less optimal in terms of specialization and adaptability. The point is illustrated by the much-discussed issue of the merits of homeothermy ("warm-bloodedness") versus poikilothermy ("cold-bloodedness"). A mixed strategy is also possible. For example, some tissues may be homeothermic and others poikilothermic (such as the extremities of some arctic birds). The obvious advantage of homeothermy is that subcompartments of the organism not involved in temperature regulation do not see variations in environmental temperature. Greater internal specialization is therefore possible as well as expansion into environments otherwise difficult to occupy. The homeothermy mechanisms and the greater specializations which they make possible are costly to maintain and construct. As the return on this specialization decreases and as the environment becomes either warmer or more equable, a point will inevitably be reached where poikilothermy has a competitive advantage.

11.9. VECTOR DIAGRAMS

There is a useful way of representing patterns of adaptability pictorially. The basic picture, which I call a vector diagram, is illustrated in Figure 11.3. The uncertainty of the environment is represented on the abscissa. The adaptability of the entire system (say population or ecosystem) is repre-

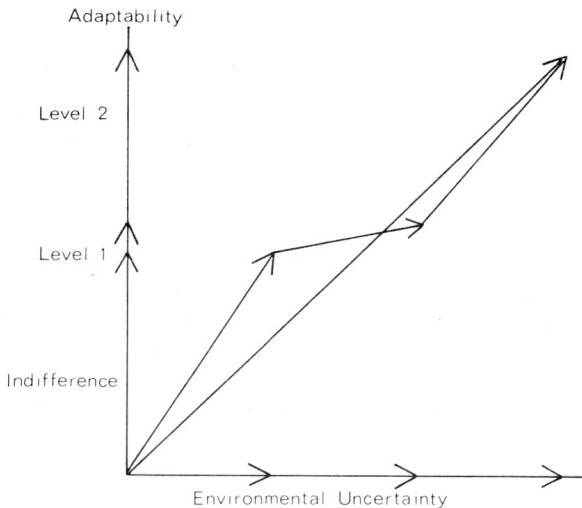

FIGURE 11.3. Vector diagram for a system with two levels of organization. The indifference component and level 2 are controlling. Level 1 is controlled. The levels might correspond to phenotypic and genetic levels, in which case the diagram would depict the adaptability structure of an organism.

sented on the ordinate. Each of the arrows may be interpreted as a vector, with the length representing a magnitude and the angle relative to the origin a direction. The arrows on the ordinate are components of the total adaptability. In Figure 11.3 these components are taken as the adaptabilities associated with each level [as defined through equation (5.24)] plus the indifference contribution. The components could also be taken as the adaptabilities associated with each compartment, again along with the indifference as a separate contribution. The arrows on the abscissa represent the fraction of environmental uncertainty which would be absorbed by the various levels (and by the indifference) if each made an equal contribution to adaptability. The sums of corresponding component vectors of adaptability and environmental uncertainty may form any angle with the coordinate axes. But according to the evolutionary tendency equation the total sum will tend to shrink and rotate clockwise until it forms a diagonal. Vector sums pointing above the diagonal line are associated with controlling levels, whereas those pointing below the diagonal are associated with more controlled levels. The controlling levels contribute to homeostasis by absorbing and dissipating a greater share of environmental disturbance.

If compartmental rather than level adaptabilities are used, the magnitudes of the environment vectors can be weighted according to biomass. The

---

FIGURE 11.4. Patterns of adaptability in populations. The diagrams picture the major patterns discussed in the text. The dotted lines in (a) indicate that among protista and other lower forms any of the modes of the component adaptabilities (populational, phenotypic, genetic, indifference) could be high. But if some are high, others must be low, as the vector sum should tend to the diagonal line. Parts (b) and (c) illustrate possible patterns in higher plants and higher animals, respectively. The indifference could in fact be higher or lower, and any of the component adaptabilities which point higher than the diagonal could be lower. But the ones which point lower are constrained to do so. Note that for higher animals the phenotype is split into two components, corresponding to the fact that special organs of adaptability must evolve to compensate for restrictions on developmental and morphological plasticity. Actually the social component of populational adaptability could also be high. The formal correspondences are: environmental uncertainty $= \Sigma H_e(\omega_{h0}^*) + \Sigma_{n \neq 0} H_e(\omega_{mn})$; adaptability $= H(\hat{\omega}_{r4}) + H(\hat{\omega}_{r4}|\Pi \hat{\omega}_{h0}^*, \Pi_{n \neq 0}\hat{\omega}_{mn}) + \langle \Sigma H_e(\hat{\omega}_{h0}^*|\hat{\omega}_{r4}) + \Sigma_{n \neq 0} H_e(\hat{\omega}_{mn}\hat{\omega}_{r4}) \rangle$, sums not taken over r and $(m, n) \neq (r, 4)$; populational adaptability $= H(\hat{\omega}_{r4}) + H(\hat{\omega}_{r4}|\Pi \hat{\omega}_{h0}^*, \Pi_{n \neq 0}\hat{\omega}_{mn})$; phenotypic adaptability $= \Sigma_{u=0}^{3} H_e(\hat{\omega}_{ru}) + \Sigma_{u=1}^{3} H_e(\hat{\omega}_{ru}|\Pi \hat{\omega}_{h0}^*, \Pi_{n \neq 0}\hat{\omega}_{mn})$, sums taken over r, with r even for all $\hat{\omega}_{r0} (= \hat{\omega}_{r0})$; genetic adaptability $= \Sigma H_e(\hat{\omega}_{r0}) + H_e(\hat{\omega}_{r0}|\Pi \hat{\omega}_{h0}^*, \Pi_{n \neq 0}\hat{\omega}_{mn})$, all r odd. Phenotypic adaptability I and II are defined in terms of different phenotypic variables, with $\hat{\omega}_{ru} = \hat{\omega}_{ru}^I \hat{\omega}_{ru}^{II}$. Indifference $= \Sigma H_e(\hat{\omega}_{h0}^*|\hat{\omega}_{r4}) + \Sigma_{n \neq 0} H_e(\hat{\omega}_{mn}|\hat{\omega}_{r4})$.

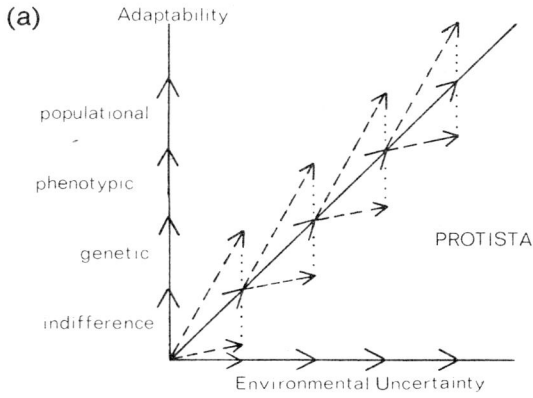

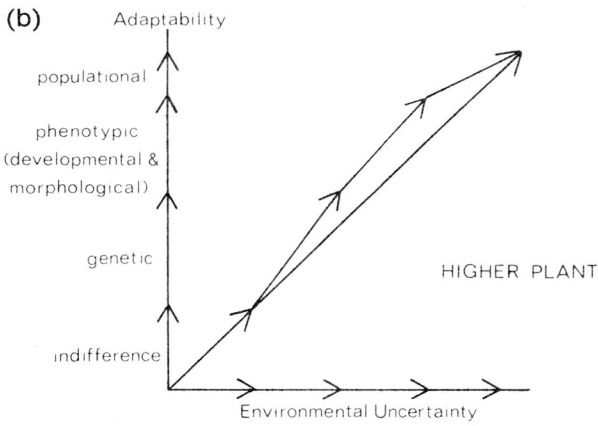

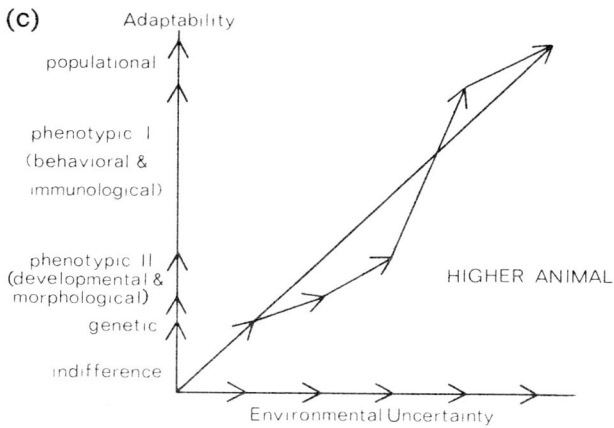

components of environmental uncertainty could also be taken as corresponding to particular environmental variables, such as temperature, light intensity, or precipitation. In that case a more detailed picture of the extent to which different levels or compartments are controlled or controlling may be obtained.

Vector diagrams directly picture the structure of a system's solution to the allocation problem. Figure 11.4a illustrates the qualitative features of the adaptability structure of a microbial population. Any of the possible forms of adaptability may in fact be high. Figure 11.4b illustrates the possible adaptability structure of a higher plant. The major restriction is that the culturability vector fall below the diagonal. Figure 11.4c illustrates the adaptability structure of a higher animal. The genetic, developmental, and culturability vectors are restricted. Special organs (brain, immune system) or new levels of organization (highly developed social organizations) assume much higher selective value than they would otherwise have.*

The important thing about vector diagrams is their covariance. There is a kind of conservation law. Given the environmental uncertainty, there is a minimal allowable value of the total adaptability. If certain forms of adaptability are restricted, they must be compensated by other forms. If other forms are not available—if the sum of all the vectors falls below the diagonal—the evolutionary tendency will be for new adaptability-supporting organs or levels of organizations. Once all the vectors but one are fixed, the evolutionary tendency will always be to fix the last so that the sum vector falls on the diagonal; but of course any evolution process is always accompanied by changes in the lengths and directions of all the vectors.

11.10. IMPLICATIONS FOR THE ETIOLOGY OF DISEASE

The vector model provides a useful framework for classifying disease processes. There are two fundamental classes. The first (class I) are the diseases of the controlled systems and the second (class II) are diseases of the controlling systems. As the controlled system becomes more complicated, the likelihood increases that any given perturbation will lead to its destabilization. As the whole organism becomes more complicated and interdependent, any such local instability is more likely to ramify and to cause an irreversible destabilization. To avoid this situation it is necessary to expend more energy on reliability and to shift more adaptability into compensating subsystems, such as the immune system and the central

---

*For an illustrative optimization calculation leading to diagram structures similar to those in Figure 11.4 see Conrad (1972). The calculation is based on hypothetical costs and advantages.

nervous system. But there is a limit to the extent to which adaptability can be shifted to compensating systems without giving rise to diseases of the second class. It is the instability and degradative power of these systems which enable them to absorb and dissipate disturbances. Increasing these capabilities means increasing the chance that the compensating system will itself serve as a source of destabilizing perturbations. There is no way of excluding this possibility in advance. As the system to be controlled becomes more complex, more situations will be perturbing to it. As the controlling systems concentrate more adaptability they decrease the likelihood that no situation will be perturbing. If the shifts in adaptability increase the complication of the system or if the increases in complexity decrease the indifference component of adaptability, these inherent problems will be further aggravated. The increased complexity is more likely to lead to diseases of complexity and the increased concentration of adaptability is more likely to give rise to diseases of adaptability.

There is therefore a tradeoff between diseases of the first class and diseases of the second class. This tradeoff is not prominent for simple systems, such as the open growth morphologies of plants. When complexity is not high the diseases of complexity are not important. Adaptability may be high even in simple systems, but in complex systems, such as the closed morphology of the vertebrate, the diseases of adaptability become more prominent due to the reallocation of adaptability. In this respect the comparative aspect of pathology, that is, the comparative aspect of undesirable variation, should mirror the comparative structure of adaptability.

The peculiar feature of the diseases of complexity is that each has many good mechanistic explanations, but no single mechanistic explanation which fits all the epidemiological data. As a system becomes more complex the number of possible causes which could give rise to any given symptoms increases, due to the fact that the class of destabilizing perturbations becomes larger. This appears to be the case with cancer, which in fact is often described phenomenologically as a failure of control. The likelihood of failure increases with the complexity of cells and organs and the likelihood of global consequences increases as the interdependence of different cells and organs increases.

Another example is the circular interaction between weakened heart cells and systemic disorders of circulation in congestive heart failure. One hypothesis is that the problem has its origin in a failure of heart cells. A second hypothesis is that it has its origin in a systemic failure connected with oxygen transport and fluid balance. In either case the systemic failures weaken heart cells, in turn aggravating these systemic failures, and so forth. According to adaptability theory it is improper to attempt to decide between these two hypotheses. As heart tissue and circulatory systems become more complex, the class of possible causes of congestive heart failure widens to the point where it is unlikely not to include both molecular and systemic events as initial causes. Still another plausible example is schizophrenia, also a disease which appears to have multiple biochemical and environmental explanations.

The peculiar feature of the diseases of adaptability is that the same basic cause can give rise to an astonishing variety of symptoms. The cause is a nonadaptive instability in one of the compensating systems. But the compensating systems are inherently instability-rich. The variety of possible perturbing situations which can arise is indefinitely large. This is the case with genetic diseases. The variety of pathological phenomena certainly cannot be more foreclosed than the variety of advantageous variants. A key feature of the argument has been that the pathologies become more frequent as the morphology becomes more complex and that the requirement for immunological and behavioral compensation is due to this. But the compensating systems suffer from the same problem, only to a lesser degree.

For the immune system the inherent problem is to distinguish self from non-self. The problem is solved by eliminating those cells which produce antibodies that recognize self at an early stage of development and by building a trigger into cells which causes them to commit suicide at a later stage if they are exposed to an antigen on a long-term basis. To the extent that this mechanism works, non-self can be converted into self. This is the phenomenon of immunological tolerance. These solutions are clever, but they can be only partial. The problem of determining whether it is functionally acceptable to incorporate a new component or produce a new antibody is effectively the self-computability problem, already shown to be logically unsolvable (in the discussion of the halting problem, Section 10.9). It is inevitable that self will sometimes be rejected and that non-self will be accepted. Self can be rejected in as many ways as there are combinations of constituents, hence the indefinite symptomatic variety of autoimmune diseases. The fundamental origin of such diseases is the ensemble of possible trajectories necessary for adaptability. In this respect the diseases of the second class are as truly diseases of entropy as the diseases of the first class are diseases of complexity.

The most prominent example is syphilis. The primary stage has its origin in the failure to reject non-self, that is, in the acceptance of the infectious agent. A secondary stage is the autoimmune response, leading to the improper rejection of self. At this stage the disease is a disease of entropy. The numerous possible modes of self-rejection mean numerous possible combinations of perturbation to the different systems of the body, hence numerous possible diseases of complexity. This is indeed the remarkable feature of tertiary syphilis. The interaction between the two classes of disease mechanisms displayed here is undoubtedly a general feature. It may reasonably be expected that diseases with an origin in complexity give rise to problems which have their origin in adaptability as well, even autoimmune problems. As a speculative example, the diseases of complexity need only lead to a failure of RNA processing which results in the production of a variant cell-surface protein.

In any complete description of an etiology the interactions between class I and class II mechanisms and the ramification of class I mechanisms can become much more elaborate. In interpreting the description it is important to remember that what constitutes a perturbing rather than a protecting variation on the part of one of the controlling systems depends on the complexity of the controlled system (including the controlled features of the controlling system). If the complexity is sufficiently high, it will become impossible to trace the origin of all cases of the disease to either a class I or a class II mechanism, due to the fact that the class of relevant perturbations will broaden to span both failures of adaptability and unsuitable adaptive states. From the standpoint of adaptability theory the question is improperly posed, just as the question as to whether congestive heart failure as a generic disease has a cellular or systemic origin was improper.

These same considerations extend to the other great compensating system, namely the brain and its system of behavioral plasticity. Here, as in the immune system, it is impossible to preclude in advance the occurrence of a situation which is perturbing rather than protecting. The reason is similar. To the extent that the behavioral plasticity of the organism is based on a process similar to evolution, it is inherently counterproductive to foreclose possibilities and impossible to filter out the undesirable ones without testing them. The test, hence the risk, is necessary. To the extent that behavioral plasticity is generated algorithmically it is logically contradictory to foreclose possibilities, else one again has a system which is not dull but which at the same time is capable of computing itself. This is sometimes called the creative feature of unsolvability. For a finite system it could be called the creative feature of intractability. Whether generated by an evolutionary mechanism or an algorithmic mechanism, behavior has a creative feature. Without this feature there is no possibility for new adaptation, but with it there is necessarily the possibility for novel maladaptations and self-perturbations. This is the reason for the great richness of human psychic and psychosomatic disorders. From the standpoint of adaptability's contribution to adaptation, narrowing the "genetic creativity" of the species is compensated by opening the mental creativity of the individual. From the standpoint of adaptability's contribution to maladaptation the costly danger of diseases of genetic origin is traded for the less costly danger of diseases of mental origin. If one takes steps to eliminate this maladaptive component of variability one will necessarily eliminate the adaptive component. The adaptive component, hence the maladaptive component, cannot be genetically determined, even though the search mechanisms and algorithms which generate it undoubtedly are. If it were, the system of behavioral plasticity would not be a compensating system. Can the resulting mental diseases be conquered—by therapy, by drugs, by social reorganization, by eugenics? The question can validly be asked whether it is possible to make too much progress in this area, since progress here would mean closing man's fundamental mode of compensating adaptability.

11.11. IMPLICATIONS FOR THE TREATMENT OF DISEASE

According to adaptability theory the attempt to eradicate disease is contradictory and can only be counterproductive. Yet the theory provides a natural framework for the design and evaluation of medical therapies. The key point is that an organism is dying (from the standpoint of adaptability theory) if the ramification of a disturbance reduces the adaptability faster than the adaptability reduces the ramification of disturbance. A treatment or regimen may have the effect of attenuating the ramification of disturbance, either by eliminating the cause or by inhibiting the effects, or it may act by enhancing adaptability. The condition for a disturbance-attenuating treatment to be effective is that it not have adaptability-reducing side effects which are eventually counterproductive. If this condition is not fulfilled the apparently beneficial effects of the treatment will be followed by regression to the original disease or by other type I disease processes. The condition for an adaptability-enhancing treatment to be effective is that it not cause counterproductive increases in the occurrence of type II diseases. If this condition is not fulfilled, new disturbances will be generated faster than they can be absorbed or dissipated.

These simple conditions are consistent with the commonly held idea that an individual's resistance or susceptibility to disease is affected by a variety of factors or actions that can be taken. But many of the actions taken by individuals and by the community are not in fact consistent with these conditions. I believe this is due to three factors. The first is a misidentification of adaptability with physical fitness (as defined in sports medicine). Cultivating the ability to cope with external physical stresses which never arise under ordinary circumstances is an energy cost to the individual which competes with modes of adaptability more fit to the real requirements. The second factor is the subtlety of the effects involved. It is much more difficult to recognize the side effects of a drug or treatment on a capacity to absorb or dissipate disturbances than it is to recognize its effects on the disturbances themselves, or on its symptomatically identifiable side effects. But it is just these side effects on adaptability which it is most important to investigate.

The third factor is the tendency to extend intuitions gained from working with machines to the problem of repairing biological organisms. To date no machine is genuinely self-repairing, whereas organisms are basically self-contained and autonomous in terms of adaptability. The machine-based intuitions are reinforced by the existence of environmentally and genetically caused diseases whose effects cannot be dissipated or absorbed by any of the organism's internal modes of adaptability. These diseases, insofar as they are treatable at all, require external mechanisms of adaptability which are ultimately vested in the behavioral plasticity of other individuals and in social institutions. The phenomenon of medicine, as an institution, can be thought of as a reallocation of adaptability to the social level of organization which compensates for either inherent or socially derived limitations on the adaptability of individuals. Historically there have been some dramatic successes based on the removal of specific causes and on the introduction of specific defense mechanisms. Famous examples are antibiotics, vaccines, and sanitation. But the program which has emerged is being unreasonably extrapolated to situations for which cultivating internal mechanisms of adaptability is much more suitable. This is particularly true for the diseases of complexity, where the class of causes becomes very large and the possibility of treating the disease without creating even more serious problems becomes slim. As the program of basing medicine on external adaptabilities makes progress it must eventually reach a point, if it has not already, where these situations become the most important. A more unfortunate consideration is that the elimination of pollutants and other environmental causes of such diseases is not as simple as the problem of sanitation. The external adaptabilities required to remove these contaminants are, like all adaptabilities, a cost, and it is not clear that societies will succeed in paying this cost.

One course of action open to the individual under these circumstances is to take steps which enhance individual adaptability. There are two simple, rather obvious courses of action. The first is for the individual to expose himself to the range of environmental circumstances he expects to meet, provided that these are of the type to which adaptation is possible. These circumstances include climatic and other physical conditions as well as diseases to which resistance can be developed. Steps to isolate the individual from diseases to which defense can be built up or to arrange matters so that the natural defense mechanisms of the body are not periodically exercised are undesirable. A highly specialized mode of life reduces adaptability and evidently renders the individual more sensitive to a range of circumstances which might reasonably be expected. The second course of action is for the individual to build up the capacity to absorb and dissipate disturbances by reducing as much as possible his exposure to circumstances which can be prevented from occurring, which are not expected to occur, or to which adaptation is impossible. In the first instance this means stresses, such as lack of sleep, which have the effect of drawing resources away from adaptability mechanisms even if some adaptation to these stresses is possible. In the second instance it means not inventing exercises which wear down the individual for the purpose of superfluously enhancing the effectiveness of particular adaptations. In the third instance it means avoiding superfluous exposure to agents which wear down adaptability mechanisms. Many such agents, such as diagnostic agents and drugs, are optional. The attempt to obtain information about an organism always causes some degradation of the structure and function of the organism (Sections 2.4 and 2.5), so it is to be expected that it will often cause some degradation of adaptability mechanisms as well. Similarly, the likelihood that a drug untested for its effect on adaptability has a positive effect is much less than the likelihood that it will have a negative effect. If these rules are taken into account it should be possible to re-internalize beneficially some of the adaptability necessary for health and to reduce the extent to which one's state of health is determined by uncontrollable factors. It is possible that some illnesses now considered to be totally outside of the individual's power to prevent can be at least to some extent moved within that power. The reader may protest that these general rules, while they have the appearance of common sense, are very unspecific. What does *superfluous* mean? What does *unnecessary exposure* mean? How does an individual determine what it means for him in particular? These are legitimate empirical questions. That there are so few data about them points to how incongruent the current data-collecting paradigm is with the adaptability theory framework.*

11.12. ADAPTABILITY VERSUS ADAPTATION

The most obvious implication of the principle of compensation is that there is an equivalence between thought and evolution. They are the

---

*The extent to which the institutionally incorporated paradigm is directed to the development of external mechanisms of adaptability rather than to the development of methods for enhancing individual adaptability is vividly illustrated by the fact that as of 1977 the referral officer at the U.S. National Institutes of Health could find no section willing to manage the peer review process for grant applications in the area of biological adaptability. To my knowledge this situation has not yet changed.

expressions of adaptability most prominently responsible for novelty. According to the principle of compensation these and other expressions of adaptability are connected by an equation. The capacity to evolve is one component in this equation. It is based on genetic mechanisms. The capacity to think, to create, and to alter behavior is another component. It may in part be based on mechanisms which utilize the same underlying principles as those which underlie evolution. It may in part be based on quite different principles, such as algorithmic principles. The equivalence between thought and evolution is one of function, not necessarily one of mechanism. The capacity to evolve through genetic mechanisms involves a cost. The capacity to evolve through thought mechanisms also involves a cost. According to the principle of compensation closing off the genetic capacity entails an opening of the thought capacity. Or it entails either the opening of some other capacity or narrowing of the environmental space occupied.

In order to appreciate the full implications of compensation it is necessary to analyze the situation more closely. Superficially it might seem that thought and other modes of adaptability exist in order to maintain or improve adaptations, such as features which allow organisms to eat and reproduce. I think it is fair to say that this statement sounds more realistic than the statement that adaptations such as features which allow organisms to eat and reproduce exist in order to maintain or improve thought and other modes of adaptability. But a deeper analysis shows that both statements, properly interpreted, are equally valid and no matter how interpreted are equally invalid if taken in isolation. Adaptability and adaptation are both essential features of a functional circle and as such neither can be more nor less fundamental than the other. It is not scientifically supportable to view adaptability as a means for achieving adaptation without at the same time viewing adaptation as a means for achieving adaptability. According to the principle of compensation this conclusion does not depend on which mode of adaptability is being considered. It is just as valid when the mode of adaptability is behavioral as when it involves the capacity to evolve through genetic mechanisms. The opening of mental adaptabilities compensates for the narrowing of the capacity to evolve through genetic mechanisms. Mental adaptabilities are just as essential to the life process as the evolutionary adaptability for which they compensate. The attitude that mental adaptabilities are properly viewed as a means for obtaining the more fundamental ends of foodgetting and offspring production is untestable and therefore unscientific, as is the attitude that genetic adaptabilities are properly viewed as a means for obtaining more fundamental ends.

The references to means and ends are misleading. Taken out of the context of a functional circle they suggest that activities described as ends have inherent value and that they therefore ought to be undertaken in preference to those activities described as means. To suppose that such value judgments and their normative implications have an empirical basis is the classical naturalist fallacy. One may choose to view thought as existing for the purpose of obtaining food and procreating. But it is a fallacy to suppose that this viewpoint is any more scientific than the view that foodgetting and procreation exist for the purpose of creating art and science. There are no conceivable empirical observations which could confirm such purposive statements. It is verifiable that man does not live by thought only. Suppose that this established thought as the means and bread as the end. It is also verifiable that man cannot live only by bread. One need only try to perform a nonroutine task without thinking and then think of how many nonroutine tasks must be performed every day. This would establish the opposite statement, that bread is the means and thought the end. With respect to scientific discourse the language of purpose reduces to absurdity unless interpreted in the limited context of a functional circle.

I have deliberately used terms such as *exists in order to, means,* and *ends.* I have considered usages such as:

P1. Thought and other modes of adaptability exist in order to maintain or improve adaptations, such as features which allow organisms to eat and reproduce.

P2. Adaptations, such as features which allow organisms to eat and reproduce, exist in order to maintain thought and other modes of adaptability.

I believe that the unscientific sound of P2 in comparison to P1 reflects an unconscious and biased value judgment in scientific usage which should not enter into the structure of scientific theories. The expression *exists in order to* should always be interpreted as *makes it possible to* rather than *serves the purpose of.* P1 and P2, read in this way, are well-formed scientific statements, as verifiable as the statements that the egg is made possible by the chicken and the chicken is made possible by the egg.

The use of the term *adaptation* is not necessary in this argument. The term is used to draw attention to the fact that features of the system of interest are being viewed from a functional standpoint, that is, from the standpoint of their contribution to the persistence of that system. Not all features make such a contribution. But one can argue that some features connected with food, reproduction, and adaptability must be necessary adaptations. Foodgetting, or metabolic adaptations, are necessary on thermodynamic grounds. Procreative adaptations are necessary, given the mechanism of variation and selection. Conceivably life could persist without this mechanism, but one can make strong informational arguments that it is necessary. Adaptability is a form of adaptation which is logically necessary since it refers to those features which enable life to persist under circumstances that are inevitably variable. One is tempted to say that the adaptabilities serve to adapt the foodgetting and procreative adaptations of the individual organism. But this is just the misleading usage into which I am here attempting not to fall. However awkward it sounds, the foodgetting and procreative adaptations must with equal necessity (all necessity is equal) serve to preserve the behavioral and other organismic adaptabilities. This follows from the principle of compensation. Unless all the behavioral adaptability is superfluous, in violation of the principle of compensation, it must enter with equal necessity into the circle of function.

The lack of normative neutrality evidenced by this asymmetry of usage has influenced the structure of important scientific theories, such as theories concerned with psychiatry and ethology. The phenomenological situation is that there are a number of different modes of behavior, such as eating behavior, sexual behavior, cleaning behavior, and exploratory behavior (including purely mental exploration). Some theories postulate that drives, or internally generated pressures, control the movement from one mode of behavior to another. Many authors have put primary emphasis on the aspects of behavior associated with sex and food and on the drives postulated to underlie sexual and foodgetting activities. The supposition is that since behavior must be structured by evolution these components must be the most fundamental. The analysis here shows that this view cannot be justified on *a priori* evolutionary grounds. According to the principle of compensation, mental adaptabilities in humans compensate for severe restrictions on evolutionary and other modes of adaptability. It is not scientific to view them as unidirectionally subserving foodgetting and sexual activities, any more than it would be scientific to regard evolutionary adaptability as unidirectionally subserving these activities. But due to compensation modes of behavior associated with mental adaptability must assume enormous importance in humans. The failure to recognize that these modes of behavior are biologically as fundamental as sexual and foodgetting modes has resulted in unnecessarily complicated theories of such phenomena as play, artistic and scientific creation, meditation, and dreaming. There is no *a priori* evolutionary reason to regard the introspectively self-evident urge to engage in these modes of behavior as illusory and properly reduced to biologically more fundamental urges. The assumption that sexual and foodgetting drives are exclusively fundamental is inconsistent with evolutionary theory. It has led to an automatic and, I believe, unhealthful undervaluing of creative and other adaptive mental activities in those societies which have accepted this assumption as a guiding axiom.

It might appear that this conclusion is contradicted by the sociobiological idea that the properties of an organism, such as its mental adaptabilities, serve to propagate patterns of genetic information, perhaps through the reproductive activities of organisms other than itself. But this just means that the circle of function threads through more than one organism. It is equally legitimate, or illegitimate, to suppose that the property which develops exists for the purpose of propagating patterns of genetic information as it is to suppose that these patterns of genetic information exist for the purpose of propagating the property. No empirical success which this inclusive fitness model could have—and it has had significant successes—could establish that the reproduction of genes is to a greater extent nature's objective than the adaptability which makes this reproduction possible.

It is interesting to consider one more area in which this asymmetry of usage has had consequences. The area is economics. There is a deep connection between the functional circle of adaptability and adaptation and the functional circle of an economic system. It is evident that inventiveness and other mental adaptabilities are necessary for technical and financial innovation and for the maintenance of economic life in the face of changed environmental conditions or disturbances emanating from the economic activities of other societies. It is evident that the production and delivery of goods and services are necessary for the maintenance of mental adaptability. The situation is summed up in the following two propositions, which correspond to P1 and P2:

P1′. Thought serves to maintain the total of goods and services maintained in an economy.

P2′. The exchange of goods and services in an economy serves to maintain thought.

According to the principle of compensation the ability to think enters into the functional circle of an economic system in the same way that the ability to evolve by genetic mechanisms enters into the functional circle of a preeconomic ecosystem. However much less compelling P2′ may sound than P1′, both are on a scientifically equal footing. It is a value judgment to suppose that the allocation of resources to food, clothing, and shelter is inherently more worthy or more self-evidently necessary than the allocation of resources to the thought process. Yet I think it fair to say that the problem for an economy is usually posed as the problem of delivering enough goods and services to satisfy criteria which the society establishes for itself rather than as the problem of delivering enough goods, services, and creative opportunities. The objection that some creative activities serve no useful purpose in terms of continuing the functional circle does not put goods and services on a higher plane. The exchange of some goods and services serves no useful purpose in this respect as well. Another possible objection made is that the creative activities should be part of the goods and services exchanged. But only the most external manifestations of the thought activity in a human society can be part of the measurable exchange of goods and services. Yet this thought activity is as essential for the regeneration of an economy as genetic mechanisms are for the perpetration of an ecosystem. To devalue it in the very posing of the economic problem is inherently necrotic.

Adaptability theory is a value-free theory. According to the adaptability theory analysis it is impossible to draw value conclusions from evolution theory. Value conclusions so drawn should not be taken as unconscious axioms in theories of human behavior and economics. Value conclusions imposed on theories of human behavior and economics should not be taken as an unconscious axiom in evolution theory. The asymmetry of usage which has established itself in these areas is not a feature of nature itself, but of the language user. It results from the use of ambiguous expressions such as *serves the purpose of*. In describing nature these are properly interpreted as *is* statements, but in describing our wishes and choices they are *ought* statements. As an individual the language user can value the different activities of life in any way he wishes. Each choice will have consequences. According to adaptability theory certain choices are viable, others are not. Which are viable depend on the external conditions and on the internal organization of the system. According to the principle of compensation, given external conditions are compatible with only certain patterns of organization and adaptability—that is, with only certain choices. There are a variety of possible choices which lead to nonchimerical forms, and an infinite variety which lead to chimera whose existence can only be short-lived. It is therefore important for scientific theories to be formulated in a way which is normatively neutral, so that it is sure that the structure of the theory will not improperly constrain the choices. This is a desirable limitation. Any scientific theory which could establish purpose in nature would serve as a guide to behavior, therefore would make it possible to derive *ought* statements from *is* statements. It would imply solutions to philosophical problems of a nonempirical and inherently arguable nature. The value-free criterion—I call it the normative razor—should be used to determine whether an interpretation of a scientific theory is acceptable or not. Evolution theory has an acceptable, nonnormative interpretation. In this interpretation activities connected with adaptability have the same status as activities connected with reproduction and metabolism, namely the status of being necessary for the continuation of the life process. There are practical reasons for using the normative razor. Any theory which purports to be descriptive of nature but which builds into its structure a constraint on the decision-making process will inevitably eventually lead to actions which are unsuitable to the conditions, therefore which will be unfavorable to the continuation of the life process. By letting language go on holiday, to use the apt phrase of Wittgenstein, I believe that this is just the situation which has established itself in commonly held interpretations of evolution theory and in putatively realistic theories of psychiatry and economics. Insofar as such theories embody the attitude that foodgetting and reproductive adaptations are self-evidently more fundamental than genetic or mental adaptabilities they are expressing the attitude of the language user rather than describing the situation in nature.

REFERENCES

Ashby, W. R. (1956) *An Introduction to Cybernetics*. Wiley, New York.

Bernard, C. (1949) *An Introduction to the Study of Experimental Medicine*. Henry Schumann, New York.

Bremermann, H. J. (1980) "Sex and Polymorphism as Strategies in Host–Pathogen Interactions," *J. Theor. Biol. 87*, 671–702.

Burnet, F. M. (1959) *The Clonal Selection Theory of Acquired Immunity*. Vanderbilt University Press, Nashville, Tennessee.

Cannon, W. B. (1939) *The Wisdom of the Body*. W. W. Norton and Company, New York.

Clausen, J. (1954) "Partial Apomixis as an Equilibrium System in Evolution," *Caryologia 1*, 469–479.

Conrad, M. (1972) "Statistical and Hierarchical Aspects of Biological Organization," pp. 189–221 in *Towards a Theoretical Biology*, Vol. 4, ed. by C. H. Waddington. Edinburgh University Press, Edinburgh.

Gardner, M. R., and W. R. Ashby (1970) "Connectance of Large Dynamical (Cybernetic) Systems: Critical Values for Stability," *Nature 228*, 784.

Grant, V. (1957) "The Plant Species in Theory and Practice," pp. 39–80 in *The Species Problem*, ed. by E. Mayr. AAAS Symposium 50, Washington, D.C.

Hutchinson, G. E. (1957) "Concluding Remarks," *Cold Spring Harbor Symp. Quant. Biol. 22*, 415–427.

May, R. M. (1973) *Stability and Complexity in Model Ecosystems*. Princeton University Press, Princeton, New Jersey.

Patrick, R. (1954) "Sexual Reproduction in Diatoms," pp. 82–99 in *Sex in Microorganisms*, ed. by D. H. Wenrich. AAAS Symposium, Washington, D.C.

Patten, B. C. (1978) "Systems Approach to the Concept of Environment," *Ohio J. Sci. 78*, 206–222.

Raper, J. R. (1954) "Life Cycles, Sexuality and Sexual Mechanisms in the Fungi," pp. 42–81 in *Sex in Microorganisms*, ed. by D. H. Wenrich. AAAS Symposium, Washington, D.C.

Ravin, A. W. (1960) "The Origin of Bacterial Species," *Bacteriol. Rev. 24*, 201–220.

Sonneborn, T. M. (1957) "Breeding Systems, Reproductive Methods, and Species Problems in Protozoa," pp. 155–324 in *The Species Problem*, ed. by E. Mayr. AAAS Symposium 50, Washington, D.C.

Stebbins, G. L. (1950) *Variation and Evolution in Plants*. Columbia University Press, New York.

Stebbins, G. L. (1960) "The Comparative Evolution of Genetic Systems," pp. 197–226 in *Evolution after Darwin*, ed. by S. Tax. Chicago University Press, Chicago.

von Uexkull, J. (1926) *Theoretical Biology*. Kegan, Paul, Trench, Tubner and Company, London.

Waddington, C. H. (1957) *The Strategy of Genes*. George Allen and Unwin, London.

Wiener, N. (1961) *Cybernetics; or, Control and Communication in the Animal and Machine*, 2nd ed. MIT Press, Cambridge, Massachusetts.

Williams, G. C. (1975) *Sex and Evolution*, Princeton University Press, Princeton, New Jersey.

Organization and Succession of Ecosystems

The starting point in the last chapter was the idea of homeostasis. Life-critical properties of the internal milieu are the objects of control. The key to the extension of the analysis to communities and ecosystems is that some features of the external milieu are drawn into this orbit of control. Organisms and communities either control their environment or are replaced by new organisms and communities, leading to a sequence of changes which can only be terminated by the appearance of a community which does control its environment. In the short run this is the ecological succession process. In time frames sufficient for genetic adaptation it is the evolution process.

But there is a difference, really a cardinal difference, between the application of adaptability theory at the organism and community levels. The organism, not the community, is the evolutionarily critical self-reproducing unit. This is why evolution can often usefully be viewed as an optimizing process, even if it is not known what is being optimized. For the community the utility of this viewpoint is very attenuated. Here the important consideration is the cyclicity of the interaction with the environment, hence the idea of environmental homeostasis. But the individual species does not know this. What is optimal from its short-sighted point of view may not be optimal for the community and therefore to its own advantage in the long run. This is the conflict between the evolutionary stability of the community and the evolutionary stability of species. The two do not always conflict, but when they do phenomena which are best thought of in terms of dual trends occur. Adaptability theory applies at both levels, but the allowable aids to inference are not the same.

12.1. MICROCOSM EXPERIMENTS AND THE REALITY
OF CONTROL

Some simple laboratory observations serve to illustrate the reality of environmental control. The observations involve aqueous microcosms (laboratory microecosystems) similar to the flask ecosystems which we have used as a gedanken device.

The experiments proceeded in the following way. Water and substrate scooped from a pond were cultured in glass tanks or jars, in this case ten-gallon tanks and completely closed sixteen-ounce jars. The composition and potentialities of this material were unknown, but the cultures were prepared as identically as possible. From time to time I removed water and substrate from tanks to make secondary jar cultures. In these cases tap water was added both to the jars and to the tanks, with no precautions taken to remove protisticides such as chlorine. These communities underwent a sequence of changes, eventually arriving at a rather slow-changing form of organization which can be viewed as a tentative climax. In some of the tank communities the final long-lasting organization observed (after about two years) was dominated by an algal species which completely carpeted the glass walls of the tank. The soil, initially on the bottom, layered the inside of this carpet. The algae had sequestered the substrate, effectively internalizing the external milieu in a way which maximized the amount of light captured by the community. The patterns which developed in some of the jars were more striking. In many an important initial process involves knitting the soil into a mat. This mat may rise due to the accumulation of gases. After about two years many of these mats had knotted themselves into a completely ball-like structure. This morphology maximizes their light-capturing ability and at the same time completely traps the substrate. This type of organization was still present at the last time of observation (after the third year).

In this simple ecosystem control of the environment is an observable fact. The morphology which develops allows for high nutrient retention and minimal loss of energy, making it resistant to replacement by any alternative community form. The connection to stability is obvious. Mechanisms which enhance nutrient retention tighten nutrient cycles, therefore contribute to environmental control. Blocking potential sources of energy which could provide a foothold for new species is also a *sine qua non* condition for stability and a form of environmental control. These features are not so obvious or easy to analyze when they are distributed in indirect ways over numerous species, but their manifest presence under conditions which are simple enough for them to be recognized argues for their generality. According to Odum (1969) the appearance of mechanisms which enhance nutrient retention and cycle tightening is in fact a general feature of the succession process.

I have done many experiments on the stability of succession in such microcosms under different conditions of environmental uncertainty and harshness. I will return to these later on. But I think the most obvious observation is the most remarkable. They persist as living systems and there is no indication that they will not persist indefinitely. I have had jar cultures which have persisted for over five years (at the time of this writing) and cultures in small test tubes which remained green for nine years. The initial medium may be very dilute in terms of biotic material, that is, with a thin layer of substrate, mostly tap water, and no life visible to the naked eye. That was the case with the ball-like communities described above. Such systems may initially be kept in a dark, cold place for two or three days. This serves to make the subsequent succession more determinate, undoubtedly by eliminating marginal potentialities. But even such initially poor systems go through a sequence of stages with different predominant species. The early communities consist of a number of floral species and some free-swimming protista. Thus, the apparent complexity increases, though one may reasonably presume that in a small, closed system the real complexity in terms of potentialities should decrease. In fact the tentative climaxes for such simple systems usually involve one to three dominant floral species, though sometimes marginal species and free-swimming protista persist or spring out of dormancy. But if one begins with a large amount of bottom matter and a fairly complex community, the complexity remains much higher and it is more common for the different interfaces to be controlled by different species. The number of species is smaller at the climax, but this is usually true for natural communities as well (Margalef, 1958).

Systems do fail when they are carried beyond the end of the physiologically tolerable. Sometimes they fail if they are exposed to major environmental shifts before they have had a chance to develop their adaptabilities. But in some cases the failures are only tentative. If one begins with a community which is very rich in terms of biotic material, an oxygen tension problem usually develops after it is sealed off. There is a crash, leading to a predominant accumulation of detritus (dead matter). This usually happens when water from a running stream is used, even if the tank or jar is left open. Such a community goes into a very dismal-looking state. But this is just a stage. If one has sufficient patience one finds that its demise is only an illusion. Some "green potentialities" have usually survived and are waiting for the long work of the less pleasant-looking decomposers.

A number of investigators dispute the conclusion that these closed systems can persist indefinitely. They argue that all such systems are slowly dying. This is a difficult question to resolve. It is possible that there is a minimum volume of life below which the loss of potentialities exceeds the origin of new potentialities. In a volume smaller than this life would slowly disappear. In a more complexly structured volume there would be more refuges for potentialities, so that the disappearance of life would be delayed. The volume would be smaller for microorganisms than for larger forms which are less numerous. Our earth is clearly above this volume. My strong suspicion is that the microcosms described here are sufficiently above this volume to survive indefinitely. When they are first moved from a large volume (nature) to a small one the number of potentialities which they can maintain becomes smaller. So at the beginning they lose potentialities more rapidly than they generate them, but eventually a point of balance is reached. At this point the basic principles are the same as for a natural community; only the details, degree of complexity, and history-dependent features differ. A completely safe conclusion is that even the simplest communities contain an enormous reservoir of potentialities and an enormous potentiality for new adaptation. These potentialities are the basis for control of the environment, which is the stabilizing factor in succession and evolution. But they are also the basis for the origin of forms which break this control, which is a major destabilizing factor. These interesting processes cannot be understood in terms of mathematical models which do not incorporate conceptual and operational connections to potentiality, as adaptability theory does.

12.2. ERGODIC ANALOGY

The microcosm experiments establish the reality of environmental control as an empirical fact. But it is interesting to consider some purely theoretical arguments. The basic idea is that the ecosystem will spend more time in a stable form of organization than in any particular unstable form. The stable form must be reachable. But it will be reachable as long as it is in principle possible for the ecosystem to assume a cyclic mode of behavior, that is, if it can cycle materials and periodically reset the environment. Under these conditions the environment is by definition controlled. The issue is one of the expected dynamical behavior of a collection of evolved units (organisms), not principally one of evolutionary competition between different ecosystems. If there is a significant level of such competition it would only strengthen the control. If an ecosystem could reach a state which is truly cyclic, this state would certainly last longer than all of the unstable states. To the extent that an ecosystem is increasingly captured by regions of this type, the treatment of the environment as increasingly controlled is interesting and justified. But in fact further evolution at the organism level can always open up leaks which eventually drain the cycle and destabilize the system.

There is an analogy between the succession of an ecosystem to a climax form of organization and the approach of a thermodynamic system to equilibrium. The analogy has weaknesses, but locating these points of weakness serves to deepen the discussion of the mechanisms by which ecological systems enter and exit situations in which features of the environment are more or less controlled.

Recall that the macroscopic properties of a thermodynamic system at equilibrium are those compatible with the largest number of microscopic complexions. The equilibrium state is the stable state because any egress from this large class of microscopic complexions is minuscule in comparison to the entrance to it. The assumption is that the system is ergodic, or statistically homogeneous.

How does one characterize a climax community? In general one specifies the species composition, the food web structure, the topographic organization, and also certain features of the environmental morphology (e.g., lakelike, pondlike, boglike, marshlike, meadowlike, forestlike). But if one could make a detailed inspection one would find that each of these gross states of organization is compatible with a large number of complexions defined in terms of more refined observations. These more refined observations could be in terms of any of the variables used so far (see Table 5.1). Or they could be defined in terms of detailed physical states of organization.

If the class of complexions associated with some form of ecosystem organization is much larger than the class of those associated with others the ecosystem, if ergodic, will spend more time in this form of organization. It will enter into larger and larger classes, until it finally enters into the largest class. Each class will be more stable than the others and the final class will be most stable. The sequence of entrees and egresses correspond to the seres in ecological succession, with the final, most stable sere corresponding to the climax form of organization.

Now I want to consider the limitations of the analogy. First it is not correct to assume that the ecosystem approximates ergodic behavior. It is a highly constrained system driven far from thermodynamic equilibrium by a constant throughflow of energy. Each organism is an astonishing complex of flexible chemical and mechanical feedbacks and ratchets, so much so that even a randomly assembled community would inevitably be highly constrained as well. This means that all regions of the phase space are not equally accessible, a feature characteristic of flexibly constrained systems. It is therefore possible for an ecosystem to be trapped in an equivalence class of complexions which is relatively small. This should be obvious since evolution produces systems which are unusual from the physical point of view, hence compatible with fewer microscopic descriptions. It can produce systems which are compatible with a smaller equivalence class of refined descriptions, whatever the level of refinement.

This smallness is the second limitation. The egress from an equivalence class of compatible complexions is not so minuscule in comparison to the entrance as is the case with ordinary thermodynamic systems. In principle fluctuations are more important and there is a chance that the system will jump or be pushed into another class of compatible complexions. In fact, ecological successions are quite deterministic. In undefined but similarly prepared laboratory ecosystems (like the flask ecosystem) the time of the moves from stage to stage is sometimes variable and the sensitivity to external conditions is often critical. But under identical conditions the very large majority of systems develop in the same way.

The third limitation of the analogy is that the averaging procedure and averages are not so clear as for thermodynamic systems. This is due both to the smallness of the classes and to the importance of constraints. The gross variables are therefore not so crisply defined. This is a key point, for it opens up the possibility that subtle macroscopic changes will occur which either are unnoted or do not appear significant. But these changes may be coordinated to the waxing or waning of a crucial constraint, hence to the dynamic opening up of a pathway to new regions of the phase space. Even an apparently extremely stable system may therefore be undergoing undetermined changes which will eventually destabilize it. But the apparent tendency most of the time will be in the direction of a situation which is ecologically stable.

12.3. EVOLUTIONARY VERSUS ECOLOGICAL STABILITY

The situation is complicated by the fact that the evolution of individual species is indifferent to the above considerations. A trait that contributes to ecological stability may not be evolutionarily stable. There are basically four possible situations, summarized in Table 12.1. The mixed situations (the feature not both ecologically and evolutionarily stable) should not be described in terms of single, simple models. The intermixing of two models is always necessary.

It might appear that evolutionary stability should always be more important than ecological stability, therefore that the former consideration should always be more important. But the relative importance of the two considerations depends on the time scales involved and on the intensity of

TABLE 12.1
Possible Fates of a Phenotypic Trait or Community Feature

Feature is ecologically stable?	Feature is evolutionarily stable?	Feature dominates or disappears
Yes	Yes	Dominates
Yes	No	Mixed
No	Yes	Mixed
No	No	Disappears

the selective forces. Some traits might only be mildly disadvantageous to the species, but so destabilizing to the community that extremely intense selective forces would soon ensue. The issue here, certainly one on which there is a divergence of opinion, has already been discussed at some length (Section 6.6). *The subtle point is that as a trait which is evolutionarily somewhat unstable but significantly ecologically stabilizing begins to disappear, more and more powerful constraints come into play. These select against the evolutionarily stable forms of the trait as well.* Under these circumstances the ecologically stabilizing trait becomes less evolutionarily unstable. Its relative occurrence revives, leading to a rehabilitation of the species and a revival in its absolute occurrence.* Alternatively, the period of decline may enable a different species, with a comparable trait, to outrace the original one. After more than two billion years of evolution one might not be surprised to find that some genetic ratchets have evolved which decrease the likelihood of ecologically destabilizing variations.

   Some work done in the context of origin of life models—as far removed from modern ecosystems in time as living systems could be—can, I believe, be interpreted in a way which sheds some light on the issue of ecological versus evolutionary stability. I refer to the hypercycle concept developed by M. Eigen (1971). The hypercycle is a set of autocatalytic components which directly or indirectly facilitate each other's production, that is, a set of cross-catalytically coupled autocatalytic elements. In Eigen's case the elements are prebiological polymers. The choice of the hypercycle structure was motivated by the fact that it is the simplest kinetic structure which (in its idealized form) allows coexistence, mutual stabilization, and coherent growth of different autocatalytic units (Eigen et al., 1980), all interesting features from the standpoint of prebiological evolution. But it might just as validly have been motivated by the fact that a hypercycle is an

*This type of selection dynamics emerged in the computational model of evolution discussed in Section 6.6.

abstraction of an ecosystem, for an ecosystem is most fundamentally a set of self-reproducing units facilitating each other's reproduction. The argument is sometimes made that selection can only act on an ideal hypercycle as a unit and that it tends to eliminate all but the most faithfully reproducing hypercycles. Alternatively, it could be stated that the most faithfully reproducing hypercycle is of necessity the strongest basin of attraction, though internal variation and selection processes involving individual species (the constituent cycles) could not realistically be eliminated as potential sources of destabilization. These two points of view reflect the dual contributions of ecological and evolutionary stability to ecosystem succession and evolution. Hypercycles, insofar as they are exemplified by ecosystems, cannot exclude selection on individual species, as sometimes claimed on the basis of studies of idealized hypercycles. But at the same time these studies suggest that the global constraints imposed on individual species in an ecosystem are much more powerful than those which could be deduced from classical population dynamics models, a consideration which should be incorporated into thinking about ecosystem organization.

12.4. ENVIRONMENTAL HOMEOSTASIS

While the naive thermodynamic analogy is inevitably somewhat wobbly, the connection to adaptability theory may be formalized in a useful way. The equivalence class of compatible descriptions corresponds to the hierarchy of adaptive, informational, macroscopic, and microscopic equivalence classes described in Table 4.1. It is connected to behavioral uncertainty. The effectiveness of the constraints is measured by the anticipation and indifference terms. The tendency of adaptability is determined by the evolutionary and dynamic considerations described above. But the minimal tendency should dominate, as evolutionary pressures would always eliminate superfluous constraint (e.g., would eliminate superfluous ability to anticipate), while dynamical considerations suggest that superfluous constraint would always be exchanged for a larger equivalence class of compatible descriptions.

The adaptability theory formalization is again suggested by considering features of the microcosm experiment. The physical environment is divided into two parts, that inside the flask and that outside. The outside part is completely outside the control of the organisms within. But the inside part is not. For any particular organism or population there is also a biotic component of the environment which it may influence. This distinction between inexorable and exorable parts of the environment is quite appropriate for the world at large, though with some roughening. Events in space, such as radiation from supernovas, influence life, but are not influenced by it; the vagaries of weather, even if influenced by the development of life on earth, are still vagaries. But the concentrations of carbon, phosphorus, water, and other materials are largely determined by the activities of organisms.

There are two forms of control which can be exerted on these features:

1. *The diversity of community states may be increased* (cf. Section 4.10). This exerts control if the increase reflects the incorporation of mechanisms which cycle materials or otherwise repair the environment. This does not reduce environmental uncertainty. Even if there is no cycle the environment may change in an entirely determinate way. *But creating cycles increases the ability of the community to anticipate the environment and decreases undesirable indifference.* It decreases the uncertainty of the environment from the point of view of the community. Cycling does not necessarily decrease the diversity of the environment states, but it generally does so. Insofar as it is controllable, the cycle of environmental states can in principle be contracted to a single state if the diversity of community states is made high enough.

2. *The community may use its adaptability to buffer the controllable environment from the uncertain perturbations of the uncontrollable environment.* Here the community may be thought of as internalizing part of the environment and then making this internalized environment an object of control. Originally its adaptability had to exceed the uncertainty of both the controllable and uncontrollable environment. But after internalization the adaptability is increased and the controllable uncertainty is decreased. This contradicts the evolutionary tendency for superfluous adaptability to decrease, but the advantage of internalization outweighs this. When the internalization phase comes to an end, the excess adaptability can decline without resulting in an increase in the uncertainty of the controllable part of the environment. There are many examples of such internalization: the environment inside the nest, the beaver's dam, the human home. At one time in the ancient past the present day internal milieu may have been an internalized environment, such as regulated sea water in a body cavity.

To make this buffering argument precise it is convenient to write

$$\hat{\omega}^* = \hat{\omega}_c^* \hat{\omega}_u^* \tag{12.1}$$

where $\hat{\omega}_c^*$ is the partial transition scheme for the controlled features of the environment and $\hat{\omega}_u^*$ is the partial transition scheme for the uncontrolled features. There are two possible points of view. In the first $\hat{\omega}_c^*$ is treated, as usual, as part of the environment. Then equation (6.4) becomes

$$H(\hat{\omega}) - H(\hat{\omega}|\hat{\omega}_c^*\hat{\omega}_u^*) + H(\hat{\omega}_u^*|\hat{\omega}) + H(\hat{\omega}_c^*|\hat{\omega}\hat{\omega}_u^*) \rightarrow H(\omega_u^*) + H(\omega_c^*|\omega_u^*)$$

$$(12.2)$$

In the second $\hat{\omega}_c^*$ is treated as an extension of the biota (in the sense that the beaver's dam is an extension of the beaver). This gives the equation of extended adaptability:

$$[H(\hat{\omega}) - H(\hat{\omega}|\hat{\omega}_u^*)] + [H(\hat{\omega}_c^*|\hat{\omega}) - H(\hat{\omega}_c^*|\hat{\omega}\omega_u^*)] + H(\hat{\omega}_u^*|\hat{\omega}\hat{\omega}_c^*) \rightarrow H(\omega_u^*)$$

$$(12.3)$$

If the features of the environment associated with $\hat{\omega}_c^*$ are an object of control, the terms $H(\hat{\omega}_c^*|\hat{\omega})$ and $H(\hat{\omega}_c^*|\hat{\omega}\hat{\omega}_u^*)$ should be small. This is possible if

$$d[H(\hat{\omega}) - H(\hat{\omega}|\hat{\omega}_u^*) + H(\hat{\omega}_u^*|\hat{\omega}\hat{\omega}_c^*)]/dt > 0 \qquad (12.4)$$

That is, an increase in extended adaptability allows an increase in the correlation between the controllable features of the environment and the biotic community proper. There is an advantage only if $H(\omega_c^*|\omega_u^*)$ in equation (12.2) becomes smaller, that is, if the controllable features become more certain but freer from the uncontrollable features. But then, according to equation (12.2),

$$d[H(\hat{\omega}) - H(\hat{\omega}|\hat{\omega}_c^*\hat{\omega}_u^*) + H(\hat{\omega}_u^*|\hat{\omega})]/dt < 0 \qquad (12.5)$$

$[H(\hat{\omega}_c^*|\hat{\omega}\hat{\omega}_u^*)$ is here omitted since it is small by assumption.] In words, the usual adaptability should decrease to its minimum possible value, as usual. Suppose that

$$dH(\hat{\omega}|\hat{\omega}_c^*\hat{\omega}_u^*)/dt \sim dH(\hat{\omega}|\hat{\omega}_u^*)/dt \qquad (12.6a)$$

and

$$dH(\hat{\omega}_u^*|\hat{\omega}\hat{\omega}_c^*)/dt \sim dH(\hat{\omega}_u^*|\hat{\omega})/dt \qquad (12.6b)$$

where similarity here means that at least the sign is the same. Then it is impossible for the extended adaptability to increase and at the same time for the usual adaptability to decrease. This is inconsistent, so one of the assumptions must be wrong. There are only two possibilities. The first is that the correlations (12.6a) and (12.6b) do not always hold. Then

$$dH(\hat{\omega}|\hat{\omega}_c^*\hat{\omega}_u^*)/dt > 0 \qquad \text{with} \qquad dH(\hat{\omega}|\hat{\omega}_u^*)/dt \leqslant 0 \qquad (12.7a)$$

and/or

$$dH(\hat{\omega}_u^*|\hat{\omega}\hat{\omega}_c^*)/dt > 0 \qquad \text{with} \qquad dH(\hat{\omega}_u^*|\hat{\omega})/dt \leqslant 0 \qquad (12.7b)$$

(The terms on the left-hand side could also equal zero, but then the terms on the right could not.) Equation (12.7a) means that the biota becomes more indifferent to the influenceable (now controlled) features of the environment, but not to the uninfluenceable (uncontrolled) features. Equation (12.7b) says that the controlled features of the environment anticipate the uncontrollable features less well, but the biota anticipates them better. Conditions (12.7a) and (12.7b) describe the final situation, but of course do not explain how the system reaches this situation.

The second possibility is that the evolutionary tendency assumption expressed by equations (12.2) and (12.5) does not always hold. This would happen if the advantage of an increasingly controlled environment exceeded the cost of an increase in superfluous adaptability, a very reasonable assumption. The development of control would then proceed in two stages:

Stage 1. The influenceable features change in a nonrepeatable, non-random way unless they are controlled. Thus stabilization of an ecosystem is necessarily accompanied by an increase in extended adaptability. This drives an increase in the usual adaptability, including an increase in $H(\hat{\omega})$.

Stage 2. Once the ecosystem is stabilized, the minimal tendency of adaptability becomes manifest again. Increased anticipation and decreased indifference allow realization of the conditions expressed in equations (12.7a) and (12.7b). The actual magnitudes [such as the magnitude of $H(\hat{\omega})$] should also decrease.

The essential feature of a controlled environment is that it is more independent of both the community and the external environment than it would be if it were uncontrolled. In the first stage the development of this control dominates; in the second stage refinement of correlation between biota and external environment dominates.

12.5. FOOD WEBS AND CONSERVATION LAWS

Now it is appropriate to classify the mechanisms by which communities either adapt to or control variations in the input of matter and energy. To do this I describe the material and energy flow in terms of a food web and then utilize the fact that these flows must be compatible with the conservation laws of energy and matter. The simple web in Figure 12.1 illustrates the notation to be used here. It splits into two parts, a grazing pathway part and a detritus pathway part. The grazing pathway is basically the sun $\rightarrow$ grass $\rightarrow$ cow $\rightarrow$ lion type series of transformations. The detritus pathway is the degradation pathway, important for cycling materials back to the environment. Real food webs are much more complicated. The distinction between grazing and detritus pathway species is not always sharp. Species

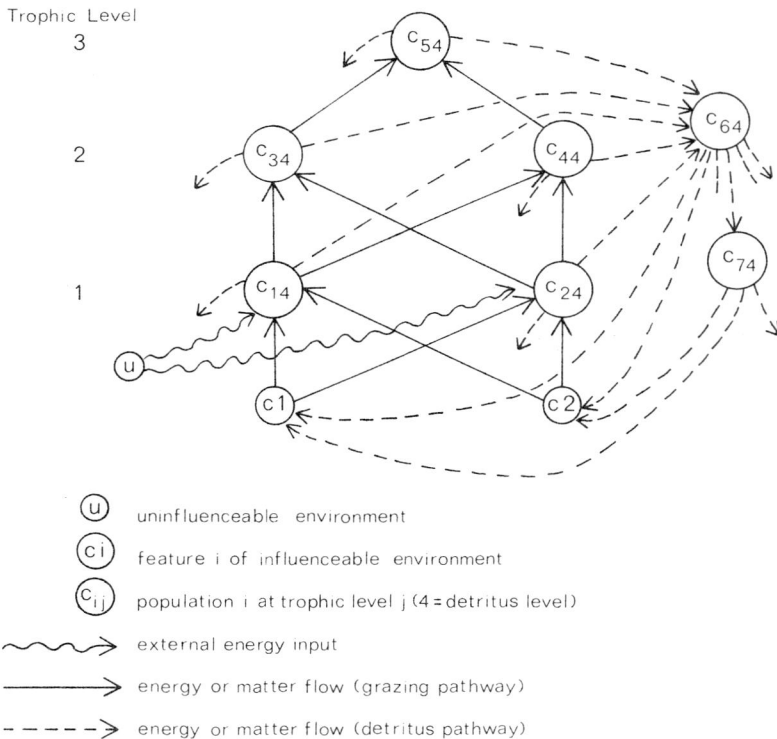

FIGURE 12.1. Illustrative food web. The uninfluenceable environment (u) includes the external energy input, usually radiant energy. Influenceable features of the environment ($c1$ and $c2$) include the sources of material input, although materials could also be injected from u. Arrows which do not lead to other units represent dissipation of energy or leakage of materials. Other possible food webs could be obtained by adding or removing arrows of any type, provided that there is no biotic unit without some arrow leading into it.

may also feed on more than one lower trophic level and may feed on themselves. But thinking in terms of a simplified web is sufficient to illustrate the major principles.

The matter and energy available to a community may vary because of variations in the input of radiant energy (an absolutely uncontrollable feature), because of variations in the input or topographic distribution of chemical matter, or because of variations in the conditions affecting the rates of processes. But the conservation laws of energy and mass, equation (2.1), impose powerful constraints on the possible reactions of the community to these variations. These laws imply that the variation in the total of the energy and mass entering a community must be equalled by variation in the total of the energy and mass of the community plus that leaving the community. *But the variations in the total of the energy and the mass do not have to equal the total variation of either the energy and mass inputs or the energy and mass residing in and leaving the community.* There are many local variations and only rarely could they all be in the same direction. Many compensate one another, so that the variation in the totals is less than the sum of all the local variations. Adaptability cannot reduce the variation in the totals, but it can exploit compensating variation to reduce the total variation or to distribute it in a less costly way. The conservation laws allow the community (also each species) five possible mechanisms for doing this:

1. Through physiological adaptability mechanisms which allow individual organisms to cope with variable conditions affecting the use of energy.
2. Through variations in the size or activity of individuals.
3. Through variations in the number of individuals in populations.
4. By organizing themselves so as to take advantage of compensating variations in sources of energy or in conditions affecting the uses of these sources.
5. By organizing themselves so as to reduce or ignore the environmental variations.

Each of these alternatives involves costs which significantly determine the allocation of control. Alternatives (1) and (2) involve the usual costs associated with physiological variability (cf. Chapter 6). Utilization of alternative (1) is allowed by the conservation laws only if the mass and energy inputs are constant. If they vary, or if alternative (1) fails, all remaining forms of physiological variability must involve variation in the energy storage or energy transformations of individuals. Ultimately this means variation in their heat export, including variation in activities associated with growth, maintenance, and reproduction. The organism must pay for this capacity to vary its level of activity whether it does so in a way which appears graceful or in a way which appears manifestly enforced (such as the capacity to suffer hunger). The cost is not so great in simpler organisms, where graceful mechanisms such as dormancy, growth plasticity, and wide variation in acceptable metabolic rate are possible. It is greater in complex forms in which the internal environment must be maintained within a much narrower range. Here growth plasticity involving storage mechanisms (such as fat deposits) is of course possible, but the extent of the variation is less and the cost in terms of required regulatory mechanisms greater.

Alternative (3) involves oscillations in population size. It is also costly since the regrowth of the population generally lags behind the reappearance of the energy supply. The growth of the population involves extra energy expenditure as well. These costs are not so important in the simpler, culturable organisms, but they are increasingly significant in the more complex forms. Lagging behind the energy supply implies a decrease in the energy content of the community.

The possibility of exploiting compensating variation (alternative 4) increases if species use more sources of energy or if they have fewer requirements. In both cases the species distinguish between fewer energy sources, therefore are less externally specialized. As discussed previously (Section 6.2) such decrease may be connected to the indifference or to the behavioral uncertainty term. In either case it is costly since the species must be able to synthesize or degrade more types of components.

Alternative (5) amounts to restoration of the physiochemical character of the environment, that is, to the condition of balance. There is no possibility for controlling variations in the radiant energy available for photosynthesis. But mass components are cycled, therefore variations in the variability can in principle be controlled. The basic requirement is that the niche structure be complete in the sense that enough chemical and structural transformations are affected to restore the chemical and morphological character of the environment, and to do so without significant time delays. If the rate of withdrawal of materials from and replenishment of materials to the environment is not as equal as possible over any given time, internal oscillations will be inevitable, implying a variable availability of materials even if the community is stable in the long run. Such completion of niche structure either requires an increase in the number of functions performed by individual organisms or an increase in the number of types of species. The cost and advantage issues are complex and will be discussed later (Section 12.10). Alternative (5) includes as a logical possibility the case in which either the community or the species refrains from fixing all the energy it could or absorbing all the mass components which it could. If the amount sacrificed were sufficient to exceed any likely downward variation, the environment would in effect be less variable. But such sacrifice is clearly costly.

12.6. FORMAL CONNECTIONS TO COMPONENTS OF ADAPTABILITY

This section presents a number of theorems which establish the formal correspondences between components of adaptability (behavioral uncertainty, anticipation, indifference) and the forms of adaptability to variations in availability allowed by the conservation laws. These theorems make it possible to assess the contribution of food web plasticity to community adaptability by considering the structural and dynamic features of the web. They do not imply that webs will always tend to become structured for maximum possible adaptability. Such structuring is always ecologically stable, but only evolutionarily stable up to a certain point.

In what follows the potential or allowable uncertainty of a food web structure is the maximum uncertainty compatible with the continued presence of all the species in the community. Variations in the availability of energy or materials include both variations in their input and variations in conditions affecting their use. Unless otherwise stated, the theorems below and in the remainder of the chapter refer to both cases.

THEOREM 1 (uncertainty of energy flow). The adaptability of a community to variability in the availability of energy or materials increases as the potential uncertainty of its food web structure increases, assuming no compensating reductions in lower-level modes of adaptability. The potential uncertainty of the food web structure increases as the proportions of energy and materials flowing between sources and species or between species become more uncertain. It also increases as the number of alternative pathways of energy and material flow increases. The uncertainty of the web is highest for a given total variation in the flow of matter and energy if the proportions flowing between its elements (sources, species) become more independent of one another and more independent of lower-level modes of adaptability.

PROOF. By definition [Section 5.2, item (1) on page 84, also Table 5.1] the partial state of the community can be specified by

$$\dot{\beta}^i_{15} = \left(tr^i_{11}, \ldots, tr^i_{mn}; v^i_1, \ldots, v^i_p \right) \tag{12.8}$$

Here tr_{hk}^i is the fraction of total energy or matter reaching species k from source h when the community is in partial state i.* All other community variables $(v_1^i, \ldots, v_p^i)$ may be ignored without loss of generality. The adaptability contributed by the food web is

$$H_e(\hat{\omega}_{15}) - H_e(\hat{\omega}_{15}|\hat{\omega}_c^* \hat{\omega}_u^*) \tag{12.9}$$

where

$$H_e(\hat{\omega}_{15}) = cH(\hat{\omega}_{15}) + \text{independence terms} \quad (0 < c < 1) \tag{12.10}$$

The term

$$H(\hat{\omega}_{15}) = \sum_{i,r,s} p\big[\beta^i(t), \varepsilon^s(t)\big] p\big[tr_{11}^i(t+\tau), \ldots, tr_{mn}^i(t+\tau)|\beta^r(t), \varepsilon^s(t)\big]$$

$$\times \log p\big[tr_{11}^i(t+\tau), \ldots, tr_{mn}^i(t+\tau)|\beta^r(t), \varepsilon^s(t)\big] \tag{12.11}$$

represents the potential uncertainty associated with the food web structure (cf. also Table 5.2). This proves the first claim, that adaptability increases as the potential uncertainty of the food web structure increases.

The second claim, that potential uncertainty increases as the transfer proportions become more uncertain, follows from the fact that

$$p\big[tr_{11}^i(t+\tau), \ldots, tr_{mn}^i(t+\tau)|\beta^r(t), \varepsilon^s(t)\big] \leqslant \prod_{h,k} p\big[tr_{hk}^i(t+\tau)|\beta^r(t), \varepsilon^s(t)\big]$$

$$\tag{12.12}$$

Thus as the uncertainty measured over the individual probabilities increases, the potential uncertainty measured over the joint probabilities increases. The third claim, that potential uncertainty increases with the number of alternative pathways, follows from the observation that the index set of i can increase if the index sets of h and k increase.

The claim concerning independence of the flows follows from the fact that the $p[tr_{11}^i(t+\tau), \ldots, tr_{mn}^i(t+\tau)|\beta^r(t), \varepsilon^s(t)]$ become more equal with less total variation in the energy and matter flow as all conditional probabilities, $p[tr_{hk}^i(t+\tau)|tr_{11}^i(t+\tau), \ldots, tr_{mn}^i(t+\tau), \beta^r(t), \varepsilon^s(t)]$, approach the unconditional probabilities, $p[tr_{hk}^i(t+\tau)|\beta^r(t), \varepsilon^s(t)]$. The claim concerning independence of levels follows from the fact that total variation in matter and energy flow is smallest when the independence terms are largest. These terms, $[H(\hat{\omega}_{15}|\hat{\omega}_{14}), H(\hat{\omega}_{15}|\hat{\omega}_{14}, \hat{\omega}_{24}) \ldots]$, are largest when the specification

---

*Recall that the superscript i is a dummy variable. Thus the $\dot{\beta}_{15}^i$ is the same as the $\dot{\beta}_{15}^r$ in Table 5.1.

of the biomass and energy transformations of each species gives the least information about the transfer proportions.

COMMENTS. The proof does not depend on the assignment of the transfer terms, tr_{mn}^i, to the community level. They could also have been assigned to organisms or populations. But the assignment to the community level corresponds to our original association between variables of a biological system and levels of organization (Table 5.1). A different choice of association would only mean a different, though less convenient, way of representing these variables in terms of levels.

MacArthur (1955) and earlier Odum (1953) suggested that the stability of a food web could be indexed by the amount of choice of pathways of energy flow. MacArthur used the index $I_s = -\Sigma_i q_i \log q_i$, where q_i is the fraction of total energy reaching to the top of each web along each distinct pathway. Thus this index is highest when the q_i are all equal and equal to the reciprocal of the number of possible pathways. It has definite limits and may be the same for food webs involving many species with few connections as for food webs involving fewer species with many connections. Theorem 1, first presented outside the formal framework of adaptability theory (Conrad, 1972), was stimulated by MacArthur's most insightful paper. But it should be noted that the index chosen by MacArthur is actually not a suitable measure of choice from the standpoint of either adaptability or stability. The adaptability contributed by the web depends on the *variability* in the proportions, $(tr_{11}^i, \ldots, tr_{mn}^i)$, not on the proportions themselves. In the absence of such variability there will be no contribution to adaptability even if there are many lines of flow. It is the potential contribution to adaptability which increases with the equitability of the proportions and with the number of lines of flow. I_s does provide an index for the potential contribution.

Another important point is that there is no contribution to adaptability unless the uncertainty is reduced. The significance of this is shown in Theorem 2. As shown in Theorem 3, indifference makes no contribution to community adaptability. But when a single population in the community is considered, with the remaining populations treated as part of its environment, some of the mechanisms of adaptability are shifted into the indifference term. These theorems show that there is no necessary connection between dynamical stability of food webs and structural features such as proportions of energy flow and diversity of species. The proper connection is to the adaptability of the web.

THEOREM 2 (contribution of anticipation). Community adaptability to variations in availability increases as the uncertainty of the food web structure is increasingly reduced by specification of the environment. It increases further as specification of the biomass and energy transformation of species further reduces the uncertainty of food web structure. Part of the reduction is due to constraints imposed by the conservation laws of energy and mass. Fixed thresholds, that is, limitations in the amount of energy and materials which species can assimilate from each source, allocate uncertainty in the web but do not contribute to the reducibility of this uncertainty. Flexible thresholds allow for routability of energy and materials and therefore can be used to increase reducibility.

PROOF. According to equation (12.9) the adaptability increases as $\langle H_e(\hat{\omega}_{15}) - H_e(\hat{\omega}_{15}|\hat{\omega}_c^*\hat{\omega}_u^*)\rangle$ decreases. The lead term, $H(\hat{\omega}_{15}|\hat{\omega}_c^*\hat{\omega}_u^*)$, expresses the first claim. The dependence terms (such as $H(\hat{\omega}_{15}|\hat{\omega}_c^*\hat{\omega}_u^*\hat{\omega}_{14})$) express the second claim. The conservation laws of energy and mass fix variation in the total energy and biomass, therefore constrain the total variation. Fixed thresholds are independent of the environment, therefore cannot be used to increase the difference between the web uncertainty and the anticipation term. They can force routing of energy and mass to different parts of the web (for example, from grazing to detritus pathways). By lowering the thresholds it is possible to reduce the variation in one part of the web, but then this variation must appear in some other part. Flexible thresholds make it possible to increase web uncertainty by increasing the uncertainty of the transfer proportions. These proportions are determined once the resource availability of each of these sources is specified, so an increase in reducibility is possible.

COMMENTS. The role of threshold behavior is illustrated in the simple webs of Figure 12.2. Fixed thresholds in upper-level grazing pathway species allocate the uncertainty to the detritus pathway. In web (c) a fixed threshold in only one of the second-level species allocates uncertainty to both the other upper-level grazing pathway species and the detritus pathway. The thresholds are flexible if different amounts of mass and energy can be withdrawn from the sources. In web (c) a compensating variation in the sources can be completely absorbed in the variability of the transfer proportions if both the species at the second trophic level have fixed total thresholds, but can assimilate variable amounts from each of their sources. Flexible thresholds without any fixed total thresholds allow for distribution of the uncertainty through the web, but not for controlled allocation of the uncertainty. Utilization of compensating variation in the sources is possible in webs (a), (c), and (d), but not in (b).

It is interesting that allocation of uncertainty to the detritus pathway through fixed thresholds and use of compensating variation both stabilize

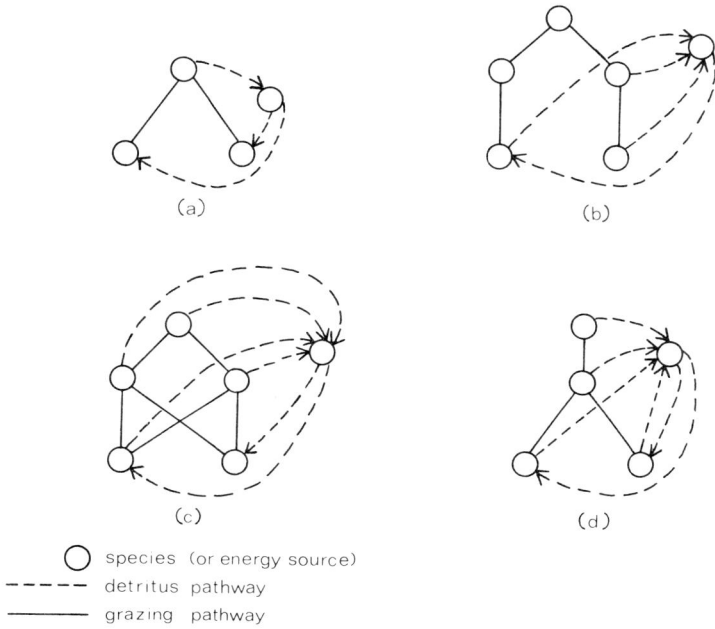

species (or energy source)
— — — — detritus pathway
———————— grazing pathway

FIGURE 12.2. Simple food webs. The maximum index of grazing pathway stability is 1 bit for webs (a), (b), and (d), and 2 bits for web (c). The actual stability may be quite different when each species is characterized by a threshold, or when variation in the total energy at the bottom level is less than the sum of the absolute values of the variations of individual species (or sources). The possibility of taking advantage of such compensating variations increases with the average number of sources per species. This is 2 for web (a), $4/3$ for web (b), 2 for web (c), and $3/2$ for web (d).

upper-level grazing pathway species. That is, the trajectories of these species appear less perturbed by environmental variation. This illustrates the point that the dynamical stability of the web is not directly related to the number of possible pathways of energy flow. Complete stability is possible even when the number of pathways is small, provided that the transfer proportions can vary. But this increased stability is compensated for by increased instability in the detritus pathway.

THEOREM 3 (indifference component). The adaptability of a community to conditions affecting the use of energy can be increased by increasing the indifference contribution. But the adaptability to variability in the input of energy and mass cannot be increased in this way. This limitation applies to total variability as well as to the variability of the total.

PROOF. The indifference component of the complete community is given by $\sum H_e(\hat{\omega}_{h0}^*|\hat{\omega}_{15}, \cap \hat{\omega}_{i4})$, where $\hat{\omega}_{h0}^*$ includes both controllable and uncontrollable features. $\hat{\omega}_{h0}^*$ can then be written as $\hat{e}_{h0}^* \hat{f}_{h0}^*$, where $\hat{e}_{h0}^*$ refers only to the energy and mass components of the state set on which $\hat{\omega}_{h0}^*$ is defined and $\hat{f}_{h0}^*$ refers to all other environmental variables. If terms such as $H_e(\hat{f}_{h0}^*|\hat{\omega}_{15}, \cap \hat{\omega}_{14})$ increase, this means an increase in community adaptability to conditions which in general would affect the use of energy and mass. But according to the conservation laws of energy and mass $H_e(\hat{e}_{h0}^*|\hat{\omega}_{15}, \cap \hat{\omega}_{14})$ cannot change.

COMMENTS. The importance of this theorem is that it shows clearly that utilization of compensating variation is represented by the anticipation, not the indifference term. The condition for compensating variation is

$$\sum H_e(\hat{\omega}_{h0}^*) > H(\cap \hat{\omega}_{h0}^*) \tag{12.13a}$$

or

$$\sum H_e(\hat{e}_{h0}^*) > (\cap \hat{e}_{h0}^*) \tag{12.13b}$$

for the energy and mass components. Since the community cannot utilize any of these correlations through indifference mechanisms, the only possibility is to utilize them through the anticipation mechanisms (as in Theorem 2).

It is interesting to compare the expressions for adaptability as written from the standpoint of an individual species. From the standpoint of the community

$$\left[\sum H_e(\hat{\omega}_{ij}) - \sum H_e(\hat{\omega}_{ij}|\cap \hat{e}_{h0}^*, \cap \hat{f}_{h0}^*)\right] + H\left(\cap \hat{e}_{h0}^*, \cap \hat{f}_{h0}^*|\cap \hat{\omega}_{ij}\right)$$
$$\rightarrow H(\cap e_{h0}^*, f_{h0}^*) \tag{12.14}$$

where $\sum H_e(\hat{\omega}_{ij}) = H(\hat{\omega}_{15})$. From the standpoint of any constituent species

$$\left[\sum_{k \neq s} H_e(\hat{\omega}_{hk}) - \sum_{k \neq s} H\left(\hat{\omega}_{hk}|\cap \hat{e}_{h0}^*, \cap \hat{f}_{h0}^*, \hat{\omega}_{15}, \bigcap_{i \neq j} \hat{\omega}_{14}\right)\right]$$

$$+ H\left(\cap \hat{e}_{h0}^*, \cap \hat{f}_{h0}^*, \hat{\omega}_{15}, \bigcap_{i \neq j} \hat{\omega}_{14}|\hat{\omega}_{j4}\right) \rightarrow H\left(\cap e_{h0}^*, \cap f_{h0}^*, \hat{\omega}_{15}, \bigcap_{i \neq j} \omega_{14}\right) \tag{12.15}$$

where $\sum_{k \neq s} H_e(\hat{\omega}_{hk}) = H(\hat{\omega}_{j4})$. In both cases the terms in the square brackets include all components of physiological and population adaptability discussed in the previous section (variability in size, activity, numbers). As usual the amount of variation required for a given degree of adaptability is least when the different forms of adaptability are most independent and when the population is best able to anticipate the environment. The important difference is in the indifference terms. According to Theorem 3 the indifference term for the community cannot handle variability in the inflow of mass and energy. But the indifference term for the individual species can. This is perfectly consistent with the conservation laws and in fact in the limit it is possible to have

$$\sum_{h,\,i\,\neq\,j} H_e\left(\hat{e}_{h0}^*, \hat{f}_{h0}^*, \hat{\omega}_{15}, \hat{\omega}_{14}|\hat{\omega}_{j4}\right) \rightarrow H\left(\cap\, \hat{e}_{h0}^*, \cap\, \hat{f}_{h0}^*, \hat{\omega}_{15}, \bigcap_{i\,\neq\,j} \hat{\omega}_{14}|\hat{\omega}_{j4}\right)$$

$$(12.16)$$

This would happen if all compensated variation were annihilated by the $\hat{\omega}_{15}$ term, that is by the variability of the transfer terms. This is the reroutability process. Thus the energy and mass content of a species, in fact of all species in the community, could be indifferent to compensated variation in the environment. When thresholds are lowered, some species become more indifferent, but at the expense of other species becoming more sensitive.

This is what has happened. From the community point of view the contribution of the transfer terms, hence reroutability, belongs partly to anticipation and partly to indifference. Anticipation involves the use of information to alter feeding patterns or clock mechanisms to adjust activity levels prior to a change in the environment. Indifference means despecialization, therefore it increases as the species can choose from among more different sources of food. The implication is that whenever such anticipation or indifference is an advantage to the species, hence evolutionarily stable, it will appear automatically as reroutability from the standpoint of the community. There are, then, conditions under which reroutability is evolutionarily as well as ecologically stable.

12.7. SPECIES DIVERSITY

THEOREM 4 (species numbers). The potential contribution of food web structure to a community's adaptability to variations in the availability of energy or materials increases as the number of species increases.

PROOF. The potential number of alternative pathways of energy and matter flow increases as the number of species increases. Thus Theorem 4 is a direct corollary of Theorem 1.

Comparison to the natural situation. Species diversity increases with number of species, but also may be defined to reflect the equitability of their occurrence [Table 3.1, also equation (11.7)]. The relation between adaptability and equitability of occurrence is complicated by the fact that a population may consist of a large number of small organisms. If the average proportions of matter and energy flowing between different elements of the web become more equal, more variability of these proportions, hence more adaptability, should in general be possible. Also the presence of only a very few individuals in a population may be incompatible with variations in the proportions which are associated with organism death.

A number of authors have reported that diversity does increase in the course of succession, particularly in the early stages (Margalef, 1958; cf. Odum, 1971). A marked increase in the early stages is often followed by a less marked decrease in the final stages, but more complicated patterns of increase and decrease have also been observed. Several factors contribute to these changes. The entry of new species by migration, the pressures for completion of the niche structure, and the selective advantage of specialization all work to increase the number of species in the early stages of succession. But the adaptability may also increase through overlapping multifunctionality, flexible thresholds, and the use of compensating variation. These despecializations also increase the number of possible pathways of energy flow and in many cases are an advantage to the individual species. Thus competitive exclusion and also elimination of marginal species in the final stages of succession should decrease the number of species.

The conclusion is that the phenomenological increase in diversity implies a potential increase in community adaptability. But increase in adaptability is not the only or even the main causative factor here. Selective despecialization in the final stages of succession also increases community adaptability and may explain some of the decrease in diversity which is often observed at this point.

12.8. DETRITUS PATHWAY AS BUFFER SYSTEM

CLAIM (detritic control). The contribution of the detritus pathway to community adaptability increases in the course of succession and that of the grazing pathway decreases. The detritus pathway populations increasingly control both the grazing pathway populations and the environment. The grazing pathway populations become increasingly controlled.

ARGUMENT. According to equation (12.14) the adaptability of the complete community can be written as

$$\left[H_e\left(\hat{\omega}_{15}\right) - H_e\left(\hat{\omega}_{15}|\cap \hat{e}_{h0}^*, \cap \hat{f}_{h0}^*\right)\right] + A_g + A_d \to \sum H_e\left(e_{h0}^*, f_{h0}^*\right)$$

$$(12.17)$$

where A_g is the adaptability of the grazing pathway populations and A_d is the adaptability of the detritus pathway populations. These are given by

$$A_u = \sum_{q<5} H_e\left(\hat{\omega}_{pq}\right) - \sum_{q<5} H_e\left(\hat{\omega}_{pq}|\cap \hat{e}^*_{h0}, \hat{f}^*_{h0}\right) \qquad (12.18)$$

where $A_u = A_g$ when (p, q) runs over the index sets of the grazing pathway populations and $A_u = A_d$ when (p, q) runs over the index sets of the detritus pathway populations. The indifference term has been omitted since it makes no contribution when the disturbances involve the inflow of mass or energy. According to Theorems 1 and 2 the routability terms [bracketed in equation (12.17)] do contribute to adaptability. But according to the laws of conservation of mass and energy, variability in the total inflow of energy and mass must appear in either A_g or A_d, that is, in terms of culturability, variation in activity, or variation in size. This establishes that small A_g requires A_d to be large. This is possible if there are limitations on grazing pathway populations which cause variations in energy and mass to be routed to the detritus pathway populations (cf. Theorem 2). Thus the successional question reduces to the ecological and evolutionary stability of such limitations.

1. *Obligatory limitations.* The advantages which large size and complexity bring to many (not all) grazers make the obligatory limitations which are concomitant to large size and high complexity evolutionarily stable. These grazers will not be replaced by forms which have a greater capacity to utilize upward variations in the availability of mass and energy. But the opposite is the case for detritics, where small size and high specialization are often an advantage. Hence natural selection acting at the organism level leads to a community structure in which the greater portion of variability in input is captured by the detritus pathway. This capture reduces variations in the size of grazing populations which would otherwise follow variations in input, so detritus buffering from upward variations implies buffering from downward variations as well. It also reduces the amount of variation generated internally by the growth dynamics of the grazers.

2. *Voluntary limitations.* Voluntary limitations are those which result from self-regulation rather than being unavoidable correlates of physiology. Density-dependent control mechanisms which hold population size below the maximum which it could reach at any point in time are an example. Here material is not so much captured by the detritus system as sacrificed to it. A species which makes this sacrifice is always evolutionarily unstable. But there are three factors which may nevertheless lead to the evolution of some

voluntary limitation. Shunting of material to the detritus pathway protects the population dynamics of the grazers from a greater range of environmental variation. In the short run a mutant form which utilizes the upward variation as much as possible will leave the most offspring, but a population of such mutants will be more decimated during periods of downward variation and will leave fewer offspring in the long run. This factor depends on interdeme and group selection processes and therefore its significance is open to question. The second factor is that in terms of community utilization of resources high A_d is more efficient than high A_g. But the significance of this also depends on higher-level selection. The third factor is that high A_d regulates the environment more than high A_g does. A high A_g/low A_d pattern would mean population oscillation in the grazing pathway rather than in the detritus pathway, hence pileups of dead matter which is not cycled back to the environment. Such a community could not be ecologically stable. It certainly would be succeeded by one with higher A_d. But then the uncertainty of the environment would go down, allowing A_g to fall as well. The existence of voluntary limitations would be favored by these powerful cyclic constraints.

Comparison to the natural situation. The increase in the amount of energy flowing over detritus pathways predicted by the above analysis agrees with the results of field studies (Odum, 1969). The detritus system does play an increasingly important role in the course of succession. But it should be recognized that this is in part due to the requirement for balance. In the initial stages of succession primary production leads to an accumulation of biomass, then an accumulation of dead matter which necessarily is followed by an increase in the biomass flowing through the detritus system. The theory here is that the detritus system serves not only to complete the cycle of materials but also to buffer both the grazing system and the environment from external and internal disturbances. The critical predictions are that the variability of flow should increase in the detritus system and decrease in the grazing system and that the amount of energy flowing into the detritus pathway should increase even after the ecosystem comes into balance. The average biomass of the detritus system should also increase as soon as the point is reached at which no further increases in the rate of decomposition processes are possible. I know of no field studies directed to the first prediction, but it appears that the shift to detritus pathway dominance is a long-term process. The analysis suggests that in the final stages of succession there should be some shift of energy flow and biomass back to the grazing pathway. In these final stages utilization of compensating variation should decrease and the uncertainty of the environment should be reduced.

There is one mechanism of energy–mass adaptability which has been ignored in the above analysis. This is variation in size. The possibility for such variation and its cost depends on structural principles, such as whether the system has an open or closed growth system and how much food it is capable of storing. But a structure-independent principle is that large size buffers the organism from a larger range of variation in food web availability. The grazing system has the advantage in this respect, implying that a community with larger grazers can have higher A_g. It is interesting that the paleontologists have in fact remarked on a tendency to increase in grazer size in the course of evolution (Simpson, 1953), though other factors may contribute to this. Many grazing pathway populations (such as plankton, microscopic herbivores, and carnivores) are also small and highly culturable. If all grazers were of this nature, the factors favoring high A_d would not be so weighty.

12.9. TROPHIC COLLAPSE AND TROPHIC ELONGATION

I have made use of the phenomenological fact that detritics are often smaller and more specialized than grazers, many of which are large and complex. But it is also possible to justify this assumption theoretically. The argument is interesting, since it shows that there are energy-independent restrictions on the number of trophic levels which greatly reduce the types of web structure. The number of levels in the grazing pathway collapses to the smallest number compatible with biochemical and morphological constraints. Detritus pathway chains tend to become as long as possible compatible with these same constraints.

Recall that plants are assigned trophic level 1, herbivores trophic level 2, and primary carnivores trophic level 3. A carnivore might actually eat both herbivores and plants, but it will be assigned trophic level 3 if it obtains a substantial amount of its energy from herbivores. More generally, a grazer which obtains a substantial amount of its energy from trophic level n will be assigned to level $n + 1$. A detritic will be assigned to level $n + 1$ if it takes its energy primarily from dead organisms at level n or from their waste products.

CLAIM I (grazing pathway collapse). Grazing pathway organisms collapse into the smallest number of trophic levels compatible with constraints associated with basic biochemical differences between the plant and animal kingdom or with size differences between predators and prey. The number of levels is at minimum three and is always independent of energy input. Except for parasites, the complexity of an organism must in general be greater if it takes more of its resources from higher trophic levels. Feeding patterns, including cross-level feeding patterns, are determined by a compromise between adaptability and efficiency, but the number of trophic levels is independent of both.

ARGUMENT.   The free energy at trophic level $n + 1$ is smaller than that at level n. This is required by the second law of thermodynamics. Suppose that no constraint prevents an $n + 1$ level species from dropping to level n and competing with its former prey for its source of food at level $n - 1$. In that case remaining at level $n + 1$ is not evolutionarily stable. Any mutant of the $n + 1$ level species which takes a larger proportion of its food from level $n - 1$ will have increased fitness, for it is always more efficient to obtain energy from level $n - 1$ than from level n. Plants, by definition, cannot drop a level. Herbivores cannot drop a level without fundamental biochemical change (sunlight is different than grass). Similarly primary carnivores cannot drop without fundamental change (plants are nutritionally different from herbivores). But herbivores are nutritionally basically the same as primary carnivores, so a secondary carnivore could drop. No fundamental biochemical changes would be necessary.

It might at first be thought that this argument leads to an unoccupied niche. But in fact this niche—the secondary carnivore niche—is evolutionarily unstable. It can only be occupied if a nonnutritional constraint, such as size incompatibility of predator and prey, prevents the $n + 1$ level species from obtaining energy from the $n - 1$ level source. The conclusion is that in the absence of nonnutritional constraints, the number of trophic levels should be three. Size incompatibility, transitional situations, and opportunistic harvesting may increase the number of levels, but there is in no case any dependence on energy-density.*

The successive decrease of free energy at successively higher trophic levels is also responsible for the trend to increased complexity. At each higher level the predator must harvest over a larger area, implying ever greater requirements for organs of perception, information processing, and motion.

---

*This argument was originally developed with H. Hastings (Hastings and Conrad, 1979). It included two technical points which should be mentioned. Let $h(t)$ be the energy gain from an effort t expended on herbivores and let $c(t)$ be the gain from an effort t on carnivores. Both $h(t)$ and $c(t)$ are assumed to be increasing functions, convex because of diminishing returns, and close to linear. The problem is to maximize the total gain $h(t) + c(1 - t)$ subject to the constraint that the competition with potential carnivore prey requires a small effort ε. The maximum occurs at $t = 1 - \varepsilon$ as long as $h'(t) > c'(1 - t)$ for $0 \leqslant t \leqslant 1 - \varepsilon$, implying that trophic collapse occurs as long as ε is small.

Parasites are the special case. A parasite may obtain its energy either from a carnivore or from a herbivore. A parasite whose host is a carnivore is in a sense a secondary carnivore. But if it dropped a level it would not fall into the same niche occupied by its former host. Rather it would become a parasite on a prey organism eaten by its former host. Since the former host continues to exist, it remains a tenable niche for some parasite, implying that the secondary carnivore niche is evolutionarily stable if it is occupied by a parasite. The trend to greater complexity obviously does not apply to parasites. The parasite lives in a locally rich thermodynamic environment, but at the expense of sacrificing as much size and complexity as possible.

A feature of this argument is that it completely divorces the issue of feeding pattern from the issue of number of trophic levels. Feeding pattern results from the tradeoff between adaptability and efficiency. The basic rule is that efficiency increases with increase in number of requirements or decrease in number of sources, whereas adaptability decreases under these circumstances (cf. Theorem 1). Thus the omnivore which uses herbivores to provide more of its food requirements gains in efficiency but loses in adaptability. The omnivore which uses herbivores to increase its number of sources loses efficiency, but increases adaptability. This case is basically equivalent to that of the herbivore which increases its number of possible plant sources, except that the requirement for an animal source prevents a trophic collapse. But an omnivore which takes both herbivores and primary carnivores can always do better by concentrating on herbivores.

Comparison to the natural situation. In nature grazing food chains rarely involve more than four or five trophic levels and often involve only three (Hutchinson, 1959; Slobodkin, 1961). The number of levels is not noticeably different in arctic or tropical ecosystems, despite the fact that primary productivity varies by a factor of 10^4 (Pimm and Lawton, 1977). The classical explanation is that the number of levels is limited by the inefficiency of energy transfer, known to range between 5% and 20%. But this is not compatible with the constancy of levels and with the absence of secondary carnivores. Other explanations have involved the idea that more levels would destabilize the population dynamics. But in general breaking up a dynamical system into more levels of organization reduces the number of interactions and increases its stability. Also arguing against dynamical explanations is the consideration that secondary carnivory would have to be ecologically very unstable in order to overbalance evolutionary stability. The argument here is that secondary carnivory is evolutionarily unstable and from the energetic point of view ecologically unstable as well.

The large size differences possible among protista should increase the number of evolutionarily stable links in the microbial part of the grazing chain.

CLAIM II (detritus pathway elongation).   Detritus pathway organisms expand into the largest number of levels compatible with biochemical constraints. Such elongation increases efficiency, decreases the cost of culturability, and does not significantly affect the cost of physiological adaptability.

ARGUMENT.   It is not thermodynamically contradictory for herbivores to be broken down and then reconstituted into nutritionally equivalent carnivores. Organisms which die, but not through predation, can be reconstituted into nutritionally equivalent scavengers if they are not too decomposed. But waste products and decomposed organisms are thermodynamically too poor to support a nutritionally equivalent species at the next level which is evolutionarily stable. Such a species would have to range over a large area in order to obtain sufficient energy. More important, few of its requirements could be satisfied by waste products; therefore it would have to possess a full complement of synthetic capabilities. A small, simpler species would always be more competitive. Ecological stability of course requires that each decomposer species modify the character of the food it produces for a subsequent decomposer; otherwise a cycle of materials would be impossible.

Now suppose that it is biochemically possible to replace decomposer species A by two species, A1 and A2. The waste product of A1 is processed by A2 and the waste product of A2 is the same as that of A. A1 and A2 are each more specialized for a particular task than A, but their external specialization is comparable to that of A since they have the same number of sources and requirements. But as they need fewer degradative enzymes than A, they are more efficient. Since the external specialization is the same and the only basic difference is in the number of degradative enzymes, the cost of physiological adaptability is the same. But the cost of culturability, the major mode of adaptability, is different. Since A1 and A2 have fewer degradative enzymes, it is less costly for them to form spores and culture rapidly when presented with a favorable source. Species A must therefore be evolutionarily unstable since A1 and A2 are both more efficient and more adaptable. This is true even if A uses its energy source more completely than A1. A cannot use the energy not used by A1 to outcompete A1 since this energy is used by A2.

The argument may be extended, splitting both A1 and A2 into more refined decomposers. The evolutionarily stable situation is the one in which the detritus chain is as long as possible. This situation should also be more ecologically stable since it is more efficient.

Comparison to the natural situation.   The highly refined horizontal partitioning of resources predicted by the above argument is a striking feature of the detritus pathway. Medium-size and small organisms which can mechanically break down grazers and waste products are necessary for rapid cycling of materials. Organisms such as centipedes and termites play this kind of a role. Once detritus is broken up it might at first appear that a single small organism with a large set of enzymes would degrade it as far as energetically possible. But such a species would not be evolutionarily stable. It is interesting that the evolutionarily stable situation, that of a maximally long detritus chain, is the one which allows the detritus pathway to buffer the grazing pathway most effectively.

12.10. NICHE DIVERGENCE AND NICHE CONVERGENCE

Now I want to return to the idea of environmental control. This occurs when the community organizes itself so as to reduce the uncertainty of the environment [alternative (5), Section 12.5]. The inflow of energy cannot be controlled. But control can be exerted over the occurrence of different chemicals in the environment and their physiochemical condition. This is possible if the community develops enough functional capabilities to complete the matter cycle. The number of functional capabilities correlates with the index of internal specialization [equation (11.9)]. But the number of enzymes in the community provides a simpler and more direct index in this case. There must be enough enzymes to support a cyclic set of reactions and to support enough different structures and modes of behavior to ensure that this full set of reactions occurs.

There are two ways to achieve this. Many enzymes can be used to complete the cycle in a highly efficient way, or fewer enzymes can be used to complete it in a less efficient but also less costly way. Furthermore there are two ways of packaging these enzymes. They can be packaged into a large number of specialized species (the niche-divergent pattern) or a small number of simple, despecialized species (the niche-convergent pattern). The niche-convergent pattern of organization uses fewer enzymes; therefore the efficiency with which energy is utilized is less. But the cost of adaptability is also less.

About these two basic forms of community organization it is possible to make the following prediction:

PREDICTION (niche divergence and convergence). Energy-rich, certain environments favor niche-divergent communities. Energy-poor, uncertain environments favor niche-convergent communities. Decrease in the energy quality or increase in the uncertainty of the environment will at a certain point cause the community to switch from a divergent to a convergent pattern, and conversely. (The energy quality of the environment decreases either as the amount of energy input decreases or as the physical conditions become less favorable for any present forms of life to utilize energy efficiently.)

ARGUMENT. I will call the energy which the community requires for construction and maintenance the investment energy. Of the energy available to the community, only some is actually converted to biomass or turnover, that is to maintenance or construction. This will be called the return energy.* All the return energy is invested in construction or maintenance. Thus, if the return energy exceeds the investment energy, the community either increases in biomass or becomes more active; if the investment energy exceeds the returned energy, it decreases in biomass or becomes less active.

The investment and return energy per unit volume are functions of the same factors as efficiency [equation (11.3)]. But here there is no assumption that the population is at a steady state, so the biomass turnover must now be included among these factors. Notationally,

$$I = I(A, C, M^*, E^*) \qquad (12.19a)$$

and

$$R = R(A, C, M^*, E^*) \qquad (12.19b)$$

where I is the investment energy per unit volume and R is the return energy per unit volume. A is adaptability, C is internal specialization, M^* is biomass turnover, and E^* is energy quality (determined by a number of environmental parameters). As internal specialization (for example, number of types of enzymes) increases, the return energy increases, though with diminishing returns. The investment energy must also increase. Notationally,

$$\partial I / \partial C > 0, \qquad \partial^2 I / \partial C^2 < 0 \qquad (12.20a)$$

*The distinction between investment and return energy earlier arose in connection with the discussion of patterns of adaptability in populations (Section 11.7).

and

$$\partial R / \partial C > 0 \qquad (12.20\text{b})$$

As internal specialization increases, the cost of adaptability in terms of return and investment energy must decrease. Again notationally,

$$\partial I / \partial A < 0 \qquad (12.21\text{a})$$

and

$$\partial R / \partial A < 0 \qquad (12.21\text{b})$$

Now suppose that the energy quality is decreased, the environmental uncertainty is increased, or both. The investment energy per unit volume must stay the same if the biomass or turnover is not to decrease. But this is not possible since the investment energy must change to equal the return energy, which must necessarily decrease. An equilibrium is possible if M^* reaches a value which satisfies

$$\partial I / \partial M^* = \partial R / \partial M^* \qquad (12.22)$$

This requires that

$$\partial I / \partial M^* < \partial R / \partial M^* \qquad (12.23)$$

for all values of M^* less than the initial equilibrium value and greater than the final one. That is, the amount of energy invested per unit volume of a community with the given amount of internal specialization must decrease faster than the energy returned per unit volume of this same community. This is not such an easy condition to meet. If it cannot be met, the community will certainly be replaced by another one with lower internal specialization. If it can be met, but only at much lower biomass, replacement by a less divergent community is still likely. A significant decrease in biomass is likely to be accompanied by the loss of some species in any case. Achieving an equilibrium by increasing internal specialization is not possible because of diminishing returns.

Will the change in internal specialization be continuous or discontinuous? The answer depends on whether the internal specialization can change on a short enough time scale. There are two possibilities, corresponding to the two ways of decreasing the number of enzymes. The first possibility is to distribute fewer enzymes over the same number of organisms, with no change in niche divergence. This is an evolutionary process and therefore clearly too slow. But it might occur on a faster time scale if it involved changes in the relative proportions of species already present in the system or replacement of species by migration from the outside. This is possible only to the extent that such changes are compatible with a coherent (ecologically stable) community and to the extent that the different types are themselves evolutionarily stable and present as potentialities.

The problem is that maintaining the potentiality for such a graduated series of communities requires adaptability which is very much higher than necessary. The second way of reducing the number of enzymes is by reducing the niche divergence. This may occur through the step-by-step removal of species. The problem now is that removal of any essential functions is incompatible with an ecologically stable community. In the absence of replacement species the community would be superseded by a different one. The conclusion is that it is in principle possible for the degree of divergence to change continuously in response to continuous changes in energy quality and environmental uncertainty. But as a practical matter the potentialities which are present at any given time make it highly likely that discontinuous (catastrophic) shifts will occur in some cases.

The argument could also be developed in terms of efficiency, which depends on the same factors as investment and return energy. When the energy quality or environmental uncertainty increases to a point where the investment energy exceeds the return energy, the efficiency must go to zero. The community is going out of existence at that point. The degree of internal specialization is at an optimum when the investment energy equals the return energy and the return energy is as high as possible. Then the efficiency is at a maximum. But most communities must be very far from this optimal situation. This is due to limits on the optimizing power of the evolutionary process. But peculiarly these limits afford important protection to the community. If the community were at the optimum, there would remain for it no option other than a catastrophic response to worsening conditions.

Formulation in terms of catastrophe framework. The civil engineer Thompson (1975) has discovered a similar phenomenon in technical structures such as bridges. Simultaneous optimization of all features leads to a catastrophe-prone structure. Here optimization of internal specialization for a high degree of required adaptability leads to a catastrophe-prone community.

Thompson's analysis is a special case of R. Thom's catastrophe framework (Thom, 1970, 1972; for a thorough review see Poston and Stewart, 1978). It is possible to picture the argument geometrically in terms of this framework. Suppose that the only significant state variable is internal

FIGURE 12.3. Catastrophe surface for niche divergence.

specialization (or niche divergence) and the only significant control variables are harshness (inverse of energy quality) and environmental uncertainty. The possible situations are summarized by Figure 12.3. This is the well-known cusp catastrophe. It allows for a continuous or a discontinuous shift in the internal specialization in response to continuous change in the control variables. If there are fewer than three variables, no discontinuity is possible; if there are more, the possible discontinuities become more complicated. The catastrophe language suggests the possibility of discontinuous shift in niche divergence. But whether or not such a shift will actually occur and whether or not optimizing internal specialization increases its likelihood can only be investigated by biological argumentation.

The reader may wonder whether it is possible to use the canonical forms of catastrophe theory to obtain a useful quantitative description of "phase transitions" in community structure. One problem is that catastrophe theory is too local. There are a number of significant variables other than the ones mentioned. Locally these variables may travel in a small set of coherent bundles, making it possible to select out principle variables which reflect the dynamics of the system. But globally there is no reason for these bundles to maintain their integrity.

Comparison to the natural situation. The contrast between the niche-divergent patterns of tropical ecosystems and the niche-convergent patterns of arctic systems is compatible with the above analysis. The passage from an energy-rich to a harsh environment forces the community pattern to switch from one which is species-rich (with many species highly specialized in terms of feeding habits) to one which is much less diverse in terms of number of species. The interpretation (undoubtedly overly simple) is that the energy-richness and absence of high uncertainty in the tropics allows for a high investment energy, hence for a situation in which many enzymes are distributed into many externally specialized packets. There is enough energy available to support the extra costs of adaptability in such a system. But in the arctic ecosystem this is not so. The result is a less diverse, patchy community in which much of the adaptability involves independent and often significant local variations in population sizes (see Dunbar, 1973). But there is a paradoxical feature of arctic systems which was noted by MacArthur (1955). The aquatic ecosystems are reported to have high biomass, possibly higher than that of their tropical counterparts. The environment is really energy-rich because of nutrients carried in by undersea currents. I have not been able to find decisive data, but it is not implausible that these currents are highly variable. The resulting high uncertainty in energy input would favor a niche-convergent pattern, as would the harsher physical conditions. If the average amount of energy input is high, the average biomass could be high enough to support high internal specialization. But during periods of downward variation the investment energy would exceed the return energy, thereby nipping any predilection to divergence.

Sharp spatial demarcations between different communities—such as the tree line on mountains or the sharp floral line which can be seen in ponds with increasing water depth—are well known. Here there is an evident discontinuous shift in the community in response to a continuous change in a feature of the environment. If the species in the two communities can survive in each other's zones, one usually finds a shared transitional band. The analysis in this section suggests that such demarcations should also occur in response to environmental uncertainty and quantity of energy, all other factors being equal. The prediction is that these are the major factors which determine the number of niches in a community. The topographical complexity of the habitat is secondary. In principle a relatively convergent community could be ecologically stable in a topologically diverse environment. It could support high biomass. But whether it would be evolutionarily stable would depend on whether the uncertainty and quantity of available energy allow for a self-supporting divergent system. If so, this divergent system will break the convergent system's controlling grip on the environment, replacing it by its more distributed, intellectually less graspable, but more defendable grip.

The divergence argument is completely independent of trophic collapse and trophic elongation. The number of grazing levels should still be independent of energy density. The number of degradative levels should still be as large as possible, but the smaller number of grazing species in a convergent system reduces the number of distinct degradative functions and therefore should reduce the diversity of detritic species.

12.11. ADAPTABILITY, COMPLEXITY, AND THE STABILITY OF HISTORY

In some cases the successional development of an ecosystem is extremely stable to variations in initial or external conditions. In other cases a continuum of community forms develops; in still others initially similar communities diverge (bifurcate) into two or more distinct forms. Three factors control these stability properties: energy input, complexity, and adaptability. The connections between these are subtle. But the analysis has now been developed to the point where they can be stated in a definite way.

The size of the ensemble of perturbations to which the successional process is stable (in the sense that its qualitative character does not change) increases with adaptability. This is true by definition. But the number of alternative patterns into which the community can diverge in response to perturbation outside this range also increases with adaptability (since the more adaptable system has more potentialities). The cost of adaptability in terms of energy increases with complexity (therefore with niche divergence). We know that increasing complexity can increase efficiency. But if the required adaptability is too big or if the available energy is insufficient, the community must switch to a simpler, niche-convergent pattern of organization. An out-of-the-range disturbance does not necessarily drive the complex community into such a form. If the adaptability is high enough, sufficient potentialities may be present to allow an alternative complex form. But if the adaptability is not high—a situation which can develop more rapidly in a complex system—replacement by a niche-convergent community is more likely. The qualitative invariance of succession in such communities is generally high. High adaptability is easy to maintain—or more difficult to discard—and a great deal of quantitative variation in response to different external conditions can be expected. But in an energy-rich environment the convergent community is evolutionarily unstable or, more to the point, unstable to the invasion of species from the outside. Then a number of lines of development are again opened up and the apparent determinism of succession decreases.

These connections can be given a dynamical formulation (using the cross-correlations between adaptability theory and dynamics developed in Chapter 8). Complexity is a precondition for stability insofar as it allows for the formation of cycles. But as shown by May and others (cf. Section 11.4), the probability of orbital stability decreases with increasing complexity. This means that of two cyclic systems the one which is more complex is less likely to forget the effects of perturbation in detail. The orbital stability goes down as complexity goes up; nevertheless, if adaptability is high, there are enough solutions which are similar in detail so that the overall dynamics appears structurally stable. As structural stability increases, adaptability increases, though not necessarily conversely. Thus the subtle connection between adaptability and the number of alternative paths of development translates to a connection between structural stability and the number of these paths. The connection between adaptability, complexity, and energy translates to a connection between structural stability, complexity, and energy.

It is interesting to compare these conclusions to observations on the stability of microcosm succession. One can set up different incubators which are more or less uncertain in terms of light and temperature changes, more or less harsh in terms of temperature variation, and more or less rich in terms of light input. By switching cultures from one incubator regime to another, one can generate a variety of histories. Classifying these cultures is a difficult pattern recognition problem, carrying with it the same type of subjective difficulties which attend the classification of species. But certain features are unambiguous. One is that in experiments cultures with similar histories often diverged—the successional determinism typical of systems cultured on the laboratory shelf was shattered. A second feature is that cultures with different histories sometimes developed morphological features which appear similar. A third feature is that cultures exposed to a richer input of energy generally exhibited more variation and were generally more difficult to classify. A fourth feature is perplexing. Some communities initially cultured in uncertain environments and moved to highly certain, mild environments flourished; others became quite barren. A similar divergence occurred with communities kept in the uncertain, harsh environment and with communities switched from the mild, highly certain environment to the harsh, uncertain one. The communities which developed most determinately were those which spent all their lives in the mild, highly certain environment and the ones which flourished most prominently were those which were moved to this environment from the uncertain, harsher one.

I believe the situation is this. In cultures which are initially biotically dilute (as these were) the environmental conditions during the first stages had the most decisive influence. Slight asymmetries in the light input—in the relative amount of fluorescent versus incandescent light impinging on the community—determined which species dominated and how fast they locked up biomass. Like a crystal defect, the resulting floral morphology (or its slow-to-degrade detritic aftermath) set the stage for all future development. Shifts from regime to regime during these early stages—when the spread in degree of development was greatest—accentuated these differences. In communities with low adaptability (few potentialities) there is little chance of repairing the resulting defect, hence divergent succession is to be expected. But unlike a crystal, communities with high adaptability (many potentialities) have some chance to repair, hence are more likely to show convergent succession. *The key point is therefore how these asymmetries and shifts at variant stages of maturity affected adaptability.*

Since all of these communities were initially dilute, they all started with low adaptability. As the biomass built up, the total adaptability could build up. But the question is, could it build up fast enough for the community to survive the uncertainty of the environment? In the uncertain regime the communities which were well placed relative to light input could build up their adaptability fast enough. Those which were not well placed, or which started with a culture which was initially too dilute, could not, hence declined to a barren state. This decline could not be reversed by moving them to the mild, highly certain environment. The communities cultured in the mild, highly certain environment all built up more biomass, hence all increased in adaptability. But the adaptability required was not as high as that required by the harsh, uncertain environment. Some of those which were shifted to this environment did not have enough potentialities to make the transition. The communities which successfully cultured in the uncertain regime had the highest adaptability. It was this extra potentiality which enabled them to develop the richest-appearing structure when moved to the highly certain, mild environment. In general these communities then developed quite determinately. But in a rich environment more of the marginal potentialities become visibly expressed, hence the greater eccentricity generally shown by communities in the richer regime. This phenomenon has also been observed by epidemiologists who have studied the effects of enrichening diet. This leads not only to a fuller development of the individual, but also to a fuller expression of any pathologic potentialities, patently a feature of richer societies as well.

12.12. SUCCESSION TO INSTABILITY

The sometimes vacillating trends exhibited by successional communities are driven by two underlying processes. One is cycle formation. The other is the time evolution of adaptability. They are linked. The two mechanisms of cycle formation are functional diversification and adaptability. Both allow for control over the environment, hence for a stabilization of whichever community features mediate this control. The dominant principle during the first stage of succession is that the community cannot stop changing until it falls into a cycle, that is, until its niche structure is initially complete. It is this falling process which drives the increase in diversification and adaptability that is characteristic of the first stage. The formation of cycles makes it easier to anticipate the environment. The accumulation of biotic adaptability makes it possible for the community to buffer controllable features of the environment from unanticipated perturbation by uncontrollable features.

The second stage of succession is dominated by increases in internal correlation which exploit this stabilization of the environment and by reallocations of community adaptability which are either ecologically or evolutionarily stabilizing. Some of the adaptability accumulated during the first stage of succession becomes superfluous due to the buffering of the environment. This superfluous adaptability begins to decline during the second stage. A major feature of the second stage is an increase in the allowable pathways of energy flow, hence an increase in adaptability to disturbances involving the flow of mass and energy. This occurs first through increased species diversity and second through increased multifunctionality. A major feature of the second stage is a shift of energy flow from the grazing to the detritus pathway. In part this is a concomitant of cycle completion. But it is also concomitant to a shift of adaptability from the grazing to the detritus pathway. The latter increasingly buffers the dynamics of the former from both internal and external perturbations.

The third stage of succession is dominated by further enhancement of internal correlation, further stabilization of the environment, and further loss of superfluous adaptability. These internal correlations allow for more effective anticipation of the environment. Some of the phenomena which should become more prominent are clocking mechanisms and both inter- and intraspecies communication. Some decline in species diversity often occurs as a result of increased multifunctionality and competitive exclusion. The presence of large organisms in the grazing pathway provides an effective adaptability mechanism which is internal to this pathway. At the end of the first stage the grazing pathway in general collapses to three levels, but some size incompatibilities may extend the number of levels in some cases. The number of links in the detritus pathway increases in the direction of the maximum possible. It is possible that some shift in energy flow back to the grazing pathway occurs due to increased utilization of compensating variations, to the appearance of flexible thresholds, or to the enhanced stabilization of the environment.

During the fourth stage of succession there is a further withering of unexercised adaptability. There is no way of preventing this atrophy. It can only be slowed down by the development—through evolution—of less costly forms of adaptability. An environmental disturbance whose effect would have been mild at an earlier stage of succession may now have significant consequences. Such a disturbance might be of external origin, or it may originate from slight leakage in the cycle. In dynamical terms, the second, third, and fourth stages of succession are characterized by a shift from structural to orbital stability. The ecosystem becomes more complex, integrated, and more efficient. It is apparently more stable. The environment is apparently more stabilized. But in reality it is more delicate. The increasingly highly tuned organization is more expensive in terms of adaptability and more likely to undergo catastrophic change in the absence of a suitable adaptive mechanism.

There are some implications for the vulnerability of a community to invasion by outside species. This is clearly greatest during the first stages of succession, when the niche structure is incomplete. The high adaptability characteristic of the first stages decreases vulnerability to invasion insofar as it increases control over the environment. But it increases vulnerability to functionally similar species which are less adaptable. Atrophying of adaptability during the final stages decreases vulnerability to invasion by less adaptable species, but it increases the vulnerability of the community as a whole to disturbance. An out-of-the-range disturbance becomes more likely, indeed eventually inevitable. Then major new opportunities are opened up for outside species or for marginal species already present.

What would happen if our ecosystem contained humans? Suppose we could culture many similarly prepared human ecosystems under controlled conditions. Our cultures would now contain new levels of organization with new forms of institutional complexity. New modes of adaptability are possible, most importantly modes which make it possible to transmit acquired information to future generations. But I think the basic principles of adaptability, complexity, stability, and instability would still obtain. The nonhuman ecosystem cannot avoid the loss of adaptability. It cannot know that it will need this adaptability in the future. An individual man can. But unfortunately this does not mean that a human society can. Foresight is expensive. The man or the institution which mutates to the nearsighted state is likely to replace his farsighted competitor. As in the humanless system, unexercised adaptability tends to wither away. The creative intelligence of man, rather than halting the progression to instability, accelerates it. As the most potent known source of novelty in the universe it is the most devilish source of disturbance. The first thought is to stabilize history, indeed to stop it. The first act is to increase the orbital stability of the society. The first illusion is the importance of the enhanced efficiency which accrues from the shift from structural to orbital stability. The second illusion is that the resulting structural instability can be avoided the second time around by doing the same thing more carefully. If one really wanted to stabilize history, the best strategy would be to preserve as many mechanisms of historical change as possible. The only society which can be stable in the long run is one which allows instability in the short run.

REFERENCES

Conrad, M. (1972) "Stability of Foodwebs and Its Relation to Species Diversity," *J. Theor. Biol. 34*, 325–335.

Dunbar, M. J. (1973) "Stability and Fragility in Arctic Ecosystems," *Arctic 26*(3), 179–185.

Eigen, M. (1971) Self-organization of Matter and the Evolution of Macromolecules, *Naturwissenschaften 58*, 465–523.

Eigen, M., P. Schuster, K. Sigmund, and R. Wolff (1980) "Elementary Step Dynamics of Catalytic Hypercycles," *BioSystems 13*(1,2), 1–22.

Hastings, H. M., and M. Conrad (1979) "Length and Evolutionary Stability of Food Chains," *Nature 282*, 838–839.

Hutchinson, G. E. (1959) "Homage to Santa Rosalia or Why are There So Many Kinds of Animals?" *Am. Nat. 93*, 145–159.

MacArthur, R. H. (1955) "Fluctuations of Animal Populations and a Measure of Community Stability," *Ecology 36*, 533–536.

Margalef, R. (1958) "Information Theory in Ecology," *Gen. Syst. 3*, 36–71.

Odum, E. P. (1969) "The Strategy of Ecosystem Development," *Science 164*, 262–267.

Odum, E. P. (1953) *Fundamentals of Ecology*, W. B. Saunders, Philadelphia.

Odum, E. P. (1971) *Fundamentals of Ecology*, 3rd ed. W. B. Saunders, Philadelphia.

Pimm, S. L., and J. H. Lawton (1977) "Number of Trophic Levels in Ecological Communities," *Nature 268*, 329–331.

Poston, T., and I. Stewart (1978) *Catastrophe Theory and Its Applications*. Pitman, London.

Simpson, G. G. (1953) *The Major Features of Evolution*. Columbia University Press, New York.

Slobodkin, L. B. (1961) *Growth and Regulation in Animal Populations*. Holt, Rinehart and Winston, New York.

Thom, R. (1970) "Topological Models in Biology," pp. 89–116 in *Towards a Theoretical Biology*, Vol. 3, ed. by C. H. Waddington. Edinburgh University Press, Edinburgh.

Thom, R. (1972) *Stabilité Structurelle et Morphogénèse*. Benjamin, New York.

Thompson, J. M. T. (1975) "Experiments in Catastrophe," *Nature 254*, 392–395.

13

Evolution and the Organization of Potentiality

There is one major difference between the succession of ecosystems and their evolution. The origin of fundamentally new potentialities—the species formation process—becomes the decisive factor. In this chapter I add this ingredient to the analysis of ecosystem development. I also review the central features of adaptability theory and consider its relation to alternative descriptions of biological nature.

13.1. ADAPTABILITY THEORY OF SPECIES FORMATION

Mechanisms of species formation are discussed in many books on evolutionary theory (e.g. Mayr, 1963; Stebbins, 1950; Dobzhansky, 1951). The ideas all revolve around the conditions under which a gene pool—a collection of genomes tied together by mechanisms of genetic interchange—either conserves or changes its structure. There are two basic possibilities (illustrated in Figure 13.1). In the first the structure of the gene pool changes, but without bifurcation into noncommunicating parts. In this case the species undergoes evolutionary change, but no new species are formed. This is phyletic evolution. In the second the gene pool bifurcates into two or more noncommunicating parts. This is the defining feature of the speciation process.

There are a number of possible gene pool situations:

1. The large, highly connected pool, emphasized by Fisher (1930) and Haldane (1932).
2. The pool broken up into weakly communicating subpools, emphasized by Haldane (1932) and Wright (cf. 1970).
3. The small pool with fluctuation phenomena, emphasized by Wright (cf. 1970).

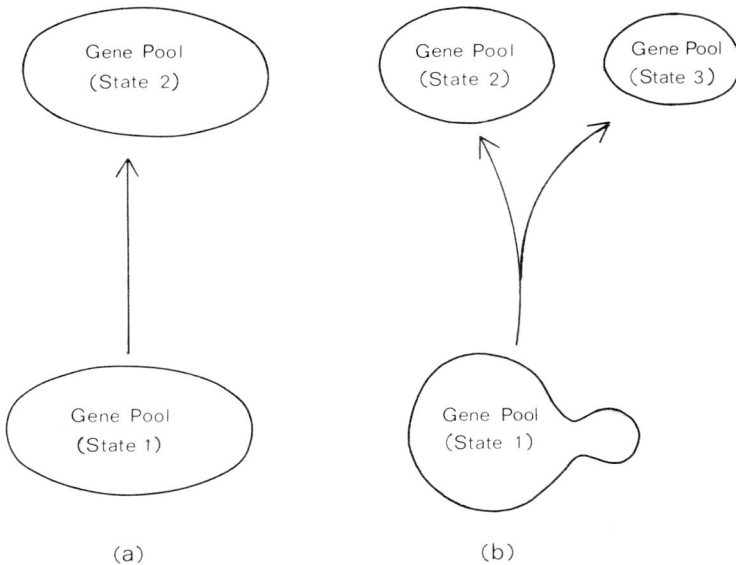

FIGURE 13.1. Phyletic (a) and speciating (b) modes of evolution.

4. The degenerate case of asexuality (effectively, the isolated genome gene pool).

There has been extensive debate on the relative importance of these different gene pool "phases" for the evolution process. In this regard two issues should be distinguished. One is the extent to which the phases actually occur in nature. The second is the extent to which they allow or prevent evolutionary change and the extent to which this change is likely to be adaptive.

PROPOSITION (gene pool structure). The potential for evolutionary adaptation is greatest in large populations which are broken up into subpopulations (demes) with high exchange of genetic material internally, but low interchange among the demes. The number of organisms in a deme which is optimal for the rate of evolution depends on the degree of genetic interchange and on the total population size. In order for divergent evolution to occur the gene pool must fission, that is, genetic interchange between demes must cease. High gene interchange serves to prevent divergence and to prevent evolutionary change. If the number of organisms becomes very low, random genetic phenomena will occur (genetic drift). This leads to a high potential for evolutionary change, but to a low potential for evolutionary adaptation. If the number of organisms is reduced to one (by the development of asexuality), the potentiality for evolutionary adaptation will be low, but the cost of maintaining adaptation will also be low.

ARGUMENT.   First recall (from Table 5.2) that gene pool uncertainty is expressed by $\Sigma H_e(\hat{\omega}_{v0})$, where v runs over the index set of some species, say c_{i4}.* The potentiality for evolutionary adaptation in response to natural selection depends on the extent to which this uncertainty is reduced by the environment, that is, on the difference $\Sigma H_e(\hat{\omega}_{v0}) - \Sigma H_e(\hat{\omega}_{v0}|\hat{\omega}^*)$. The condition for this change to be adaptive rather than maladaptive is that the indifference term stay small. Thus the potentiality for evolutionary adaptation increases as the gene pool plasticity increases, provided that the increase in the conditional plasticity (anticipation) and indifference terms is less.

The argument is based on the following two properties of the gene pool plasticity terms:

1. The modifiability part of the gene pool plasticity term increases as genetic interchange increases. This follows from biological considerations. If genetic interchange increases, the number of different base sequences which can occur with nonnegligible probability increases, implying that each of the $H(\hat{\omega}_{v0})$ increases. The modifiability component of the gene pool is proportional to the sum of all these terms taken over all the organisms in the population. The contribution of the modifiability must increase as the number of organisms increases, except in the unbiological case in which the genome of an additional organism is completely determined by specifying the genomes of all other organisms. For given mechanisms of gene interchange between organisms, the modifiability should be a function of the total number of organisms, K, and the number of organisms in each deme, N. However, it is sufficient and simpler to use the average fraction of the total number of organisms per deme, denoted by $f = \langle N \rangle / K$. The addition of an organism to a deme increases each of the modifiability terms, $H(\hat{\omega}_{v0})$, for that deme. But the addition of a second organism increases each of these terms less, since the chance of an exchange of genetic material between any two organisms decreases as the number in the deme is increased. Symbolically,

$$\partial M / \partial f > 0 \tag{13.1a}$$

$$\partial^2 M / \partial f^2 < 0 \tag{13.1b}$$

where $M = M(f, K)$ denotes the modifiability component of gene pool

*According to the reference structure conventions used in Fig. 5.3 genomes are denoted by $\beta_{(u-1)0}$, $u = 2, 4, 6, \ldots$. I am here taking $\beta_{v0} = \beta_{(u-1)0}$.

plasticity. As the size of a deme goes from one to more than one organism, the increase in the modifiability must be very large—since this is the condition for sexual reproduction. As K becomes larger, the increase in modifiability with f becomes smaller. Again symbolically,

$$\lim_{f_0 \to 1/K} \partial M / \partial f |_{f_0} \gg 0 \qquad (13.2a)$$

$$\partial^2 M / \partial f \partial K < 0 \qquad (13.2b)$$

2. The independence part of gene pool adaptability decreases as genetic interchange increases. At least some of the independence terms, $H(\hat{\omega}_{m0}|\hat{\omega}_{n0}), \ldots, H(\hat{\omega}_{m0}| \cap_{m \neq n} {}_n \hat{\omega}_{n0})$, must decrease as genetic material is exchanged. In the absence of genetic interchange the modifiability will be low, therefore the independence must be low. For given mechanisms of gene interchange between organisms the independence is a function of the same variables as the modifiability, that is, of K and f. The independence terms equal the modifiability terms for $f = 1/K$ since in this case there are no correlations between the organisms. Now suppose that the number of demes for given K decreases by one. The contribution of the independence to gene pool plasticity must decrease since there are fewer uncorrelated groups of organisms. If the number of demes is again decreased, the contribution should decrease by the same amount. Symbolically

$$\partial I / \partial f < 0 \qquad (13.3a)$$

$$\partial^2 I / \partial f^2 \approx 0 \qquad (13.3b)$$

where $I = I(f, K)$ denotes the independence component of gene pool plasticity. But of course it should be realized that N cannot assume a continuum of values for integer numbers of demes once K is given.

The problem is to find the value of f which gives a maximum of the total gene pool plasticity, expressed by the sum

$$P = M + I \qquad (13.4)$$

The situation is illustrated in Figure 13.2. The maximum will occur for $f < 1$ as long as

$$\partial M / \partial f |_{f_0} = - \partial I / \partial f |_{f_0} \qquad (13.5a)$$

$$\partial^2 M / \partial f^2 |_{f_0} < - \partial^2 I / \partial f^2 |_{f_0} \qquad (13.5b)$$

for some f_0. That condition (13.5b) is satisfied follows directly from equa-

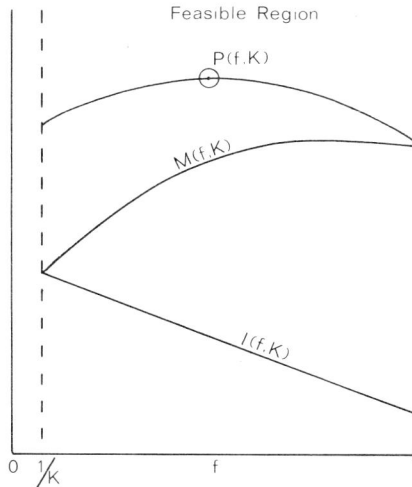

FIGURE 13.2. Deme structure of population with maximum potential for evolutionary adaptability. The maximum value of f (average fraction of total number of organisms per deme) is less than 1 under the assumptions described in the the text and decreases as the total number of organisms, K, increases. The value of f cannot be smaller than $1/K$, otherwise the average number of organisms per deme would be less than one.

tions (13.1b) and (13.3b). In order for the maximum not to occur for $f < 1$ one of the following conditions must therefore hold:

$$\partial M / \partial f|_{f=1/K} < - \partial I / \partial f|_{f=1/K} \qquad (13.6a)$$

or

$$\partial M / \partial f|_{f=1} > - \partial I / \partial f|_{f=1} \qquad (13.6b)$$

But condition (13.6a) is difficult to fulfill because of restriction (13.2a) and condition (13.6b) difficult to fulfill because of restriction (13.2b), at least for large K.

Restriction (13.2b) implies that f_0 becomes smaller as K becomes larger, therefore that the number of demes which give the maximum potential rate of evolution increases as the total size of the population increases. Since M increases with K and I does not decrease with K for constant f, the maximum potential rate of evolution increases as the size of the total population increases.

Fissioning of the gene pool is connected to the conditional plasticity and indifference terms, which have so far been ignored. The simplest way to picture this connection is to imagine the gene pool as a cloud of points in a molecular genetic space with four axes for each possible base in DNA (see Section 10.5). The coordinates of genome v at time t are given by the vector

$$\vec{r}_v(t) = r\left[\beta_{v0}^i(t)\right] \tag{13.7}$$

where $\beta_{v0}^i(t)$ is a nucleotide sequence and r is the function which maps this sequence into a point in the molecular space (by mapping each base into a one for one of the four axes for each position and into zeros for each of the other three axes). As the genomes in the population diverge, this cloud spreads out. Now suppose that genomes located far from each other in this space are highly correlated to different environments. Clearly, interchange of genetic material will reduce this correlation, thereby increasing $\Sigma H_e(\hat{\omega}_{v0}|\hat{\omega}^*)$ more than $\Sigma H_e(\hat{\omega}_{v0})$. This means that the potential for evolutionary change decreases even though the gene pool plasticity increases. The indifference term also increases due to the breakdown of correlations, so the cost of the evolutionary adaptability in terms of less fit individuals will be higher despite the decrease in the evolutionary adaptability. This would be true even under the implausible circumstance that interchange between genomes adapted to different environments reduces the fitness, but not the statistical correlation. The conclusion is that in order for a deme to maintain the specificity of its adaptation to a particular environment the gene flow between it and other demes must be small. In order for divergent evolution (bona fide speciation) to occur the flow of genes must cease altogether.

If the pool reanneals after the split, the decorrelation terms will become high, but the plasticity term will also increase. This is the case of hybridization. Opening up channels for hybridization can increase the potential for evolutionary adaptability, but only if the plasticity increases more than the conditional plasticity. This is likely to occur (and only likely to occur) if the correlations between gene pool and environment are already low. The conclusion is that high gene exchange serves to hold a condensed gene pool in a condensed state. If the parents are highly adapted, the offspring are more likely to be as well. But if the parents are not well-adapted, the chance of improvement is less (unless a deme structure develops or is imposed). Offspring emanating from a divergent but sexually communicating gene pool are less likely to be adapted, except to environments to which the parents are not adapted.*

---

*These advantages and disadvantages of hybridizing divergent gene pools should not be confused with the (rather vague) notion of hybrid vigor. This phenomenon is not based on mixing divergent gene pools, but rather on masking genetic defects which are unmasked by inbreeding.

As K becomes small the situation changes radically due to the occurrence of fluctuations in the structure of the gene pool. The expected and average actual structures at time t correspond to the expected and actually occurring center of mass points in the molecular genetic space. The locations of these points are given by

$$\vec{r}(t)_{\text{obs}} = \sum_{v=1}^{K} \vec{r}_v(t)/K \qquad (13.8)$$

and

$$\vec{r}(t)_{\text{avj}} = \sum_{v=1}^{K} p\left[\beta_{v0}^i(t)|\beta^r(t-\tau), \varepsilon^s(t-\tau)\right]\vec{r}_v(t)/K \qquad (13.9)$$

where the summation is a vector operation and $\beta^r(t-\tau)$ includes $\beta_{v0}^i(t)$. According to Chebyshev's inequality the probability that $|\vec{r}(t)_{\text{obs}} - \vec{r}(t)_{\text{avj}}|$ is greater than any given positive value d is given by

$$p\left(|\vec{r}(t)_{\text{obs}} - \vec{r}(t)_{\text{avj}}| > d\right) \leqslant q^2/d^2 K \qquad (13.10)$$

where q is the standard deviation of the points in the cloud and $q/\sqrt{K}$ is the standard deviation of the center of mass points. The distance is given by

$$|\vec{r}(t)_{\text{obs}} - \vec{r}(t)_{\text{avj}}| = \sum_{j}\left(r(t)_{\text{obs}(j)} - r(t)_{\text{avj}(j)}\right)^2 \qquad (13.11)$$

where j runs over all axes in the space. The point to note is that the standard deviation expected of the average genetic structures (the centers of mass) goes up by a factor of $1/\sqrt{K}$ as K increases. The chance of a major deviation also increases as genetic interchange increases, since then the standard deviation of the points in the cloud, hence q, increases. The probability that at time $t + n\tau$ the deviation will be greater than nd is given by

$$p\left(|\vec{r}_{\text{obs}}(t + n\tau) - \vec{r}_{\text{avj}}(t)| > nd\right) = \prod_{j=0}^{n} p\left(|\vec{r}_{\text{obs}}(t + j\tau) - \vec{r}_{\text{avj}}(t + j\tau)| > d\right)$$

$$\leqslant \left(q^2/d^2 K\right)^n \qquad (13.12)$$

Thus

$$p\left(|\vec{r}_{\text{obs}}(t + n\tau) - \vec{r}_{\text{avj}}(t)| > nd\right) \leqslant 1 \qquad \text{when} \qquad q^2/d^2 K \geqslant 1$$

$$(13.13)$$

implying that K must be large ($> q^2/d^2$) in order to ensure that the gene pool does not drift a genetic distance nd in time $n\tau$. Since d can always be made smaller and n can always be made larger, drift will always occur, but it can be reduced to an arbitrarily slow rate by increasing K. Selection cannot prevent this drift from occurring. It could always push the pool back in the direction of the original center of mass at each generational step, but for a small population there would be a significant chance of extinction in this case.

The conclusion is that the potential for evolutionary change becomes very high as K becomes small. But the critical question is, what happens to the potential for evolutionary adaptability? The gene pool plasticity—the indeterminateness in the cloud of points—must decrease as long as the amount of genetic interchange does not increase. The potential for evolutionary adaptability depends on the difference between this plasticity (defined over the transition probabilities $p[\cap \beta_{v0}^i(t + \tau)|\beta^r(t), \varepsilon^s(t)])$ and the conditional plasticity (defined over the $p[\cap \beta_{v0}^i(t + \tau)|\beta^r(t), \varepsilon^s(t), \varepsilon^u(t + \tau)])$, subject to the condition that the indifference (defined over the $p[\varepsilon^u(t + \tau)|\beta^r(t), \varepsilon^s(t), \cap \beta_{v0}^i(t + \tau)])$ does not increase. The indifference must increase with decreasing K since the correlation among the genomes does not change, but there are fewer of them to provide information about the selecting environment. Some which persist with high probability in a particular environment do not occur and some which ordinarily persist with low probability might now persist in a variety of environments due to the lack of competition. In both cases the correlations between environment and gene pool weaken, causing the conditional plasticity to decrease more slowly with decreasing K than does the plasticity. This justifies the claim that the potential for evolutionary adaptability decreases despite the fact that the potential for evolutionary change increases.

The situation again changes when $K = 1$. This is the case of asexuality. The standard deviation q (defined on an ensemble of initially identical organisms) becomes very small, implying [from equation (13.12)] that $p(|\bar{r}_{obs}(t + n\tau) - \bar{r}_{avj}(t)|) > nd$ is less than a small number. This means that the cost of maintaining an adaptation becomes very much lower than in a big population with heavy exchange of genes (also a conservative system), but the potentiality for evolutionary adaptation is negligible. Of course facultative switching between different degrees of genetic interchange can serve to switch on different degrees of evolutionary potentiality.

Further comments on drift and inertia. I want again to draw attention to the distinction, automatically explicit in the adaptability theory analysis, between evolutionary change and evolutionary adaptation. The important result on this point is that the acceleration of evolutionary change in small populations by random genetic walk does not correspond to an acceleration of evolutionary adaptation. Such random walking provides many more possibilities for losing desirable gene combinations than for acquiring new ones. According to adaptability theory the chance that a fitness-increasing trait arose through drift is *always* less than the chance that it arose through usual mechanisms of variation in a large population. But this does not mean that drift processes have not played a fundamental role in the course of evolution. Such a conclusion would entail the unwarranted assumption that the only changes which affect the course of evolution are fitness-increasing. But circumstances often occur in which population size is small and random genetic walk cannot be avoided. This is the general circumstance on a small, isolated island or when a species migrates to a new area (called the founder principle by Mayr, 1963). It is also the circumstance when environmental disturbances have a decimating effect on the number of organisms in a population. Many peculiarities exhibited by present-day organisms may trace their origin to genetic fluctuation phenomena under such conditions, but adaptive breakthroughs, whether major or minor, could only rarely have their basis in such processes.

It is interesting to look at the general statements which have been made in this section in terms of the growth dynamics of an unusually fit gene or gene combination. As the population size increases, the extent to which it can raise the large mass of less useful combinations to its own fitness level decreases. In effect, the genetic inertia exhibited by the population in response to natural selection will be higher. From this point of view the smaller population has the advantage. But it is much less likely that a new useful genome will appear in a small population, so from this point of view the large population has the advantage. This leads to the compromise situation of a large population consisting of many weakly interacting subpopulations. This intuitive argument is actually very similar to the formal adaptability theory argument presented in this section. The genetic inertia is connected with the dependence terms and the chance for a new useful combination with the modifiability terms. In a large panmictic population the independence terms will be relatively low. But unless the fitness difference between the new useful gene structure and the other gene structures in the population is very large, the development of correlation (required by low independence) will restructure the smaller number of more fit genomes more than the larger number of less fit ones.

The argument in this section should be distinguished from the earlier argument (Chapter 10) that high genetic variability structures the adaptive landscape into a more climbable form. In that case the modifiability terms are increased by increasing the likelihood of near-neutral variation.

13.2. LACK OF ADAPTABILITY IN THE SEXUAL MECHANISM

There is an interesting connection between adaptability and how bifurcation-prone the gene pool is. The most important mechanism is separation of the population into two sexually noncommunicating parts by a spatial barrier, followed by a morphological, physiological, or behavioral change which works against a reversal of the isolation (allopatric speciation). It is possible that in some cases a gene pool becomes so extended that it spontaneously fissions due to the unviability of intermediate forms (sympatric speciation). But again this must be followed by the development of a phenotypic isolating mechanism. This means that a complex sexual mechanism which is phenotypically nonadaptable favors bifurcation of the gene pool. A complex mechanism is more likely to change significantly in response to genetic change (cf. Section 11.5). This change is most likely to be isolating in the absence of adaptability. This developmental instability and lack of adaptability of the sexual apparatus are the basis for the instability of the gene pool, hence for its adaptability. Possibly there is a connection to the intricacy of courtship systems, which are sometimes much more intricate than would appear to be required for species specificity, and to the repugnance engendered in many people by alternate human courtship systems.

One further mechanism which is now of interest is the transport of genes between different species by viruses. Whether this occurs on a significant scale in nature is unknown. It may have been important earlier in the history of life, during the great age of biochemical radiation. Such a mechanism would introduce an element of vagueness into the concept of a gene pool. But it would not alter the mechanics of speciation.

13.3. BIOGEOGRAPHIC RADIATION AND VOLUME OF LIFE

Recall (from the argument in Section 13.1) that the potentiality for evolutionary adaptation increases with population size, assuming that the population is deme-structured (either by geographical constraints or by behavioral features) and fission-prone. The implication is that more species should originate in a larger volume of space than in a smaller one, assuming that the number of organisms is proportional to the volume (or at least increases with the volume). Increases in fitness, insofar as they are possible or required by fitness-reducing environmental changes, should also occur at a higher rate in the larger volume. As a consequence there should be a net tendency for species to move from larger defined regions of space to smaller ones. In effect the larger regions should provide a source for species and the smaller ones a sink. This agrees with the empirical situation in biogeography. According to Darlington (1957) the major sources of new species have been the old and the new world tropics. When species radiating from these large land masses collided due to the formation of the land bridge between

North and South America, the old world forms—the forms from the larger area—dominated. A similar phenomenon has been observed in island biogeography, in which the tendency is for large islands to be sources of species which invade smaller islands, either adding to their species list or excluding competitor species. The explanations which have been developed for these phenomena are basically similar to the one given above. The well known model of MacArthur and Wilson (1967) for island biogeography involves an equilibrium of migration and extinction rates, but if I understand this model correctly, the driving force is the higher potentiality for speciation on the larger islands. It is therefore of some interest that the assumptions on which these explanations are based have an independent and rather robust basis in the adaptability theoretic analysis of the speciation process.

The conclusion that the number of species should increase with volume is consistent with field data, which may often be expressed in terms of the empirical proportionality

$$N_s \propto V^z \tag{13.14}$$

where N_s is the number of species, V is taken as an area, and z is usually between 0.2 and 0.3 (cf. May, 1975). But there are a number of complicating factors which underlie this rough generalization, including species saturation as the niche structure becomes complete, increasing the likelihood for the occurrence of genetically isolated populations and increasing the likelihood for the occurrence of ecological analogs which are genetically isolated even though they occupy physically equivalent habitats.

There is a connection between these biogeographical phenomena and the idea that there is a critical volume below which community self-maintenance is impossible. This idea was previously discussed in the context of simplifications in community structure which are sometimes observed in experiments on laboratory microcosms. If the rate of speciation decreases as the habitable volume decreases, the equilibrium number of species should decrease as well. In essence the equilibrium between the birth of new potentialities and the death of old ones should be lower. The rate of speciation of course also depends on the number of species present as well. Reduction in the number of species with reduction in volume should not be extrapolated to the conclusion that the community is dying out. But at some critical volume the rate of origin of new species or genetic variants not at the moment present should decline to a level which is always below the extinction rate. This is the critical volume below which the self-maintenance of life would be impossible.

13.4. SPECIES SENESCENCE AND EVOLUTION TO INSTABILITY

Some of the early students of evolution thought that species undergo an aging process similar to that of the individual organism. Freud thought that the species itself had a drive to extinction analogous to the drive to death which is manifested in the physiological aging process and which he called (in its psychological dimension) the death wish. Later evolutionists have disputed this idea of a drive, undoubtedly correctly. As Simpson (1949) points out, it is inconsistent with the nearly random declines and rebounds seen in the fossil records of many organisms. But the evolutionary tendency in the direction of minimum allowable adaptability does suggest that species are subject to a form of senescence—not a drive to death, but an atrophying of the potential for evolutionary change and adaptation.

The consideration is simply that the features which increase evolutionary adaptability—like all adaptability-conferring features—are costs to the individual. High amenability to evolution requires high replaceability of amino acids and a high degree of polygenic control, both costly (cf. Sections 10.5 and 10.6). High recombination or mutation rates, necessary for high modifiability, mean a higher genetic load. Mechanisms which mask this load, such as chromosome inversions, or mechanisms which can turn it up or down, such as facultative apomixis, are also costs, though lesser ones. Deme structuring, necessary for high independence, may often involve mechanisms which bind organisms to smaller regions of space than would otherwise be necessary. Bona fide species formation is facilitated by complications in the courtship mechanism which may significantly increase the energy expended on reproduction.

If the uncertainty of the environment decreases, these speciation-enhancing mechanisms will diminish along with other forms of adaptability. The uncertainty may later increase, reversing the decline in adaptability. But any species whose adaptive mechanisms had fallen below the requirements imposed by this later increase would fail to survive it. Such a species would have improved its efficiency, but the drive to efficiency would have reduced its adaptive capability. This aging of the species is not inevitable, but there are circumstances under which its nonoccurrence would appear to be unlikely.

It is interesting to consider these circumstances—and to consider the evolution process—in terms of the dynamics of aging. The situation is basically similar to succession, except that the time scale is long enough for speciation to occur.

As in succession the niche structure changes until cycles are formed and the environment is stabilized. This stabilization facilitates anticipation and to some extent decreases the statistical uncertainty of the environment. The basic completion of the niche structure reduces the possibilities for speciation. The appearance of new species which can fit into this structure becomes increasingly unlikely. From a statistical point of view this saturation of the species structure is accompanied by the disappearance of an internal source of uncertainty. The environment of each species and of the entire community becomes still less uncertain and easier to anticipate. The advantage increasingly shifts from species which can develop new adaptations to those which can maintain old ones. This shift is favored by both ecological and evolutionary stability.

But new forms of external disturbance or the effects of slow leaks in the cycle eventually reverse the decrease in environmental uncertainty. Some species go extinct or are seriously reduced in numbers. Dominant and established species are especially likely to be damaged since these are the ones most likely to have fallen below the critical level of adaptability. In many cases they may be replaced by congener species present in neighboring communities whose speciation mechanisms and other forms of adaptability are less frozen. But after a very long time there may be no unfrozen congeners, or the trauma to the biota may be so great that it is reset to what is effectively an early stage of succession. Forms hitherto inhibited from radiating by competition from already well adapted species are now at much less of a disadvantage. If these forms belong to the same genera, a basically similar flora and fauna may develop. A well known example are the many radiations, extinctions, and parallel reradiations of the cephalopods (before their final extinction). But if some really new potentialities have reached the threshhold of existence, the possibility for their springing into dominance is open. The best known example is undoubtedly the rise of the mammals after the great extinction of the dinosaurs.

The succession paradigm clearly favors the view that these dramatic events are principally due to factors intrinsic to the ecosystem rather than to external factors. Even if external factors served as the immediate cause in most of the major extinctions, the ultimate cause is the decrease in adaptability implicit in the imperative to efficiency.

There are also many forms which have persisted without visible (at least morphological) change for an enormous span of time. A well known example is *Limulus*, the horseshoe crab, which does not appear any different today than it did 200 million years ago. Such a form occupies an invariant niche in the ecosystem and therefore is always subject to stabilizing rather than moving selection.* Presumably no alternative potentiality has appeared which is more efficient. Under these special circumstances the decline in evolutionary adaptability is always advantageous. Other forms of adapt-

---

*I believe that the terminology of stabilizing and moving selection was first used by Schmalhausen (1949).

ability will always be the important ones. These tend to decline as well, but even if a decimating event occurs, no competitor is available which can outrace either regrowth or reimmigration of the original form. A parasite or predator may evolve which decimates this stable, evolutionarily frozen form. But if this form plays a linchpin role in the ecosystem, significant destabilizing leaks will be opened up. This is the kind of situation that may lead to the fall of dominant species which may be outraced by low-lying potentialities. But it is also likely to lead to the fall of the predator or parasite, or to a sequence of changes which stops only when this predator or parasite is controlled by some third species. The stable niche is then reoccupied by its best tenant.

In some of the older literature and in some of the museums of natural history it is stated that one cause of extinction is the evolution of maladaptive features, such as awkward antlers. Presumably the idea is that such awkward features arise through sexual selection or that they are an allometric byproduct of selection on another feature. The argument against this point of view is that traits which on balance are selectively disadvantageous cannot evolve through natural selection, that they can only become on balance disadvantageous if the environmental conditions change. This counterargument is basically correct. But it is interesting that traits which develop when the species is young in terms of its potentiality for evolutionary adaptation may be very resistant to changing in the reverse direction when it is old. From this point of view it seems legitimate to explain some extinctions in terms of the increasing difficulty of rectifying traits which become liabilities.

13.5. BIOLOGICAL ORGANIZATION AND THE HISTORY OF UNCERTAINTY (REVIEW OF THE THEORY)

Now I would like to review the most important adaptability theoretic principles of biological organization and to reconsider the overall process of evolution in their light. The main idea is that for every biological system there is an ensemble of modes of allowable behavior which is consistent with its functional integrity and which underlies fundamental biological phenomena: reliability, transformability, stability and instability, and, in different disguises, all the different modes of genetic, phenotypic, populational and community adaptability. The larger this ensemble, the greater the potential for adaptability. I have used an entropy-type measure, $H(\hat{\omega})$, defined over a notational transition scheme, as a measure of ensemble size and have suggested an operational definition in terms of a preparation procedure. But even if the real transition scheme does not fulfill all the technical assumptions of the notational scheme, the idea that $H(\hat{\omega})$ exists and obeys the usual rules of addition and multiplication of probabilities is quite reasonable. There should be an ensemble of possible trajectories which correspond to the living state just as there is an ensemble of complexions of a gas compatible with its macroscopic state, independent of the theoretical or experimental method used to assess the size of this ensemble.

At bottom adaptability is the use of information to handle environmental uncertainty. How effective this use of information is is reflected in the anticipation term. If information is used ineffectively the ensemble of allowable trajectories must be larger. This is a cost in the thermodynamic language which has been used to discuss costs and advantages. But effective information processing is also a cost.

To some environmental events the biological system is indifferent. This is the basis of the third component of adaptability, the indifference component, also a cost. All the components of adaptability are thermodynamic costs, leading naturally to the idea that unexercised adaptability tends to atrophy in the course of evolution. There are two possible mechanisms of atrophy. Either the ensemble of allowable trajectories becomes smaller or some of the allowable trajectories become less efficient. I can think of no ratchet which could permanently prevent one or both of these processes, even in the case of the most inexpensive forms of adaptability. Only the incomplete efficiency of the evolution process as an optimization process could slow the decline of adaptability.

There is another use for an ensemble of trajectories. By using redundancy it is possible to generate similar trajectories which form natural bundles, hence which allow for reliability. But, even more important, it is possible to achieve transformability, a fundamental requirement for processes such as evolution and learning. The simplest and most important example is the gene. The basic idea is to treat the mutability of the gene as a source of uncertainty and to ask how the protein must be organized so that this uncertainty is buffered into gradual transformation of structure and function. The point of view extends to the entire genome. Much of the apparently neutral variation exhibited by single genes emerges as both a precondition for and a result of Darwinian evolution. Evolution facilitates itself through the development of such neutralism. This is the bootstrapping process. It extends to the genome and therefore to the developmental process as a whole. A great deal of the redundancy in biological systems—both structural and functional—arises through this bootstrapping process. The paradoxical consequence is that only those organisms which are far from the optimum of efficiency and difficult to describe have any chance of arising and increasing their efficiency through the evolution process. Peculiarly this does not contradict the efficiency-driven trend to minimal adaptability. Reliability and transformability are hidden in the actual magnitudes of the entropies. They involve a shift from anticipation to indifference which has its basis in a blowup in the size of the ensemble of possible trajectories.

The next idea is that biological systems have a hierarchical and compartmental organization, so that it is possible to consider how the uncertainty of the trajectory is allocated to these different levels and compartments. The assumption of hierarchy is self-justifying. Adaptability is most efficient, in terms of its thermodynamic cost, in systems with this type of structure. The main result is the *principle of compensation*. If one mode of adaptability is contracted, due to the presence of a constraint, another must expand or the different modes of adaptability must become more independent. Alternatively, the niche can become smaller (the indifference term) or anticipation can improve. Constraints include the time scale of the disturbance and the complexity of the organization. The more complex organizations are less likely to be compatible with as many variant organizations which are viable. At least it is more costly for them to have as many. Some complex systems—such as the brain—allow for an enormous ensemble of trajectories. Here high internal constraint opens up new degrees of freedom. This is not unusual. The same is true for the genetic system, in which the extremely high internal correlations underlying the transcription and translation mechanisms allow for a large ensemble of variants. It is the variation of these internal correlations which is costly and increasingly costly as the degree of correlation increases.

A corollary idea is that, in order for a level or compartment to be predictable using a law expressed in terms of its own state variables, it must be protected from disturbances by adaptability at other levels or in other compartments. Dynamical descriptions of a particular biological system—or of a particular level of a particular biological system—can only be successful to the extent that other systems or other levels provide crucial regulation. If the adaptability of these other systems is inadequate, disturbances will reach the system of interest through them, destroying its apparent autonomy. The system of interest will undergo either adaptive or maladaptive changes, breaking the original law used to describe it and calling for a larger law which involves the unrepresented variables. There is a paradoxical consequence for the structure of dynamical biological models. To the extent that these models are successful at isolating and predicting the behavior of particular systems or levels, they fail to incorporate all the really interesting biological processes which support these dynamics. But in fact such highly abstracted dynamical models only rarely have powerful predictive capabilities. The support—the bath of unrepresented adaptabilities—is never so perfect.

Adaptability theory provides a better way of looking at biological dynamics and for interpreting dynamical biological models in a useful way. The system with high predictability must be controlled by a system with high unpredictability. But from a dynamical point of view the unpredictable system can be viewed as an unstable system, say a multistable system. The instability correlates to the modifiability component of adaptability. The predictable system can be viewed as stable. At least its dynamics must fulfill some stability criteria. The system or level may have dissipative dynamics. This is the usual situation in biology. In this usual situation one of the systems which is unrepresented is the heat bath. Its weak stability serves as the ultimate controlling mechanism. The dissipative dynamics may be structurally stable, also a common situation. Such structural stability correlates to the independence component of adaptability and to transformability. So, corresponding to the principle of compensation of adaptabilities, there is a principle of compensation for stabilities and instabilities. As the dynamics of one subsystem is required to be more orbitally stable in the course of evolution (or as stable, despite increasing complexity), the dynamics of some other subsystem must become less stable, or more multistable, or the dynamics of the original subsystem must become structurally more stable. The dynamics must itself adapt to the requirements of adaptability.

Now I want to return to the discussion of organic evolution, but in the light of the adaptability theoretic principles of biological organization. Usually when thinking about the history of life one thinks about the plant and animal forms suggested by the fossil record. One is especially inclined to think of the sequence of forms which involve increases in complexity and which lead to our own species. But complexity is only a part of the story. The other, unseen part is the evolution of uncertainty and of the structure of uncertainty.

There are four major processes which contribute to this evolution. Their interlinking dynamics are pictured in Figure 13.3. The first is the evolution of transformability. This is the bootstrapping process. The first event is the origin of a set of constraints capable of supporting evolution. One can imagine that during the first great era of primitive life the most significant events involved the evolution of these constraints—the precursors of the present day genotype–phenotype relationship—in a way which further facilitated the evolution process. This means a coding mechanism which supports high replaceability of codons and amino acids and gene structures with mechanistically superfluous amino acids. These mechanisms increase neutralism, hence increase the ensemble size. The increased ensemble size hitchhikes along with the advantageous specializations and mechanisms of adaptability whose evolution it facilitates. During this pre-

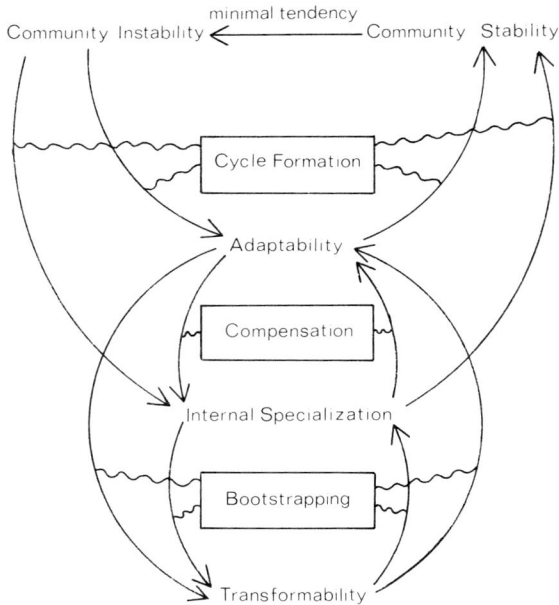

FIGURE 13.3. Interlinking of major processes in evolution. Transformability is bootstrapped by the improvements in internal specialization and adaptability which it stimulates. Restructuring of adaptability is required by changes in internal specialization and changes in internal specialization may be stimulated by new constraints required for the restructuring of adaptability. Community stability is based on the development of adaptabilities and internal specializations which allow for cyclicity of ecosystem processes. If the community is unstable, it will necessarily change, giving rise to new structures of adaptability and new internal specializations. But even a stable community will eventually destabilize due to the minimal tendency of adaptability, thus reinitiating the compensation cycle and potentially opening the possibility for a change in the dominant biota. If the compensation cycle leads to new forms of biological organization, it will reinitiate the bootstrapping cycle.

liminary stage of evolution the visible structural change must have been slow. But once the transformability reaches the critical level at which at least one of the possible mutations increases fitness, the adaptive development and divergence of the system should accelerate prominently.

This development and divergence are controlled by the second of the major processes, that is, the process of compensation. Compensation is governed by the costs and advantages of different forms of adaptability. These costs and advantages are governed by the interdependence structure of a system, correlated to its internal specialization. More internal specialization does not necessarily mean more efficiency. It does only when it brings an ability to utilize resources which cannot be effectively utilized by any simpler form. The problem is that the increase in efficiency in this case is accompanied by a decrement in efficiency due to the increasing cost of adaptability. So the selective force which favors the increase in internal specialization automatically induces a selective force which favors the restructuring of adaptability. This restructuring may involve reallocation to already existing subunits or levels. But a net increase in efficiency may only be possible if new subunits or levels, therefore new constraints, are introduced. This will happen if there is no mode of adaptability inexpensive enough or if there is no way of coupling an inexpensive mode of adaptability to the original interdependence structure. In this case the selective force favoring less expensive adaptability induces a selective force which favors further internal specialization. Now the possibilities are open for efficiency-increasing increases in internal specialization in these new levels or compartments, thereby reinducing the selective force which favors restructuring of adaptability. It is this recursive interaction between complexity and adaptability which I believe is principally responsible for such major evolutionary developments as cellular organization, multicellular organization, societal organizations, and special organs of adaptability such as the brain and immune system. According to this point of view, the evolution of human intelligence is due as much to the high cost of alternative forms of adaptability as to the direct advantages of intelligence.

The new forms of organization arising as a result of the compensation process cannot immediately undergo adaptive divergence. This requires amenability to evolution, so the bootstrapping of amenability again becomes the prominent but barely visible process. It must have been especially prominent subsequent to the formation of the first cellular organizations and subsequent to the formation of the first multicellular organizations. Important present-day features such as highly transformable regulatory mechanisms, polygenic control, and highly transformable control systems such as the cyclic nucleotide system must have appeared during these periods. It is possible that some physically ambiguous features of communication systems and language also arose during periods of bootstrapping. Such bootstrapped features are not mechanistically necessary and are energetically disadvantageous for the individual, but they are evolution-enhancing. Some organizations are both evolution-enhancing and advantageous to the individual in terms of adaptability. Buffer mechanisms which decrease the interdependence of different parts of the phenotype fall into this category.

The third process is cycle formation. If the community is not stable the constituent adaptabilities and internal specializations necessarily change. In order to be stable it must stabilize the environment (the cyclicity feature). The cycling mechanisms must have enough adaptability to tolerate variations in the environment, including variations which affect the input and flow of mass and energy. A high diversity of states increases the efficiency of the cycle. A large ensemble of pathways of mass and energy flow stabilizes this cycle. Some of the features which increase this ensemble are species diversity and multifunctionality in the grazing pathway. An especially important mechanism is the use of high culturability in the detritus pathway. A key point is that the adaptations and adaptabilities which contribute to cycle formation and buffering should satisfy pressures of evolutionary as well as ecological stability. The pressure for evolutionary stability leads to trophic collapse in the grazing pathway and trophic elongation in the detritus pathway.

Why are the dominant flora and fauna of communities so often replaced in the course of evolution, even when these replacements open up no new resource dimension of the environment? One explanation is that the newer forms are more efficient than the older ones. Ultimately this may often turn out to be the case. But I think the more immediate cause is that the older flora and fauna have become too efficient. The advantages of efficiency wring out unexercised adaptability. The advantages of conserving effective adaptations work against the maintenance of high transformability and of the plasticity of the speciation mechanism. At first the community appears more capable of conserving itself because of more effective cycling and because of the quenching of less fit variations among the dominant forms. But in reality it has traded potentiality for efficiency. This fourth process, the minimal tendency of adaptability, is implicit in the process of compensation. It inevitably leads to an instability which reinitiates the interlinking processes of compensation and cycle formation. It may open up the opportunity for novel, highly transformable forms to outrace older ones.

13.6. THE LIMITS OF PREDICTABILITY

Surprise is sometimes expressed that I would construct a theory to deal with biological phenomena of such a highly statistical nature. It would appear at first that a deeper dynamical theory is possible. As a practical matter it might be that we do not know the transition functions, but in principle it might be thought that they could be known and used to obtain more detailed answers to all the questions with which I have dealt in this book. On this view my approach is a first approximation. Careful empirical studies should allow these equations to be solved, leading to detailed predictions about development, learning, succession, evolution, and the other processes considered. It is usually understood by the exponents of this view that there may be both practical and computational obstacles to the completion of this program, therefore that an approach such as mine might practically speaking be a reasonable one. The situation is sometimes viewed as analogous to that of macroscopic thermodynamics and mechanics. In principle one ought to be able to answer all questions with mechanics. But ignorance intervenes. So one assumes only the general form of the microscopic laws and then makes some statistical assumptions, hard to justify theoretically, but practically speaking most useful.

If this analogy were correct I would certainly take it as an adequate justification for the theory presented here. But I think that the analogy is seriously flawed and that in principle there is no deep dynamical model which underlies the theory. The subtle point is that the laws of physics currently used for statistical mechanics are assumed to be of a fixed nature, allowing variation only in the initial and boundary conditions and in the forces of constraint. The dynamical models used in biology, even though they may assume a form which is superficially similar to the equations of physics, are of a very different nature. They are themselves the products of evolution, capable of changing further through adaptation, learning, and evolution. If we formulated these laws in terms of the fundamental principles of physics, this would not be the case. But this is not the nature of dynamical models in present-day biology. The dynamical models of present-day biology are abstractions which isolate systems and aspects of these systems which could never be isolated in reality. The proper question cannot be, given these dynamical laws, what is the spectrum of stabilities and instabilities? This question assumes the immutability of these tentative laws. The requirements of adaptability can change, leading to the evolution of new laws. However, these requirements have the basic aspect of real laws, which is that they are fixed. The structure of adaptability changes, just like the state of a statistical mechanical system changes. But the basic principles of adaptability, which at bottom reduce to the algebra of entropies, hold throughout, just as the basic principles of statistical mechanics, which at bottom are a logic of probabilities, hold throughout.

There are situations in which a nonphysical dynamical model can serve as a good predictor for an isolated aspect of an isolated system. This occurs when this aspect is highly regulated by the bath of unrepresented adaptabilities. This is why valid dynamical models of an empirical nature are possible for a number of physiological processes. But this just means that it is less possible to make useful models of a similar nature for some other, unrepresented system. It might be thought that this limitation could be avoided by taking more variables into account and by using time-dependent dynamical laws. It should then be possible to map the biological reality more completely. Only the tractability of the equations would be in question. But this returns us to the original problem. If a nonphysical mathematical description can predict the future of a system, then it must be concluded either that no physical processes are relevant (as in the computational behavior of a digital computer) or that the nonphysical description is equivalent to a physical description, which is evidently not the case. The real situation is that as one attempts to map more and more of the reality with a nonphysical description, more and more of the physically relevant processes are omitted. The addition of more and more complexity to dynamical biological models can only cause the predictions elicited from them to recede further and further from the biological reality.

There is a sense in which it can also be said that no dynamical physical law underlies the theory. By a dynamical physical law I here mean an equation of motion based on fundamental principles (superposition principle, symmetries of time and space, positivity of energy, exclusion principle, equivalence principle, microscopic reversibility). For biological systems it is almost certainly impossible to determine the initial conditions, boundary conditions, and constraints which would be required to write such a law without modifying the system to the point where these data are no longer applicable. Practically speaking the law, even if written, could not be computed. In principle it could not be computed unless a computing system could be built whose computational capabilities exceeded those of the biological system whose behavior it computes. Even if it could be computed, the problem of interpreting the results of the computation in terms of identifiable biological variables would introduce a requirement for auxiliary methods of computation. So reductionism to dynamical physical law is impossible, at least from the practical point of view, and in this sense the idea that a physical law of this nature underlies the theory loses clarity. This is of course the reason for making nonphysical dynamical models in the first place. But the great difficulty I want to underline is that neither nonphysical nor physical dynamical models can deal with novelty in the living world, the former because the physics should not be irrelevant and the latter because the physics is too complicated to compute.

In principle, the process of computation is itself a fundamental source of novelty. A system which executes programs, whether as simple as a computer or as complex as a biological system, can always show novel behavior which could never be predicted without actually executing these programs. If this were possible, it would be possible to determine in general whether such a system is caught in an infinite loop, therefore to solve a problem, namely the halting problem, which is unsolvable on logical grounds. Due to the physical limits of computation and due to the limited efficiency with which effectively prescribed computations can be performed, there is a point at which assuming the existence of a computer powerful enough to predict all such behavior becomes in principle untenable. Could we, using our intuition, go beyond this point? This is a dubious proposal, for it would imply that humans possess a new computing primitive which enables them to turn classes of problems which computers can only solve by enumeration into classes of problems which can be solved without enumeration. We deceive ourselves if we suppose that we can prognosticate to a level of resolution where even the heaviest computation must fail.

What then is the status of dynamical models in biology? I believe that it is the same as (but complementary to) that of adaptability theory. It provides organizational principles. The difference is that dynamical methods deal with $\hat{\omega}$ rather than $H(\hat{\omega})$. The analysis of $\hat{\omega}$ provides principles which concern the connection between system structure and stability. These principles are valid even if $\hat{\omega}$ changes in an unpredictable way. The analysis of $H(\hat{\omega})$ provides principles for predicting which patterns of stability and instability are possible, therefore which patterns of system structure are chimerical and which are not. The possible directions which the spectrum of uncertainties can take in the course of evolution are therefore limited. The possible directions which the corresponding organizational structures can take are limited as well. But the limitations are like thermodynamic limitations. They concern the spectrum of uncertainties, not the detailed course of events. The language of adaptability theory provides no means for predicting this detailed course, nor can the language of dynamics be properly construed as providing this means.

REFERENCES

Darlington, P. J. (1957) *Zoogeography: The Geographical Distribution of Animals.* John Wiley & Sons, New York.

Dobzhansky, T. (1951) *Genetics and the Origin of Species.* Columbia University Press, New York.

Fisher, R. A. (1930) *The Genetical Theory of Natural Selection.* Clarendon Press, Oxford.

Haldane, J. B. S. (1932) *The Causes of Evolution.* Harper and Bros., London.

MacArthur, R. H., and E. O. Wilson (1967) *The Theory of Island Biogeography.* Princeton University Press, Princeton, New Jersey.

May, R. M. (1975) "Patterns of Species Abundance and Diversity," pp. 81–120 in *Ecology and Evolution in Communities,* ed. by M. L. Cody and J. M. Diamond. Belknap Press of Harvard University Press, Cambridge, Massachusetts.

Mayr, E. (1963) *Animal Species and Evolution.* Harvard University Press, Cambridge, Massachusetts.

Schmalhausen, I. I. (1949) *Factors of Evolution.* Blakeston, Philadelphia.

Simpson, G. G. (1949) *The Meaning of Evolution.* Yale University Press, New Haven, Connecticut.

Stebbins, G. L. (1950) *Variation and Evolution in Plants.* Columbia University Press, New York.

Wright, S. (1970) "Random Drift and the Shifting Balance in Theory of Evolution," pp. 1–31 in *Mathematical Topics in Population Genetics,* ed. by K. Kojima. Springer-Verlag, Berlin.

14

The Age of Design

In this final chapter I want to turn to human ecosystems and to those aspects of evolution which man now can take into his own hands. The issue of predictability becomes one of practical importance. I shall argue that the basic trends which obtain in prehuman ecosystems hold as well in ecosystems dominated by humans, and that they hold with greater force due to guiding ideas about prediction, planning, and efficiency which are unsound from the standpoint of adaptability. In considering this question I draw out some lines of contact between the theory already developed and economics. It is evident that the problem of economic adaptability has many special features to which a single section of a single chapter could never do justice. But I believe that viewing an economy as a part of the ecosystem process and treating the problem of economic adaptability as a component of ecosystem adaptability are most suitably begun here.

14.1. THE LIMITATIONS OF PREDICTION, EFFICIENCY, AND PLANNING

It is possible to consider some of the limitations on the predictivity of models from the standpoint of the formalism of the theory itself. Activities such as modeling, predicting, controlling, and designing are themselves part of the ecosystem process. They are forms of information processing and adaptability, therefore automatically built into the structure of adaptability theory. It is only necessary to look at the principle of compensation from a human-centered viewpoint. The argument can be formulated in terms of the compensation equation (6.5) by separating all terms which concern the human system from terms connected with the remainder of the community. But it is simpler to use the mnemonic (two-compartment) equation

$$\{H(\hat{\omega}_P) + H(\hat{\omega}_R|\hat{\omega}_P)\} - \{H(\hat{\omega}_P|\hat{\omega}^*) + H(\hat{\omega}_R|\hat{\omega}^*\hat{\omega}_P)\}$$
$$+ H(\hat{\omega}^*|\hat{\omega}_R\hat{\omega}_P) \to H(\omega^*) \tag{14.1}$$

where $\hat{\omega}_P$ stands for the joint transition scheme of the compartments involved in modeling, predicting, and designing, and $\hat{\omega}_R$ stands for the joint transition scheme of the remaining compartments.

In order for there to be any adaptability at all (apart from indifference), it is necessary for the community to anticipate the environment, otherwise the terms in the two sets of curly brackets cancel each other out. This is a general fact, preexisting the evolution of human modelmakers and designers. The major difference is that with human beings the models and structures are developed through the mechanisms of the brain and cultural inheritance. The conditions which must be met in order for these models and structures to contribute to adaptability are the same as those which must be met by anticipation mechanisms which arise through natural selection. The most important point is that the model predict the future of the environment and that constraints exist which couple these predictions to suitable changes in the state of P or R. The second possibility is that the model helps to predict the state of R given the environment [$H(\hat{\omega}_R|\hat{\omega}^*\hat{\omega}_P)$ smaller] and that constraints exist which couple these environment-dependent predictions to suitable changes in the state of R. These constraints (also the constraints required to actually calculate the model) increase the complexity of P, therefore increase the cost of modifiability in this part of the community. But the required amount of modifiability also decreases. The appearance of new constraints through the physiological mechanisms of modeling and constructive manipulation of the environment changes the transition scheme of the ecosystem [equation (4.2)], just as new constraints that appear through the evolutionary mechanism do. Here it is exceptionally clear that these changes in dynamics are not predictable. In principle P could use all available computational resources to compute R and the environment; but then none would be left to compute P. But the problem of computing a computing process without duplicating it is even more fundamental than this.

What happens if P attempts to design R so that its dynamics are more predictable? P sees R as part of his environment, so it appears reasonable to him that if he can modify R to obey more autonomously predictable dynamics, he should be able to decrease his quota of modifiability. P will also be motivated by the fact that an increase in the predictability of R will make it a more efficient source of energy for him. The best case (based on P's assumptions) occurs when the dynamics are completely predictable. In that case $H(\hat{\omega}_R|\hat{\omega}^*\omega_P) \to H(\hat{\omega}_R|\hat{\omega}_P)$. But this means that all of the adaptability is in the difference $H(\hat{\omega}_P) - H(\hat{\omega}_P|\hat{\omega}^*)$. Rather than creating a more predictable environment, P has completely destroyed the bath of adaptabilities which protects him from the uncertainty of the external environment. What has really happened is that the human population ends up absorbing all the environmental disturbance. The biotic environment (R) does become more efficient, since it has less adaptability, but this efficiency remains high only so long as it is successfully buffered by humans or human constructions. R might include phenotypic or genetic properties of humans or features of human social organizations. The situation is the same. The compartment responsible for the modeling and designing activities, by attempting to design an environment with autonomous and efficient dynamics, will end up reducing the extent to which it is protected from environmental disturbance and increasing the extent to which it must protect delicate features on which it depends for a livelihood. The behavioral uncertainty of the modeling system, necessarily high in the stage of model development and implementation, cannot by itself provide this buffer. Constraints are necessary which actually enable it to reroute the disturbance.

P's mistake here was to misidentify the source of disturbance. He identified it with a feature of the environment which he was unable to predict in practice due to his lack of knowledge of an appropriate predictor and, in this case, due in principle to his failure to include the external environment. As a consequence he identified a source of adaptability with a source of disturbance. By eliminating this source of adaptability he at first enhanced the productive efficiency of his biotic environment, but by not compensating this loss of adaptability he ensured the eventuality of a traumatic event. If P wants to increase the efficiency of his biotic environment by modifying some components in this environment to obey more autonomously predictable dynamics, he must compensate this by modifying some other system to obey less autonomously predictable dynamics.

Suppose that P correctly identifies the source of disturbance to him. But he wishes to modify R so as to increase its environment-dependent predictability. His reasoning is that in this way he can properly increase adaptability by increasing the difference $H(\hat{\omega}_R|\hat{\omega}_P) - H(\hat{\omega}_R|\hat{\omega}^*\hat{\omega}_P)$. Also he will decrease the cost of this adaptability, therefore increase the efficiency of R as a source of resources. The best case (based on P's assumptions) occurs when the dynamics are predictable in as much detail as possible. The problem is that in the information transfer picture (Section 4.7) $H(\hat{\omega}_R|\hat{\omega}^*\hat{\omega}_P)$ is replaced by $H(\underline{\hat{\omega}}_R|\underline{\hat{\omega}}^*\underline{\hat{\omega}}_P)$, where $\underline{\hat{\omega}}_R$ and $\underline{\hat{\omega}}_P$ are the transition schemes in terms of the finer states. But $H(\underline{\hat{\omega}}_R|\hat{\omega}^*\underline{\hat{\omega}}_P)$ is in general very much larger than $H(\hat{\omega}_R|\hat{\omega}^*\hat{\omega}_P)$. If in making the system more predictable P decreases the former, he will also decrease reliability and transformability. As a consequence $H(\hat{\omega}_R|\hat{\omega}^*\hat{\omega}_P)$ will increase rather than decrease. The thermodynamic cost of behavioral uncertainty and environment-dependent behavioral uncertainty decreases, but the thermodynamic costs of evolution and other processes involving transformation of function increase. The costs arising from failures of information processing also increase. So by trying to improve anticipation P has actually degraded the underlying processes on which anticipation is based. P's mistake here was to misidentify the level at which he made the environment-dependent dynamics more predictable. He eliminated, for example, genetic loads which are necessarily concomitant to transformability and redundancies which serve as reliability-increasing noise absorbers. As a consequence he killed R's potentiality for adaptive change and information processing. P will again have to compensate by increasing his adaptability or by increasing the adaptability of some other part of R. Otherwise the undesirable indifference to the environment will increase. If P wants his surroundings to handle these problems naturally and to reduce the burden on himself, he should attempt to keep $H(\hat{\omega}_R|\hat{\omega}^*\underline{\hat{\omega}}_P)$ high. This is the necessary condition for introducing modifications in R which reduce $H(\hat{\omega}_R|\hat{\omega}^*\hat{\omega}_P)$.

Suppose that R exhibits either environment-independent or environment-dependent unpredictability, but that this actually is a source of disturbance to P. This is the situation of competition-generated uncertainty within the system. R is generating unexpected behavior which reduces the effectiveness with which it can be exploited by P rather than increasing the extent or effectiveness with which it buffers P from the environment. P recognizes that such internally generated uncertainty is evolutionarily stable but that it is not in his interest. He reasons that in this case it is entirely safe to modify R so that its dynamics are more predictable. He attempts to exercise care that it is really only the competition-generated uncertainty that he is eliminating, therefore that his design modifications only reduce the magnitudes of $H(\hat{\omega}_R|\hat{\omega}_P)$ and $H(\hat{\omega}_R|\hat{\omega}^*\hat{\omega}_P)$, not the difference between them. P is correct that by itself this would increase efficiency without decreasing adaptability. But in order to maintain these design modifications he must introduce constraints which prevent R from acting in an evolutionarily stable way. The entire community becomes more complicated and interdependent, therefore the cost of adaptability increases. The new constraints must be protected, so P must introduce mechanisms of buffering them from disturbance, further increasing the cost of adaptability. The total amount of modifiability which must be exhibited by the entire community in order to achieve the same amount of adaptability increases. There is a second, longer-range problem. Competition-generated uncertainty provides a natural reservoir of variabilities which often converts into real adaptability when the environment changes. Suppressing this variability reduces the capacity to cope with out-of-range disturbances. But even more important, introducing constraints which prevent the occurrence of competitive adaptations in R will necessarily also prevent the appearance of adaptability-increasing adaptations, thereby again increasing the burden on P.

Suppose that P identifies itself (the human community) as a source of disturbance. P knows that every time it introduces a modification into R it changes the transition scheme in an unpredictable way. The first result is always to decrease the already developed mechanisms of anticipation, whether developed through natural selection or through intelligent design on its own part. P may even recognize that its ability to anticipate the environment and to anticipate R will decrease, so that the local improvement may lead to a global increase in undesirable indifference unless many other compensating adjustments occur. This is not a novel situation. It is the usual situation in evolution. The compensating adjustments occur naturally because R has suitable adaptability and transformability. But in those parts of R which have been designed for efficiency and predictability by P, this adaptability and transformability is reduced. Unless there are compensating increases elsewhere (seen by P as a thermodynamic load) the compensating increases must be in P itself. Now P is in an obviously contradictory situation. If P is highly transformable it is more likely to be a source of novelty, hence disturbance; but if P is not highly transformable, it cannot compensate for the transformability it has wrung out of R. P can be adaptable, provided that the adaptability consists of a fixed set of responses. But adaptability that allows new adaptation (that is, adaptability based on transformability) is dangerous. There are only two possible solutions. One is to pump new adaptability into R; the other is to wring transformability out of itself. The latter solution is favored by efficiency—since transformability is a thermodynamic load. Adopting it, though, means taking steps to stop evolution, therefore steps to suppress novelty-creating processes which might later lead to real gains in efficiency. Adopting it also means suppressing the capacity to adapt smoothly to out-of-range disturbances which eventually inevitably occur, due to either internal failures of cyclicity or extrinsic events.

The fallacy in each of these situations is the attempt to design systems which satisfy unsuitable standards of predictability, hence systems with an unsuitably collapsed structure of uncertainty. This structure of uncertainty is real, not a matter of ignorance. There is no way of eliminating it without eliminating adaptability. This is true whether the system is a product of natural selection or a product of intelligent design. There is an obvious dual implication. P should not be able to construct models more predictive than these limitations allow for any system R whose structure of adaptability is adequate to ensure its survival. These limitations are summarized in the following statement, which, except for the first (physically motivated) sentence, has the status of a theorem.

THEOREM (limitations on dynamical predictability). New adaptations change the dynamics in a way that depends on physical processes which cannot be described by a nonphysical description (by definition) and cannot be computed by a physical one. If the possibility of new adaptation could be turned off (in principle allowable only if the ensemble of environmental disturbances is fixed) environment-dependent stochastic models would become possible. The two sources of uncertainty (aside from the uncertainty of the environment) are internally generated uncertainty and random processes at the finer (information transfer) level of description. To the extent that the former can be turned off and the latter ignored, environment-dependent models as deterministically predictable as the environment would be possible (in the limit of maximum possible anticipation). Environment-independent deterministic models are possible for subsystems (ignoring the information transfer picture), but only if these are compensated by other subsystems for which the best models are less predictive.

If models can be constructed which are more predictive than allowed by these limitations, this is a symptom of faulty design and eventual serious pathology. One way of creating such a faulted design is to continually take steps to increase efficiency by reducing the adaptability of functionally specific subsystems, but without compensating with increases in the adaptability of others. Increases in efficiency should in general be based on real adaptive advances, of either a biological or a technical nature. If these advances involve increases in complexity, the cost of adaptability increases, so it is especially important that the requirements for compensation be observed.

This result—a straightforward deduction from the very general principles of adaptability theory—is, unfortunately, much less intuitive than it may at the moment seem. Our experience in daily affairs is that we usually do better when we think a situation out carefully. Sometimes this thinking is an intuitive or judgmental process. But often we can improve a process—a machine, a particular agricultural process, the performance of an athlete, or the function of a particular institution—by trying to design or control it in detail. So it is quite natural to extrapolate this paradigm of modeling, prediction, control, and design from local situations, where it is often successful, to global situations. We arrange these local situations in the most efficient way for some particular purpose and so the tendency is to think that it should be possible to do the same for the global situation. The usual thought is that the only difference is that optimization is harder for a big problem. Fortunately this is true, for the real problem is that a global system which is forced to be predictable is a badly thought out system. Design in these global cases should concern itself with the proper organization of potentiality, not with its suppression in order to obtain efficiency or predictability.

The problems are clear in agricultural ecosystems. Here the society regards it as particularly desirable to have a model which is as predictive as possible and a system which is as efficient as possible. The modelmaker designs such a model. The agriculture is so designed that features which would prevent it from realizing the computable dynamics are suppressed as nearly as possible. In fact the suppression of such features is efficiency-increasing, so it probably began prior to the development of the model. Monocultures are used. Strains are as pure as possible. The whole economy may be based on one crop. A complicated system of harvesting is developed, also precise and efficiency-increasing. Steps are taken to reduce weeds, pests, diseases, and all other disturbances of a biological origin. Resistant strains could be used, but less-resistant strains are more efficient, so the strategy of controlling contaminants by other means is adopted. Domestic animals are kept under artificial conditions which are optimal for productivity, but narrower than the conditions which would occur if a failure in the controlling systems occurred. In fact all this is not necessarily a bad idea. But in order for it not to be a bad idea the reduction in adaptability and the increase in mechanistic complexity must be compensated by an increase in adaptability elsewhere in the system. Adaptability might include alternate ways of routing food and energy when failures occur in a particular country. But such compensating adaptabilities are costly in terms of efficiency, awkward to construct, and inherently unpredictable in detail. So failures occasionally occur, due to an unexpected disease, a breakdown in the control of predator dynamics, or unexpected weather. Stronger measures are taken to eliminate the source of the disturbance. The model is enlarged to include, say, the offending predator. The model may in fact lead to measures which successfully control its dynamics. Now the whole system is more complex, therefore more sensitive. But no compensating adaptabilities are developed elsewhere. In the long run some sacrifice of efficiency for an increase in adaptability, either elsewhere or preferably within the system, would give an enhanced long-term efficiency. Many agricultural ecosystems really do incorporate such adaptabilities. They are not in fact so naively constructed as portrayed here. But can we be sure that the pressure for increased yields—an urgent pressure in some parts of the world—is not slowly, in imperceptible steps, leading to the same exchange of adaptability for efficiency which is the basis of the periodic instabilities which occurred in prehuman evolution? I think the facts which are generally available are quite clear. They point to increasingly monocultural systems, increasing use of efficient, unadaptable strains, increasing attempts to control contaminating flora and fauna, increasing complexity of the support systems, and no increase in compensating adaptabilities.

But this tendency is not simply restricted to agricultural systems. Many institutions which might serve as compensating systems are also undergoing analogous processes. This tendency is not monodirectional. Every time there is a crisis new sources of potentiality are released. But to an increasing extent—either through planning or through the drive for efficiency—adaptability is being hocked for immediate gains which are urgently needed. The prediction of adaptability theory is that humanity itself, under this paradigm, must increasingly take on the burden of compensation, not through the pleasant processes of novelty and invention, but through unpleasant phenotypic and demographic processes. Unfortunately this prediction corresponds to the facts in many parts of the world.

It might at first be thought that these facts are caused by purely Malthusian factors. Certainly the carrying capacity of the world is limited, so it is inevitable that resources will become scarce. Given a sufficiently large population, hunger and other kinds of deprivation are inevitable. But I think this traditional explanation does not fit all the phenomena. It does not explain the phenomenon of breakdowns and massive starvation. These could not occur if the resource availability and utilization efficiency were constant. Furthermore these breakdowns sometimes occur far below the carrying capacity, as evidenced by subsequent general increases in population size. The adaptability theoretic interpretation, which I believe accords better with the facts, is that the instability of the population dynamics is serving to buffer efficient and predictable technologies and social organizations and at this instability is forced into existence by the requirement for compensation. Essentially culturability, a mode of adaptability suitable for bacteria, is becoming increasingly important as a mode of human adaptability, despite its inefficiency in a complex organism and its manifest undesirability.

14.2. THE DESIGN OF ECONOMIC SOCIETIES

To what extent are the trends which adaptability theory implies about preeconomic ecosystems also exhibited by economic ecosystems? By an economic ecosystem I mean an ecosystem to which the special constraints of an economy are added. The important special constraints are language and human decision-making, financial structures, and beliefs which influence decision-making. These constraints greatly amplify the possibilities for design. There is an enormous development of technologies and societal organs which extend food-getting, communication, and other functional capabilities of organisms. As in preeconomic ecosystems, cycling resources and balancing population growth is necessary for stabilizing the environment. Due to the development of technologies this may be achieved to a lesser extent than in preeconomic ecosystems, leading to a speedup of the developmental process. The financial system allows for shifts of money and resources between specialized classes within the society, which leads to a cycle of economic activity, totally apart from the instabilities connected with the cycling of resources between society and the surrounding ecosystem. Certain of the resulting distributions may be politically stable, others not, hence this new source of internal disturbance can assume great importance. The financial system and the amplified transportation systems allow for exchange of money and goods between different societies, thereby greatly increasing the extent to which a society's environment can be rapidly degraded by improvements or other events distant in space. As in preeconomic ecosystems early events usually are more critical for determining the organizational character of the system than the later ones. Due to language and the consequent increased capability of transmitting information from generation to generation the memory of early events can be especially strong, yet not particularly visible.

The four major processes—innovation, shifts in the distribution of resources due to mechanisms within the society, disturbances originating from other societies, and the influence of early events—are all present in preeconomic ecosystems. The new features—human language and decision-making, money, and beliefs—make them especially important in economic societies. I believe it fair to say that they are generally considered to be important for the development of these societies. Does this amplified importance argue that the minimal tendency of adaptability is less important, relatively speaking, than in preeconomic ecosystems? I think not. The considerations in the previous sections argue against this. Due to the high degree of institutional and political organization in human societies it is possible not only for them to lose adaptability by indiscernable steps, but also to undergo sharp, virtually phaselike transitions into adaptability-poor states. A society which adopts regimentation is an example.

To understand the contribution of adaptability it is necessary to recall its fundamental role in the functional circle of an economic ecosystem and to recall how the fundamental problem of an economy should be posed (Section 11.12). One problem for an economy is to produce and deliver enough goods and services to satisfy criteria established by the society. Some of these goods and services are connected with activities necessary for life, such as food consumption and procreation. Some are connected with adaptabilities necessary for life, such as those which involve the use of clothing and shelter. These are the nonevolutionary adaptabilities. They enable goods and services to be produced in a variety of environments and under a variety of environmental conditions. A second problem is to continue to generate enough goods and services to meet the criteria established despite long-term changes in conditions. These may involve changes in population and resource availability resulting from the society's own activities or they may involve changes in its trade situation resulting from the economic activities of other societies. The solution of this problem requires evolutionary adaptabilities, that is, adaptabilities which can lead to the discovery of new resources or to the discovery of new technological and organizational adaptations. But clearly these adaptations cannot arise through genetic evolution. The regenerative evolutionary adaptability can only be based on the compensating mental adaptabilities, that is, on creative thought. According to the principle of compensation, creative thought is as necessary for the continuation of economic life as genetically based evolutionary adaptability is for preeconomic life.

The effort expended on such creative activities is not less necessary for the continuation of economic life than the effort expended on food production and delivery, on the production of a next generation, or on the nonevolutionary adaptabilities. It is part of a coherent cycle in which creative adaptabilities serve to maintain the total of goods and services exchanged in an economy and the exchange of goods and services serves to maintain creative adaptabilities. This functional circle cannot persist when broken and cannot be adequately described by any theory which breaks it conceptually for the purpose of description. This is why it is necessary to pose the problem of an economy not simply as the problem of delivering enough goods and services, but as the problem of delivering enough goods, services, and creative opportunities. To pose the problem simply in terms of goods and services is to pose it in a value-biased form. It may appear self-evident that the allocation of resources to the so-called necessities of life—food, clothing, and shelter—should have priority over the allocation of resources to mental life, especially when times are unfavorable, as they usually are. But mentally based adaptabilities are just as necessary and indeed make their greatest contribution in difficult times. The mentally based adaptabilities which form the basis of innovation have the same claim to inclusion in a proper posing of the economic problem as do the production and delivery of goods and services, even goods and services required for subsistence. The mental adaptabilities are also required for subsistence. The failure to give equal *a priori* weight to all elements of this functional circle reflects, I believe, an unrecognized value assumption which is deeply embedded in many human communities. Due to this the minimal tendency of adaptability plays an even more important role in the development of human ecosystems than in the development of preeconomic ecosystems. In preeconomic ecosystems adaptability is lost only when it falls into desuetude. In human ecosystems it can be purposefully thrown into desuetude.

It is necessary to state the situation in the formal framework of adaptability theory. $H(\omega^*)$ is, as usual, the uncertainty of the environment. $H(\hat{\omega})$ is the behavioral uncertainty of the economic society. This uncertainty is the ensemble of possible modes of behavior of the complete society, including all levels of organization, down to the genetic and physiological. $H(\hat{\omega})$ includes, as one component, the uncertainty in the flow of goods and services. Except for barter, this is the monetarily measurable component of uncertainty. The uncertainty in the flow of goods and services includes, as one component, the uncertainty of investment behavior, including both its innovative and noninnovative components. The difference, $H(\hat{\omega}) - H(\hat{\omega}|\hat{\omega}^*)$, now includes economic as well as noneconomic modes of adaptability. The economic modes are connected with the information which the behavior of the environment provides about the flow of goods and services. The noninvestment component is connected with rerouting goods and services, excluding rerouting which involves decisions about their production or availability. This is the wound-healing form of economic adaptability. It may involve financial mechanisms of rerouting such as insurance. For example, if an entity in an economy is damaged, money flows into it from an insurance fund which can be used to draw enough goods and services from other entities to dissipate the wound. But it has previously paid for the insurance. The investment component of economic adaptability differs in a fundamental respect. It is connected with the choices which determine the availability of goods and services, therefore with the routing structure itself. For example, whether a farmer chooses to plant corn or wheat sets the constraints on the availability of these products, hence sets the limits on the possibilities for rerouting. $H(\hat{\omega}|\hat{\omega}^*)$ includes the uncertainty with which goods and services are routed as reduced by the specification of the environment.

The return on investment is determined by the particular routing which actually occurs, since specification of a particular routing specifies both the costs of production and the earnings. This return is necessarily uncertain in an uncertain environment. The uncertainty of the return is proportional to the adaptability. If the difference, $H(\hat{\omega}) - H(\hat{\omega}|\hat{\omega}^*)$, became small, the uncertainty of return could become small, but clearly uncertainty in investment behavior would make no contribution to adaptability in this case. If the magnitudes of the investment components of $H(\hat{\omega}^*)$ and $H(\hat{\omega}|\hat{\omega}^*)$ are high due to internal sources of uncertainty, the uncertainty of return on investment will be higher. The noninnovative component of investment adaptability involves uncertainty as to choices which concern the production of already existing types of goods and services. The innovative component involves the production of novel types of goods and services. It may also be associated with exploration. In this case it is connected with niche expansion, hence with a relative increase in the indifference term, $H(\hat{\omega}|\hat{\omega}^*)$.* It should be noted that the magnitude of the return on investment is not necessarily proportional to the uncertainty of the return. This proportionality is not implied by the theory. The noninvestment mode of adaptability can increase the return on investment only by preventing the ramification of disturbance suffered by entities in the economy, since it does this by redistributing the resulting losses. It can lead to no relative gain or loss on the part of these entities in the long run unless the environment or internal structure of the economy is changing. Nor can the activities associated with this mode of adaptability, taken alone, lead to a net return to the society. The extra resources which become available must be sufficient to cover the cost of the specializations (e.g., equipment) required.

As is usual, the actual magnitudes of these entropies may be large or small. So far I have not considered their significance, apart from the fact that they are increased by internal sources of disturbance. In preeconomic ecosystems they are associated with reliability and transformability. As redundancy increases, function-preserving reliability increases. Under suitable conditions function-altering transformability increases. This is the situation in genetic systems. The increase in the actual magnitudes increases the effectiveness of genetically based evolutionary adaptability. The actual magnitudes play an analogous role with respect to the investment component of adaptability, except that the mechanisms involve mental rather than genetic processes.

It is evident that mental processes are involved since the effectiveness of an investment depends on how well thought out it is. The individuals in a society may consider an enormous number of scenarios, or only a small number. The society may contemplate a large fraction of these, or only a small fraction. The investment behavior is much more likely to be effective if a large number of scenarios are considered. If a large number are considered, the magnitudes of $H(\hat{\omega})$ and $H(\hat{\omega}|\hat{\omega}^*)$ are both increased, but the difference between them also increases. The effectiveness of investment—its contribution to an economy's adaptation to changing circumstances—depends on this difference. The increase in the magnitudes of

---

*When new areas or features of the environment are opened up by exploration, the environmental uncertainty, $H(\hat{\omega}^*)$, increases and the indifference term, $H(\hat{\omega}^*|\hat{\omega})$, becomes relatively smaller. The behavior of the society gives more information about regions of space or features of the environment about which it previously provided no information. For example, previous to the discovery of an oil field the behavior of a society gives no indication of events which might occur in this field and indeed the state of the field may for all practical purposes be constant. As soon as oil wells are drilled the changes in the state of the field become uncertain, viewed in isolation, but much less uncertain when the behavior of the society is considered. The opening of new features of the environment requires an increase in the total adaptability of the society.

$H(\hat{\omega})$ and $H(\hat{\omega}|\hat{\omega}^*)$ may have to be large in order to increase their difference even by a small amount. This is due to the fact that a great deal of thinking is in general necessary to increase the return on investment. The increase in the magnitudes is not for the most part monetarily measurable. It is impossible and counterproductive to measure the imaginative activities taking place in a society in terms of earnings for services rendered. In many cases these activities do not even lead to macroscopically visible behavior. Yet it is the uncertainty of these monetarily invisible thought activities which underlies the effectiveness of the smaller, monetarily measurable uncertainties of investment behavior, hence underlies the regenerative adaptability of economic systems.

I want to underline the importance of this observation about the contribution of the absolute magnitudes of the uncertainties to the effectiveness of investment. In the adaptability theory analysis of an economic system terms automatically appear which connect to the mental processes underlying the effectiveness of investment. To my knowledge no other formal tool presently used in economics incorporates this feature, nor do informal discussions of economic systems pay a great deal of explicit attention to it. It is interesting to consider some of the phenomena which contribute to the magnitudes of these terms. Art, philosophy, and other imaginative and inventive activities contribute to it. These are explorations of alternative possibilities and intersubjective expressions of these alternatives. They are part of the ensemble of scenarios which underlie effective investment behavior on the part of individuals and societies. I certainly do not propose that artistic and speculative activities do not perform other functions, such as providing fulfillment and pleasure or providing support or criticism for a social order. But they can now be understood as having a fundamental role in the functional circle of an economy. They are not luxuries which follow the successful development of an economy. They are necessities for the success of an economy. The failure to appreciate this has, I believe, been a major source of economic declines.

In theories of economics there is in fact a construct which to some extent plays the role of evolutionary adaptability. This is *risk*. But according to adaptability theory risk, if it is to play this role, should not be narrowly defined as uncertainty of return on investment.* It should be identified with the total evolutionary adaptability of a society. I will call risk, so identified, total risk, to distinguish it from investment risk. The latter is the monetarily measurable component of total risk. Another major component is nonmonetary speculation. A play written on speculation, tinkering in a garage, or

---

*A distinction should be made between insurable and uninsurable risk. Uncertainty of return on investment is associated only with uninsured risk (Knight, 1921).

drawing up a proposal for the consideration of an investor are examples. Such activities are often described by the individuals involved as being done on speculation. It is evident that the amount of effort going into these activities is not to any significant extent monetarily measurable. It is monetarily measurable only insofar as it leads to monetarily measurable investments which are ultimately more effective. These nonmonetary speculations contribute to the magnitudes of the entropies associated with the economic component of adaptability. *Total risk consists of both investment risk and nonmonetary speculation. It is thus constituted not just of observable investment behavior, but more importantly of the total of the creative and inventive activities of individuals, their abstract and empirical exploratory activities, and their fantasies.* Total risk, as the economic form of evolutionary adaptability, serves to maintain the total of goods and services exchanged in any economy, that is, the gross national product (GNP). But with equal validity it can be stated that the exchange of goods and services serves to maintain total risk. Also it should not be forgotten that the economic modes of adaptability are not the only ones necessary for the perpetuation of an economic ecosystem. The noneconomic, biological modes are the *sine qua non* for the perpetuation of the system. In the development of human ecosystems these modes have become more and more drawn into the arena of design, hence placed more and more under the influence of economic decision-making.

Now it is possible to consider the compensating changes which occur in an economic society in the course of its development. As the society develops more efficiency-increasing internal specializations, constraints are placed on the adaptability of its citizens. The degree to which the adaptability of most of its citizens can ever be effectively coupled to concrete innovation decreases, as does the likelihood that innovation will have undesirable effects. As always, the restricted modes of adaptability must be compensated by the appearance of others. The increasing internal interdependencies of the society thus inevitably bring with them the requirement for compensating organs of adaptability such as academic institutions, research institutes, or other classes of individuals specialized for creative, experimental, and innovative activities. From the dynamical point of view the behavior of an economy becomes less stable as it becomes more complex and more internally specialized, all other things equal. The cost of instability becomes higher as well. The society may attempt to introduce control mechanisms to stabilize its dynamics and to regulate the environment. This is the process of trading structural for orbital stability, that is, of trading high qualitative sensitivity to a wide range of disturbances for high quantitative stability to a narrow range of disturbances. To compensate for the increase in orbital stability and to regain as much structural stability as possible it is necessary to compensate with special organs of instability. These are the research institutes, the academic institutions, the intelligentsia, and so forth.

The only alternative to this development is for the society to develop a wall of indifference, that is, to attempt to isolate its economy from external influences and to prevent internal changes which could destabilize it. But this line of development is incompatible with the development of efficiency-increasing internal specializations. The society must pay for its indifference by being forced to fabricate more of its own requirements. It sacrifices efficiency by failing to utilize comparative advantage, that is, by failing to take advantage of the relative differences in the efficiencies with which different goods and services are produced in different societies. The society becomes less externally specialized, to use the terminology previously used for preeconomic populations. Under these circumstances fewer resources are available for investment in and maintenance of technology. The level of internal specialization which the society can maintain must fall. It is unlikely that such a society could, in the long run, succeed in isolating itself from disturbance and innovation emanating from other societies. It is inevitable that its indifference will eventually break down and that it will suffer a crisis due to inadequate adaptability.

On the other hand, the society which moves in the opposite direction of developing more external trade becomes more externally specialized, that is, more dependent on events in other societies. Efficiency increases due to a fuller utilization of comparative advantage. The environment which the society sees appears more uncertain to it. As a consequence the requirements on all other modes of adaptability (aside from indifference) increase, as do the pressures for compensating changes in the society's structure of adaptability. If any society develops along this line, that is, the line of increasing internal specialization and compensating changes in the structure of adaptability, all other societies will be forced either to do likewise or to develop a wall of indifference which will eventually inevitably be broken down. Only under very special, inherently protected circumstances will a society be able to remain static in terms of its internal organization or structure of adaptability. Of course it is perfectly possible for a society not to succeed in developing either an increased level of internal specialization or a protective wall. Such a society will be in perpetual unpleasant crisis. The ensemble of deteriorated conditions which it runs through is the ensemble of states which provides its adaptability. These three possible modes of development—development in the direction of increasing complexity and interdependence, development in the direction of increasing

isolation, and development in the direction of increasing disorganization—
are perhaps more starkly represented in the modern world structure than
they have ever been in the past.

The support for special organs of adaptability in those societies which
do develop in the direction of greater internal specialization and interdependence, and even in those which are developing in the alternative directions,
shows that many societies explicitly recognize the importance of creative
and innovative activities. Superficially it would appear that this support
argues against a trend to declining adaptability. However, there are a
number of circumstances in which this support is ineffective or inadequate.
When it is present the question must be asked, is the organ being supported
in fact an organ of adaptability? In economic terms, is it an organ of risk, in
both its nonmonetary and its monetary forms? The society may develop
structures which have the form of academies, research institutes, or classes
of individuals specialized for mental work. But this does not mean that these
structures necessarily serve to compensate for restrictions on other modes of
adaptability.

It is interesting to consider the most important of these risk-reducing
circumstances. The first is the attempt to include the opportunities for
creativity and innovative thought, that is, the speculative part of total risk,
in the GNP. Investment risk can obviously be included since by definition
this involves monetarily measurable behavior. It is impossible to measure
the thought processes occurring in the individual members of a society by
the exchange of goods and services. However, it is possible to support
special research institutes or special individuals for the purpose of engaging
in the thinking that ultimately underlies investment risk. To this extent
thought processes do become monetarily measurable and it appears that
innovative adaptability becomes a bona fide component of GNP, indeed a
component of investment risk. But to this extent purposes and agreements
are established which reduce the uncertainty with which individuals behave,
thereby reducing the adaptability. The evolutionary adaptability is highest
when it is least preordained. Universities and research institutes can be
effective organs of compensation. They can provide mechanisms for exchange of ideas and for teamwork which can make them especially effective.
But whether or not they are effective depends less on the amount of
financial support they receive than on whether this financial support actually serves to enhance or diminish total risk. The real safeguard to the
effectiveness of compensation is the values held by the society. If it does not
value freedom of inquiry the support will merely serve to constrain inquiry,
in which case any steps taken to increase innovative adaptability by drawing
it into the framework of GNP will actually decrease it.

The second important risk-reducing circumstance is the attempt to increase GNP by increasing productivity without making supporting technological or organizational improvements. Such increases must be based on increases in the amount of energy and time which individuals in the society allocate to monetarily measurable activities, that is, to activities associated with the production and delivery of goods and services. The increase in GNP must then entail a decrease in energy and time allocated to thought, inventiveness, and other forms of creative and speculative activity which make investment adaptability effective. All other things constant, this must entail a decrease in total risk. It is possible that the increase in monetarily measurable activities on the part of one class of individuals will allow another class to perform a more effective compensating role. But in order for this to happen the compensating class must not itself be subject to increased constraints. It must instead either be relieved of some constraints or have more resources for exploration put at its disposal. These are not conditions which it is always possible to meet, or desirable to meet from the standpoint of equity. Indeed, it is in general the societies least in need of increasing productivity which most emphasize efficiency and which are most likely to suffer a decrease in innovative adaptability in consequence. In such societies increases in GNP should occur in consequence of creative activity, not at the expense of it.

The third important risk-reducing circumstance is the failure to couple the monetarily unmeasurable thought component of risk to the monetarily measurable investment component. In this case the universities, research institutes, and classes of individuals dedicated to compensation become organs of speculation, rather than organs of adaptability. This decoupling of nonmonetary speculation from monetary investment becomes more and more likely as the society becomes more internally specialized and as its internal and external interdependences increase, that is, just as the requirement for effective compensation becomes the greatest. This is due to the fact that the more complex economy is more likely to be destabilized by perturbation, including internal perturbations having their origin in inventiveness. This is especially true in organizations which are guided by plans, which are organized so that they can be predicted by a tractable model, or which develop definite procedures either to guide their increasingly complex operations or to maintain a definite societal structure. If rigidly adhered to, such plans and procedures are, like computer programs, inamenable to graceful change. In each of these cases it is necessary to buffer the effects of new ideas on the structure and dynamics of the society, just as it is necessary to buffer the effects of genetic variation on the phenotype of a complex organism. Such buffering requires the presence of redundant entities which make failure in a single entity tolerable. More importantly, it requires the participation of individuals who can exert sensitive judgment and alter the plan, procedure, or model accordingly. Fulfilling these requirements is costly. Decoupling the research effort from the investment effort is not costly. Bringing this research effort increasingly into the framework of GNP, but constraining it to product improvement rather than coupling it to innovation of consequence is less costly and may initially even be profitable. Both decoupling and constraining have an immediate effect which has the appearance of buffering. But they are not buffering and they are incompatible with the regenerative adaptability of an economic system.

The fourth important error is the failure to recognize conditions which require increased economic adaptability. This is particularly likely to occur when the society avails itself of comparative advantage, that is, increases its dependences on other societies. The resources available to the society are augmented, but so is the uncertainty of the environment. If all of these resources are allocated to improve the standard of living, the society will have in effect lost adaptability. The development of more extensive trade relations is thus associated with a cost in terms of adaptability as well as a gain due to comparative advantage. Clearly the cost should not exceed the gain. But more important, this cost should be covered by utilizing some of the gain to increase adaptability, not by simply absorbing it as loss of adaptability. If it is not covered, the society will eventually suffer a disturbance which will outweigh any short-run gain in standard of living. A related problem arises when steps are taken to eliminate an inefficient unit or disemploy a segment of the population. The disturbances generated must be absorbed by buffering mechanisms or they will ramify to other systems of the society. This is especially true if the internal interdependence of the society is high. If the society does not pay for these buffer mechanisms the effectiveness of its adaptability may be compromised and the cost of the ramified disturbances may in fact be greater than the cost of the buffering mechanisms. By default the systems to which they accidentally ramify, such as the justice system, become the compensating systems.

The final adaptability-reducing process which I shall consider is connected with communication. Interference with communication among the organs of adaptability, whether intra- or intersocietal, is adaptability-reducing. It is interesting that the situation is analogous to that of the exchange of genes among demes, discussed in Chapter 13. Evolutionary adaptability increases as the size of the gene pool increases and as the pool is broken up into more intercommunicating subpools, each capable of responding to selection independently. Innovative adaptability increases as the size of the idea pool increases and as this pool is broken up into more intercommunicating subpools, each capable of independent directions of investigation. If a society isolates itself from the international communication system of science, it will not only reduce the total innovative adaptability of the world, it will reduce its own innovative adaptability. It will be at a disadvantage relative to an otherwise equal society which engages more fully in this communication system. If the society or organization chooses to import ideas but to interfere with their export, the situation is more subtle. Superficially it would appear that worldwide adaptability would suffer, but that the society would be at an advantage relative to an equal society which both imported and exported ideas. However, the social form which refuses to export its ideas, like the deme which refuses to export its genes, cannot propagate itself other than by growth of its own population. In the absence of such internal growth the scientific circles in a country or organization which developed such a restrictive policy would necessarily become smaller and, as a consequence, less innovative. I believe this is one reason for the great desire among productive scientists to communicate their ideas and for the tremendous resilience of the international communication system of science, despite the absence of international support. However, there are also countervailing pressures, such as the advantages of maintaining military and production secrets. One group in a society may feel that it is to its advantage to keep another group in an uninformed state, a form of buffering which can only work if the informed group can control a large enough share of the societal resources so that it can comfortably absorb disturbances before they ramify to the uninformed group. The all too well known consequence is that societies and organizations impose upon themselves, to a lesser or greater extent, adaptability-reducing constraints such as secrecy and censorship. As is all too well known, a society's belief in the necessity for these constraints can grow out of proportion, swamping the intrinsic tendencies for communication among innovators and creating a deficit in innovative as well as other forms of adaptability.

My thesis is that due to these processes there is the same propensity to lose unexercised adaptability in economic ecosystems as in preeconomic ecosystems. Due to events of either internal or external origin, a disturbance will eventually occur which causes a crisis, that is, which causes the system to change its form. The crisis may simply lead to the release of constraints which prevent variabilities from being coupled to bona fide adaptabilities or it may release new variabilities, in which case the system regenerates its old form. Or it may assume a slightly altered form, either because of minor structural changes which persist or because of permanent changes in the external conditions. But the adaptability may also fall below a level which allows regeneration. A sequence of form changes occurs until a stable but radically different form is found. As an ecosystem, whether preeconomic or economic, increasingly regulates its environment, the range of disturbances with which it must cope becomes increasingly narrower, leading to an irreversible loss of adaptability, hence inevitably to a major form-changing event.

This aging and refurbishing process is inevitable in preeconomic ecosystems and a major determinant of both their successional development and evolution. Such systems may develop mechanisms which delay the loss of adaptability, but they have no way of preventing the loss from eventually occurring. Both the nonevolutionary and evolutionary adaptabilities tend to be lost. But the loss of the evolutionary adaptabilities is especially significant since this is the ultimate source of rejuvenation. Due to compensation the creative component of mentally based adaptability replaces genetically based adaptability as the source of rejuvenation in the economic sector of an economic ecosystem. The situation in economic ecosystems differs in two other respects. One is that a new set of processes comes into play which can contribute to aging. These include constraints on adaptability which arise from inappropriately structured efforts to support it, from slowdown of monetarily nonmeasurable investment activities caused by speedups of monetarily measurable activities, from failure to couple these monetarily nonmeasurable activities to monetarily measurable ones, from failure to couple increases in international trade relations to increases in adaptability, and from the imposition of constraints on the communication of ideas. The second difference is that conscious guidance is to some extent possible. It is possible to recognize the presence of each of the above adaptability-reducing processes and to take steps to reverse it. It is due to this possibility that the succession to instability which results from the loss of adaptability is more, not less, pronounced in economic ecosystems than in preeconomic ones. The beliefs which predominate favor adaptability-decreasing steps, not because individuals or societies purposely want to reduce their adaptability, but because the connections between adaptability and policies pursued are not recognized or because the importance of adaptability is systematically undervalued.

It is important to consider to what extent beliefs which a society has about the amount of adaptability it needs are free and to what extent they are constrained. The point to recognize is that the problem of deciding how a society ought to be organized in order to effectively convert its variability into adaptability is not the same as the problem of deciding how much adaptability is needed. The decision about how much adaptability is needed should depend on the uncertainty which the society judges appropriate to associate with the future. There is no objective procedure for arriving at such a judgment. In this respect the decision is free. But in retrospect it is highly constrained. Different societies may have different beliefs concerning the desired distribution of goods, services, and creative opportunities. Societies very different in this respect may be able to maintain their identity and continuity in competition with one another, provided that the choice of distribution is not incompatible with the dynamic stability of the society and with regulation of the environment. But if a society has too little adaptability, most especially too little creative capacity, it will to a greater and greater extent fail to meet whatever reasonable criteria it otherwise establishes for itself. Like fire protection, adaptability prolongs the probable lifetime of whatever organizational structure incorporates it, whether or not this structure is desirable from the aesthetic or ethical points of view and indeed whether or not the structure is potentially stable on other grounds. Like fire protection, the addition of a small amount of adaptability affords a small amount of protection and the addition of a large amount affords a large amount of protection. A society which has too much adaptability does not, taken in isolation, face the certainty of failure. The only intrinsic danger is that there is always a small chance that even a well-buffered creative capacity will create an undesirable instability. But competition from other societies would in the long run cause such a society to decline or cause it to forego some of its adaptability. Since it is always so easy to forego adaptability, it can with reasonable safety be said that the danger of too much adaptability is much less than the danger of too little. The danger of allocating resources superfluously to specific, noncreative activities is much greater and more irreversible. A competing society with the same productive and creative capacity could then allocate more of its resources to military or trade competition, thereby degrading the environment of its competitor society. Or it could degrade this environment by using these otherwise superfluous resources to enhance its innovative capacity.

The assertion that the maintenance of a healthy economy depends on man's creative activities and on the maintenance of other forms of adaptability sounds sensible. Nevertheless, many of the actions taken by governments and international institutions are not consistent with this. The usual idea is that, when a nation's economy is deteriorating, urgent steps must be taken to secure the immediate necessities and to improve its monetarily measurable performance. It may appear perverse and unrealistic to assert that it should expend more of its efforts at cultivating creative pursuits which are normally viewed as being of only cultural value. But this appearance of unreasonableness is based on the ill-founded belief that adaptability is not a necessity and on the failure to recognize that a breakdown of cultural integrity signifies that adaptability has become inadequate. This does not mean that inadequate adaptability is always the problem. In many cases steps are called for which may bring with them some reduction in adaptability. It also does not mean that taking steps to raise adaptability will in general be sufficient to achieve desired goals. But not taking these steps will ensure that the deterioration will proceed further. It is certain to prolong the sequence of subfunctional organizations which the society must endure before it discovers a viable one. The suggestion of adaptability theory is that the deteriorating economic situations which exist in many parts of the world today would be most ameliorated by steps just the opposite of those being taken, that is, by steps to reverse the minimal tendency of adaptability.

14.3. *BLIND CHANCE AND BETTER CHANCE*

This extension of the theory to the economic domain must not obscure the fact that an economic ecosystem is still an ecosystem, with all the hierarchical and compartmental structure of an ecosystem. From the standpoint of adaptability theory the purely economic concerns are only a small part of the design concerns. Early in human history they were a very small part. In that early stage the capacity to design was modest. With the development of economic society the effects of human decision-making became very much greater. Agricultural decisions and decisions about economic structures became part of man's design prerogatives. Decisions which might affect most levels of biological organization did not. Now the situation is changing again. New tools have become available, such as recombinant DNA and cloning techniques, which extend again the domain which can be influenced by human decision-making. Once again the proportion of man's design concerns is changing.

There is increasing worldwide pressure to develop models which can help to determine how this influence can be exerted on agricultural systems, on human genetic and physiological systems, and on social institutions. This is understandable given the stresses which exist and given the interest in utilizing resources efficiently. But the situation of these models should be appreciated. According to adaptability theory the classical design paradigm of modeling, planning, predicting, and constructing is inappropriate. Indeed, the word *design* is itself inappropriate. It may appear that we are undergoing a fundamental revolution and that we can now take our destiny into our own hands, in the same way that the artist takes a piece of clay into his own hands. It may seem that in the past Darwinian evolution provided the mechanism through which design developed, but that now we are at the threshold of an era in which the blind chance of evolution can be replaced by modeling, planning, calculating, predicting, and constructing. According to adaptability theory it is impossible to enter into such an era. Yet it is at the same time evident that the possibilities for analyzing and modifying biological processes and social institutions are greater than ever before. The problem is that if we avail ourselves of these possibilities using the classical design paradigm we will surely create organizations too poverty-stricken in terms of adaptability to survive.

It is useful to put the situation into historical perspective. Before the appearance of humans and human societies life developed under the aegis of natural selection. The element of intelligent design on the part of constituent species played at most a minor role. But with the evolution of human beings the possibility of cultural transmission of acquired information became orders of magnitude more significant than at any previous time. Linguistic mechanisms, methods of training, and use of information processing methodologies are some of the tools which make human and social model development and problem-solving increasingly important after this point. It is sometimes stated that at this point the new constraints and adaptabilities created by the Darwinian mechanism freed one species, namely ourselves, from decisive control by this mechanism and opened up the possibility for exerting decisive influence on it and on the structure of human ecological systems generally. But some degree of freedom from the Darwinian mechanism does not mean freedom from the principles of adaptability. It is one thing to use models to propose organizational structures which have desirable features, quite another to depend on models for predicting the future or to use them for designing predictable systems. The key desirable feature is that the system require the scientist to think in terms of potentialities, that is, in terms of its dispositions to respond. These dispositions would be much less rich than they easily could be if it were possible to predict them without reference to the fundamental principles of physics or to compute them from these principles. Organizations which are constructed to be predictable on either of these bases will have an unsuitable structure of adaptability—unless they are immersed in a larger, unpredictable system. They will fail, in fact in an unpredictable way.

The situation is this. Darwinian evolution is a matter of blind chance. But according to adaptability theory there is a significant amendment. Blind chance can be better chance or worse chance. It is possible for the genotype–phenotype relationship to evolve in such a way that the chance becomes better. The chance of discovering how to use subtle features of the physical world for more powerful information processing improves. It is as if blind men can move from one work room to another. They can never see, but some rooms are so constructed that they are easier to work in. If the blind succeed in making a new tool in a new room, more of them move into it, and more of them explore neighboring rooms. They discover more and more powerful mechanisms of information processing. They can use these mechanisms to take the construction of rooms into their own hands. They can build rooms even better suited for the work of the blind. They have entered a new age, but it is not an age of sight. They can never develop sight. This is the situation in evolution. We can discover more and more powerful mechanisms of information processing. The revolution is that we can use these to turn blind chance into better chance or into worse chance. At least the search for better chance can be less blind. But even better chance is still blind chance. Our intelligence and computational capabilities can never escape this. If we attempt to use these capabilities to construct a world—whether biotic or social—which is more predictable than could be found through the best blind search, we will certainly achieve just the opposite. Such a world trades away transformability and adaptability, then, a step later, efficiency and predictability. It is a world which is built for worse chance, not better chance. It is as though the blind men used their newfound capabilities to build rooms too simple and economical for the work of the blind. The only way in which intelligence can reasonably be used to design living systems and social institutions is to ensure that the conditions for better search are fulfilled and that the structure of adaptability is suitable. But this means accepting a paradoxical toll, so hard to accept in a world of scarcity, that takes away efficiency and provides in return fewer levers for predictability and control.

REFERENCE

Knight, F. H. (1921) *Risk, Uncertainty, and Profit.* Houghton Mifflin, New York.

Index

Accidents
   and acausal jumps, 21
   and mutation, 31
   See also Fluctuations; Measurement
Adaptability
   versus adaptation, 274–280
   as adaptation to environmental
      uncertainty, 10, 277
   of agricultural systems, 252–254
   allocation of, 247, 249, 255, 268
   components of
      anticipation, 57, 116–118
      behavioral uncertainty, 57, 110–113
      indifference, 57
   of computer-based societies, 234
   correlation to stability and instability,
      164–169
   cost of, 105–140, 149–150, 250–256
   definition of, 7, 56, 97, 138–139
   and dissipation, 7
   economic, 278–279, 354–368
   evolutionary, 111–112, 189, 196, 206, 335,
      356
   extended, 292
   and independence, 95–96, 101–102
   mechanisms of
      constitutive, 110–114
      indifference, 113–116
      inducible, 111–113
      selective, 111–113
   and medicine, 268–274
   modes of
      behavioral, 97–101
      culturability, 7, 97–101
      developmental, 7, 97–101
      genetic, 7, 97–101, 112
      learning, 100, 112

Adaptability, **modes of** (*cont.*)
   mental, 275–276, 356
   phenotypic, 100–101
   physiological, 7, 97–101
   populational, 7, 97–101
   routability, 7, 97–101
   social, 97–101
   topographic, 97–101
   unnamed, 123–125
   and modifiability, 95–96, 101–102
   more fundamental than dynamics, 173
   patterns of, 256–262
   and planning, 347–354
   and potential entropies, 56–57
   and reliability, 188–189
   reserves of, 120, 350
   unrepresented, 162–164, 167, 343
   upper bound of, 132–133
   and variability, 7
   withering of, 125–132, 320–321, 365–366
   See also Adaptability theory; Anticipation;
      Behavioral uncertainty; Biological
      compensation; Homeostasis;
      Indifference; Minimal tendency of
      adaptability
Adaptability theory
   contrasted to statistical mechanics, 342–343
   fundamental identity of, 55, 75–76
   fundamental inequality of, 55–56, 96
   and human decision-making, 347–354,
      368–370
   review of, 336–342
   self-consistency of hierarchical, 133–135
   See also Biological compensation
Adaptation
   definition of, 9, 276
   See also Specialization